James Cornelius Wilson

A Treatise on the Continued Fevers

James Cornelius Wilson

A Treatise on the Continued Fevers

ISBN/EAN: 9783337418762

Printed in Europe, USA, Canada, Australia, Japan

Cover: Foto ©berggeist007 / pixelio.de

More available books at **www.hansebooks.com**

A TREATISE

ON

THE CONTINUED FEVERS

BY

JAMES C. WILSON, M.D.,

PHYSICIAN TO THE PHILADELPHIA HOSPITAL AND TO THE HOSPITAL OF THE JEFFERSON MEDICAL COL-
LEGE, AND LECTURER ON PHYSICAL DIAGNOSIS AT THE JEFFERSON MEDICAL COLLEGE,
FELLOW OF THE COLLEGE OF PHYSICIANS OF PHILADELPHIA, ETC.

WITH AN INTRODUCTION BY

J. M. DA COSTA, M.D.,

PROFESSOR OF THE PRACTICE OF MEDICINE AND CLINICAL MEDICINE AT THE JEFFERSON MEDICAL COL-
LEGE, PHYSICIAN TO THE PENNSYLVANIA HOSPITAL, FELLOW OF THE COLLEGE
OF PHYSICIANS, PHILADELPHIA, ETC.

NEW YORK
WILLIAM WOOD & COMPANY
27 GREAT JONES STREET
1881

COPYRIGHT
WILLIAM WOOD & COMPANY
1881

TROW'S
PRINTING AND BOOKBINDING COMPANY
201-213 *East 12th Street*
NEW YORK

To the Memory

OF

Dr. WILLIAM W. GERHARD,

WHO

FIRST IN AMERICA APPLIED TO THE STUDY OF THE FEVERS THE METHODS OF MODERN SCIENTIFIC RESEARCH;

AND TO WHOM IS DUE

THE CREDIT OF HAVING FIRST CLEARLY ESTABLISHED SEVERAL OF THE MOST IMPORTANT POINTS OF DISTINCTION

BETWEEN

TYPHUS AND TYPHOID FEVERS.

PREFACE.

The diseases considered in the following pages constitute a group with most of which the general practitioner is more or less familiar; it has, therefore, been my aim to describe them at greater fulness than is usual in the text-books, yet without the extreme elaboration that mars the usefulness of some of the special treatises. Brief historical sketches have been introduced, and considerable attention has been given to the subject of the special causes of particular diseases, as well as to their clinical phenomena and their anatomical lesions. Purely theoretical considerations have been, as a rule, omitted, and all controversial matters have been disregarded. The sections upon treatment are designed to represent recent practical knowledge, rather than to do justice to the changing opinions of which that knowledge is the outgrowth.

With reference to the title of the book, it is to be admitted that, despite general usage and the highest modern authority, the classification of the infectious diseases, and in particular of those commonly known as the Fevers, is unscientific and provisional. Diseases being processes and not entities, are properly to be classified upon an etiological basis. Our knowledge of the exciting causes of the Fevers does not as yet admit of the employment of such a principle of classification. Much, however, has been learned within recent years, and new facts are from day to day being brought to light: the expectation that more exact and definite knowledge of the special causes of the Fevers will, in the near future, lead to a more satisfactory nosological system, is not without warrant.

Meanwhile, we must content ourselves with groupings based upon the broad clinical aspects of diseases. From this point of view, the affections treated of in this volume constitute a group sufficiently well defined. They are characterized by notable, persistent elevation of temperature, and steady continuance to a definite termination.

The group might have been made larger or smaller, but the time has not yet come, it seems to me, to include pneumonia, diphtheria, and acute rheumatism, among the Fevers, and I can find no reason, seeing that the eruptions of dengue are variable and inconstant, for classing it among the exanthematous diseases.

I desire to express my thanks to my friend, Prof. Wm. H. Greene, for assistance in reading and correcting the proofs, and for several important suggestions as to the arrangement of the topics.

JAMES C. WILSON.

1437 Walnut st., Philadelphia.
28th March, 1881.

CONTENTS.

Introduction by Professor Da Costa,

I.—SIMPLE CONTINUED FEVER.

Definition,
Synonyms,
Etiology,
Clinical History,
Analysis of the Symptoms,
Duration,
Diagnosis,
Prognosis and Mortality,
Treatment,

II.—INFLUENZA.

Definition,
Synonyms,
Historical Sketch,
Etiology,
    I.—Predisposing Causes,
    II.—The Exciting Cause,
Clinical History,
Analysis of the Symptoms,
    The Fever,
    The Catarrh,
    Symptoms Referable to the Nervous System,
Complications and Sequels,
Pathology,

viii   CONTENTS.

	PAGE
Diagnosis,	37
Prognosis and Mortality,	38
Treatment,	39

III.—CEREBRO-SPINAL FEVER.

Definition,	46
Synonyms,	46
Historical Sketch,	47
Etiology,	56
Clinical History,	64
Analysis of the Symptoms,	72
Symptoms Pertaining to the Nervous System,	72
Symptoms Referable to the Skin,	77
The Phenomena of the Fever,	78
Symptoms Referable to the Organs of Respiration,	84
Disturbances of the Organs of the Special Senses,	85
Complications and Sequels,	87
Pathology, Morbid Anatomy,	89
Diagnosis,	94
Prognosis and Mortality,	97
Treatment,	98

IV.—ENTERIC OR TYPHOID FEVER.

Definition,	107
Synonyms,	107
Historical Sketch,	108
Etiology,	116
I.—Predisposing Causes,	116
II.—The Exciting Cause,	120
Clinical History,	147
Analysis of the Principal Symptoms,	153
The Phenomena of the Fever,	153
Symptoms Referable to the Circulatory System,	161
Symptoms Referable to the Nervous System,	163
The Skin,	167
Symptoms Referable to the Digestive Tract,	170
Symptoms Referable to the Organs of Respiration,	176
The Urine,	177
Complications and Sequels,	178
Varieties,	192
Relapses,	196
Anatomical Lesions,	202

CONTENTS.

	PAGE
Diagnosis,	210
Prognosis and Mortality,	213
Treatment,	221
I.—Prophylaxis,	221
II.—The General Management of the Patient and Dietetics,	222
III.—Special Forms of Treatment,	227
IV.—The Expectant Treatment,	234
V.—The Treatment of Special Symptoms, Complications, and Sequels,	235
VI.—The Management of the Patient during Convalescence,	240

V.—TYPHUS FEVER.

Definition,	241
Synonyms,	241
Historical Sketch,	242
Etiology,	251
I.—Predisposing Causes,	251
II.—The Exciting Cause,	256
Clinical History,	260
Analysis of the Principal Symptoms,	264
Symptoms Referable to the Nervous System,	264
The Phenomena of the Fever,	269
Symptoms Manifested by the Skin,	277
Symptoms Referable to the Respiratory System,	281
Symptoms Referable to the Digestive System,	281
Complications and Sequels,	284
Varieties,	288
Prognosis and Mortality,	290
Anatomical Lesions,	293
Diagnosis,	295
Treatment,	297

VI.—RELAPSING FEVER.

Definition,	302
Synonyms,	302
Historical Sketch,	303
Etiology,	309
I.—Predisposing Causes,	309
II.—The Exciting Cause,	312
Clinical History,	320
Analysis of the Principal Symptoms,	324
Symptoms Referable to the Nervous System,	324
The Phenomena of the Fever,	326
Symptoms due to Disturbance of the Digestive Organs,	331

	PAGE
Complications and Sequels,	332
Prognosis and Mortality,	335
Anatomical Lesions,	336
Diagnosis,	337
Treatment,	340

VII.—DENGUE.

Definition,	344
Synonyms,	344
Historical Sketch,	345
Etiology,	348
I.—Predisposing Causes,	348
II.—The Exciting Cause,	349
Clinical History,	350
Diagnosis,	356
Treatment,	356

THE CONTINUED FEVERS.

INTRODUCTION.

I have been asked to write an introduction to Dr. Wilson's work on Fevers, and I shall choose for my subject that most important one, the management of Fever. For what is the study of its causation, what the care in its discrimination, what the close pursuit of the lesions in solids and in blood, unless we are thus to be led to a more clearly conceived, more thoughtful, more successful management of the Fever? As the scope of this work is limited to the Continued Fevers, so my remarks will chiefly refer to them. But there is little that I shall say that in the main would not be applicable to the other members of the great family of Fevers.

We naturally have to consider, first, the general management of the fever, as it is influenced by the arrangements of the sick-room, and by the attendance to the wants of the patient—those things which imply his nursing. Now, we all agree that good nursing is essential; but do we all enforce it, and continue to superintend it? The physician who lays aside his watchfulness on these points, finds at any moment that he is combating with one of his chief weapons broken in his hand. Reports from the nurse—written when practicable, inspection of the arrangements for ventilation, for destroying the discharges, for insuring the cleanliness of the patient, should form part of the occupation of at least one of the daily visits.

So much has been said of late years of the functions of the nurse, and there are now so many more well-trained nurses, that it will be quite unnecessary here to go into any details of the nursing of fever-patients. But the last word on this subject can never be spoken. It never can be too strongly enforced that cleanliness, cheerfulness, and regularity, are the three great qualities needed in the sick-room.

The cleanliness consists in keeping him personally clean—in sponging him with cool or tepid water, or with vinegar, or bay-rum and water, morning and evening, only parts of the body at a time, if more fatigue him; in seeing to it that his linen is unsoiled; and that the room is not encumbered with anything useless, and that all objects are free from stain and in good order. The cheerfulness sustains his spirits, and, until his nervous system is stricken with obtuseness, is a vast comfort and aid during his dreary, restless hours.

The regularity is indispensable; everything must be given at hours arranged by the physician. Well-meant but injudicious kindness may give food and medicine oftener, or fail to give them, fearing to disturb. But well-meant though injudicious kindness may thus hasten or cause death. Except under the most potent of causes, the schedule arranged by the physician must not be departed from. Of course, in these directions some latitude will be left as to how long the patient may be allowed to sleep, or under what circumstances a dose may be repeated or be omitted. But a careful physician indicates this latitude with his directions.

Besides these points essential to good nursing, there are others— some quite, others almost equally important. Equally important certainly is ventilation, admitting light and air both, not excluding them as if they were poisonous. To admit light is to influence the nervous system favorably, to keep the half-dreamy, wandering attention aroused, to procure better sleep by marking the alternations between day and night and invoking the force of habit, to moderate often a delirium. To admit pure air is to give the respiratory functions their full play and to furnish the changed blood with the means requisite for its revival. Moreover, it cools the atmosphere, which indeed, even in winter, should be kept at a very moderate degree of heat; and this, to the patient consumed with fever, is both grateful and salutary. We see what a calamity a hot atmosphere is if we are obliged to treat severe cases of typhoid fever in our cities during the summer months. They are likely to do badly—the heat adds to their gravity and prostrates the nervous system. I have often attempted to cool the atmosphere by artificial means, and have used, with at least partial success, cloths wrung out in ice-water, and hung up near an open window; I have resorted to blocks of ice that are allowed to melt in the room, and to the hand-ball atomizer charged with ice-water or cologne and water, so as to fill the room with the spray. But, with all, the torrid weather of our heated term is a terrible drawback in the treatment of grave fevers.

Another important point in the care of the sick person is that he should not be needlessly disturbed. And here it is where the well-trained professional nurse is such an advantage. Fussiness is a destructive quality; and ignorance is always fussy. Nurses who know their business but imperfectly are apt to be always in motion, always addressing the unfortunate patient, keeping him awake when he wants to sleep, constantly forcing drinks on him, never resting themselves or letting him rest. And it must be said that the overanxious eye and hand of affection are sometimes as injurious as the annoyance of the well-meaning, meddlesome nurse. The solitude which implies severance from loving watchfulness is very bad; but it may be better than the unrest which loving watchfulness misdirected occasions.

A fever-patient should be put to bed early; it saves his strength. We see what comes of not doing it, in the result of the so-called walking cases, especially in typhoid fever and in yellow fever; they are very apt to do badly. But how long should the patient be kept in bed? I think not too long. To put him to bed early and to let him up early is my rule. If the thermometer for three or four days have marked a normal evening temperature, I allow him to get up, at first for half an hour or less, and then daily more and more. I have known this plan succeed admirably in what seemed a protracted convalescence, and put a stop to night-sweats and to temperature-rises to 100°; for I think we may keep the temperature at that, or let it go back to that, by allowing the patient to stay too long in bed.

The diet varies, of course, with the character of the fever. The typhoid fever patient, with his ulcerating intestine, will not bear the same diet as the typhus fever or catarrhal fever patient. Yet in the main there is such a thing as a fever-diet, and that is, a restricted diet of bland, easily digested substances. The coated state of the mucous membrane, the difficulty of digestion, the lack of appetite, make the sick man turn almost with disgust from other food. Hence, broths, and milk, and farinaceous food form the staple of the diet, whether he have an intestinal lesion or not; and by the loathing for food Nature restricts the diet, whether we restrict it or not. Indeed, since the memorable words which Graves chose for his epitaph—"He fed fevers"—it cannot be said that the English-speaking races, at least, attempt to curtail the diet much. Our error, I think, is now in the other direction; in the earlier stages of the fever we do not curtail it enough. As regards the character of food, while the articles mentioned are those generally most acceptable, they need not, except in the case of typhoid fever, be as rigidly adhered to as is the wont. If the patient crave other food—crave solid food not actually indigestible, he may have it. Later in the febrile malady assuredly the tone of the stomach may be better sustained by some solid than by so much liquid nourishment.

From the cruel practice of refusing water to the fever-stricken patient, there has been, we all know, a strong reaction. And it is one of the doctrines now unreservedly taught, to allow the patient an unlimited supply of pure water or of other bland fluid. That in the main this is right, there can be no question. It is not simply a gratification to quench the burning thirst, but it means to get rid of the poison and of broken-down tissues by keeping skin and kidneys active. Yet, is it proper that the supply should be unlimited? I think we have gone too far in saying that it shall be. Very large supplies of water mean that the vessels of the stomach are constantly full, that the process of taking up liquid food is retarded—nay, that the desire for the really essential nourishment is greatly lessened or is changed to repugnance.

Closely connected with the subject of food and drink is that of stimulants. I cannot here go into the question of giving alcohol in fevers, because the propriety or impropriety has to be judged in each fever, and general statements are apt to be misleading. We cannot make hard-and-fast rules that will apply equally in typhoid and in influenza, in cerebro-spinal fever and in relapsing fever. Still there are, besides many special indications, some comprehensive ones which, though in different degree, turn up in all fevers, and are to be met in the same way. Whenever there are signs of failing circulation, whenever the action of the heart becomes enfeebled, stimulants are demanded. And we have no better guide in this than the law Stokes enunciated long since in typhus fever, and which more recent observations have applied to typhoid—the state of the first sound of the heart. This, indeed, can be made use of with advantage in all fevers. Let the first sound become short, indistinct, almost suppressed, and we have a certain indication for alcohol; the fainter the first sound, the more urgent is the stimulus required. Now, the pulse aids also in determining the question; yet it is not so certain. But both pulse and heart-sounds are much more available than the sphygmograph, which, though employed by some, is quite unsuited to the exigencies of professional life in framing the treatment of fever-cases.

Tremor and delirium are other signs which call for stimulants; they are mostly the result of failing nervous power. Yet, certainly with reference to delirium, we cannot make our rule too absolute. Delirium may be due, not to defective nerve-energy and poisoned blood, but to intracranial mischief, though, excepting cerebro-spinal fever, such is rarely the case.

We now arrive at the treatment of fever by strictly medicinal means. At the very threshold we come across the inquiry: Are there special plans of treatment for these fevers of the Continued Type—plans of treatment approaching to specifics, leading rapidly to cure—having, in other words, the power or something like the power, which the preparations of bark exert over the fevers of the Periodical Type? Or are we in the main still forced to treat the fevers on what is called the rational plan—to treat, therefore, chiefly the symptoms until the poison is eliminated or its results disappear? It may be a humiliating statement, but it is true that in the main such is the case. There are in some fevers special plans of treatment which aim at modifying the poison or the disease itself, which are, I believe, worthy of confidence, and are better than the so-called rational treatment; such I hold to be, for instance, the treatment of the typh-fevers by the mineral acids, of cerebro-spinal fever by opium. But these plans of treatment are few, and are not pre-eminent and striking. The result, on the whole, is better when they are employed, yet they are not curative in the highest degree; and under any circumstances, there are but a scanty number which have stood the test at all. Most of the special plans proposed are mixed up with a quantity of unmistakable rubbish, and have been cleared away; the accumulated experience of many minds acting as an ultimate court of appeal has given judgment in favor of very few, and among these have not been any based on remedies of extraordinary kind or preconceived action. The best high road to success is still the high road of the commonplace.

But it would be as illogical as absurd to suppose that we shall never possess the coveted means really to cure the continued fevers. Doubtless, to the physician of the time of Charles V. the radical and specific treatment of the malarial fevers appeared as hopeless and remote as the radical and specific treatment of the continued fevers appears to the scientific inquirer of our day.

If, then, we are still obliged to treat the fevers of continued type rather on general principles than by remedies that are specific, we have to look to those indications, and to depend largely on those agents which enable us to control the fever-process. Among these indications there are a few of paramount importance.

One, certainly, is to watch and to keep up the secretions. It is better for every fever that the skin should be moist, than that it should be harsh and dry. It is better that the urine should be abundant, than that it should be scanty and thick with tissue-waste. And it is not enough to judge by the rough tests with which the older physicians were familiar—we must resort to the more accurate chemical means. Testing the urine for albumen has much significance. It is some kind of guide to the depth of the impression the fever-poison has made on blood and nervous system. The abnormal ingredient is not present in light cases; it is rarely absent in grave ones. Watching the stools, too, and seeing that they are regular, is of value. Leaving out the special character which comes from the lesion in typhoid fever, in all fevers we can judge by them whether the food taken, be it solid or milk, is being digested; whether, therefore, it had better be continued or changed.

To reduce the elevated temperature of the fever is to all a most important, and to some the most important indication. This is accomplished by sponging the skin with cool water, by seeing that it acts freely, and by the use of drugs which lower temperature. But the most potent agent undoubtedly is the cold bath, and the treatment of fevers, especially of typhoid fever, by cold baths, is one which is now being strongly urged on professional attention. Some employ it in all cases, others only in those in which the temperature exceeds 103°. To carry out this apyretic treatment effectually requires, however, such constant repetition of the bath, such extreme care in the assistants, such facilities for resorting to it without fatigue, and such implicit obedience on the part of the patient, or rather of his friends, that in private practice, at least, it is in this country impracticable. And it is not settled that for ordinary cases it is so superior to other plans that we are bound to insist upon the discomfort and annoyance which attend it. Still, for cases of very high temperature, cases of about 105° or upwards, unless extreme exhaustion or some other contra-indication forbids, it is right to resort to it. High temperature then becomes in itself dangerous to life, and we try to subdue the bad symptom to preserve life. I have several times in the last few years made use of the cold-water treatment under these circumstances, and seen it act well. I have also known the cold-water bath to overcome that bad and destructive symptom of fever, sleeplessness, where anodynes had failed.

Other means to reduce the temperature are quinine and the salicylate of sodium. Quinine in large doses has, on the whole, proved its power to do so, certainly in typhoid and typhus fevers. Yet it is sometimes disappointing, particularly in this, that the effect gained is not at all permanent. Moreover, we must be careful not to infer that sulphate of quinia is to be employed in all the fevers of the continued type, to bring about the results mentioned. Clinical experience will have to be recorded for its use in each fever. Granting that it always has the same effect, such large doses cannot be given with impunity in various and dissimilar pathological conditions; and it is very possible that, while they reduce the temperature, they may aggravate the disease or some of the lesions. That this does not happen in typhoid and typhus fever, has, I think, been proved; for the other members of the group the problem has not, from this point of view, been worked out. Salicylate of sodium is too new a remedy to have been fairly tested; that it reduces temperature we know. But it is more apt to disturb digestion than quinia, and acting, as it often does, as a depressant to the heart, its use in low fevers will require considerable caution.

To control and influence the circulation is an indication second to none in the treatment of the continued fevers. In those, far rarer instances, in which the circulation is too active and the powers of the heart increased—in fevers, therefore, of what were formerly called ardent or inflammatory type, there is in my experience no remedy equal to aconite in quieting heart and pulse. In the much more usual indication of defective cardiac action which sooner or later is apt to show itself in the course of most fevers, alcohol steadies the feeble heart more certainly than anything else. Quinia in small doses, or strychnia, aids; but alcohol exerts by far the most influence. Digitalis, from which we might expect so much, has disappointed me; at least it has done so repeatedly in typhoid fever.

There is another point connected with the management of fevers as important as any that has been stated. It refers not to the sick-room, nor to the sick man, but to the doctor: he must manage himself. Nothing is worse than a vacillating physician, whom each motion, each wish of the patient, each suggestion of the nurse or of the family, affects. Blown hither and thither by every breath, incapable of taking a broad view of the case, his treatment soon becomes as irresolute as himself, and directions and bottles accumulate with bewildering rapidity. The fewer drugs that are used, the better; the greater the decision with which the drugs are used, the better. To do this effectually the physician must understand the mode of onset of the fever, its probable length, its natural course, the succession and duration of each symptom, its dependence or non-dependence upon a fixed lesion, the kind of complication likely to arise and the time at which it is apt to set in; he must, in one word, be pathologist as well as physician. He then knows when to act and when not to act. And in so doing we have learned equally from men and nature. From men we have learned what agents to employ when we wish to make strong impressions; from nature the uselessness or folly of such attempts when the fever is pursuing an even course.

Yet, to treat a case with the best chance of success, still something else is required—the practical skill which takes note of the epidemic influence prevailing; which recognizes that all cases are not alike because they bear the same name; which does not overlook that in the same disease apparently the brunt may fall primarily on this organ or on that organ, that the nervous system or the circulation may suffer disproportionately and exceptionally from the onset, or, as in fevers of the worst form, be overwhelmed together; which lays stress on peculiarity of causation, of temperament, of constitution; which sees, therefore, not only the disease in the sick man, but the sick man in the disease. And another quality enters into the achievement of greatest success—the tenacity which never abandons a case while there is life. In diseases that are self-limited, to continue to sustain to the last is to give nature the chance of exerting a power of recuperation which art cannot gauge.

J. M. DA COSTA.

I.

SIMPLE CONTINUED FEVER.

DEFINITION.—A continued fever, not due to specific **cause, usually of** short duration, lacking the distinguishing characteristics of **the other** fevers, rarely fatal in temperate **climates, and showing, when death** occurs, no characteristic lesion.

SYNONYMS.—Febricula; Ephemeral fever; Common continued fever; Sun and Heat **fever; Ardent continued fever;** Febris **continua simplex;** Synocha.

MUCH confusion has arisen in consequence of **the use of the term simple** continued **fever,** by different authors, **to** designate **several distinct** affections. It has been a category for many cases of uncertain character. It has been made to include cases arising during epidemics **of fever, as** for example, typhus or yellow **fever, that have lacked the** distinguishing features of the prevalent disease on the one hand **and the** traits of the other essential **fevers on the other.** It has been applied where enteric fever and remittent were endemic, to cases **of fever** occurring side by side with these maladies, yet not showing **the** typical pathological **events** which attend them, or showing them to so faint a degree as to baffle the judgment of the **observer.** Further observations, conducted with great care and analyzed in sufficient **numbers, are** needed to determine the nosological relations of such cases. It is probable that they are not of nonspecific origin in most instances, but that, by reason of the smallness of the dose of the fever-producing poison, **or an** imperfect susceptibility on the part of the individual, the specific fever is of such mildness that its characteristic phenomena are not made manifest—it is, in other words, "abortive." Cases of this kind do not correspond to the definition of simple continued fever. These views have led some **observers to deny** even the existence of the fever under consideration as a **distinct affection.** Practitioners of medicine in every clime are familiar with the fever called, from its transient character, ephemera and febricula. From a duration of one, two, or three days, which is common, this fever may, without other modification, in rare examples, be extended over a period of **ten or** twelve days. It is therefore proper to include under the heading, simple fever or simple continued fever, all essential continued fevers that are clearly of non-specific origin, whether they be in the strict sense of the term ephemeral, or be prolonged through several days. As Flint[1] has pointed out, the diminutive term febricula has relation to the duration of the fever rather than to its intensity. In many instances the fever is intense.

While the affection known as sunstroke is properly referred in systematic treatises to the diseases of the nervous system, Professor H. C. Wood[2] has shown that its phenomena are those of fever of great intensity, and that the continued fever following exposure to the sun or to a prolonged high temperature, differs from sunstroke in degree and not in kind. The terms sun fever and heat fever, are therefore properly applicable to the febrile affections brought about by the action of these causes.

Etiology.

It results, from what has been already said, that only those fevers can be regarded as simple that are due to non-specific causes—that are, in fact, neither contagious nor miasmatic. It is also important, theoretically and in practice, to exclude all symptomatic fevers, such as the fever which follows traumatism and surgical procedures, the formation of abscesses, other local inflammations, and hectic fever. It is indeed the more important because in frequent instances the symptomatic fever bears a strong clinical resemblance to ephemera. This discrimination is important on theoretical grounds, because the one is secondary to and dependent upon a primary disorder of which it is no more than a symptom, the constitutional disturbance resulting from local irritation; whilst the other is in itself the primary disease and the result of causes affecting the nervous system at large without determinable local lesion. It is important in practice, by reason of its obvious influence upon diagnosis and treatment.

Many different causes are known to be capable of producing the train of febrile phenomena which constitutes simple continued fever. Among them may be named exposure to great heat or cold, excesses in eating and drinking, mental and bodily fatigue, excitement and violent emotions. Children, by reason of the relative instability of their nervous organization, are much more prone to this form of fever than adults. It is a malady more frequently encountered in summer than at other seasons of the year, and is often produced by the fatigues of travel and unwonted exposure to the sun. It is not unfrequently due to the combined influence of the excitement, the physical exhaustion, and the exposure to the

---

[1] Clinical Medicine. 1879.
[2] Sunstroke and Thermic Fever. Boylston Prize Essay, 1871.

direct rays of a mid-day sun, which are attendant upon surf-bathing. Many cases of simple continued fever occurred in Philadelphia at the time of the Centennial Exhibition in 1876. Citizens and strangers were alike exposed to the action of some of the most powerful causes of non-specific fever. The summer was unusually hot, the distances to be traversed were considerable, the fatigue of several successive days spent on the grounds of the exhibition was often beyond the sight-seer's powers of endurance. Add the excitement of mingling with vast throngs of enthusiastic people, and at evening a hunger out of proportion to the digestive powers of jaded men and women, and it seems remarkable not that cases of fever occurred as often as they did, but that they were so comparatively rare.

Clinical History.

Simple fever begins abruptly. Prodromes are absent. Lassitude, a chill or chilliness, and a sudden rise in temperature, mark the onset of the disease. All the phenomena of fever are rapidly established. Hot skin, rapid pulse, thirst, headache, pain in the back and limbs, harass the patient from the beginning. The bowels are constipated, the urine diminished in quantity, and of high specific gravity. Except in cases due to excesses at table, and the like, vomiting is uncommon. There is loss of appetite. The tongue is white and coated. The rise in temperature is not only very rapid, it is also in many cases very great. In the course of a few hours it may reach 39.4° C. (103° F.) or even 40.5° C. (105° F.). The abruptness of the temperature rise, and the rapidity with which the maximum is reached are characteristic of this fever as compared with the other continued fevers, with the exception of relapsing fever. They are only shared in by some forms of malarious fever (intermittent), variola, measles, and pneumonia.[1] The continuance of the fever is usually of short duration. In a few hours, or a single day, defervescence sets in and the temperature speedily falls to the standard of health—an instance of *crisis*. On the other hand, the fever may be prolonged through two, three, four, or, as has been pointed out above, to ten or twelve days, the normal body-heat being regained by several days of gradual defervescence—*lysis*.

The defervescence is often marked by copious perspiration, but this is not always the case. It is sometimes attended by vomiting or diarrhœa, by a copious deposit of lithates in the urine, or by epistaxis, or hemorrhage from the uterus or rectum.[2]

Simple continued fever is attended by no constant or characteristic eruption.

---

[1] Aitken: Science and Practice of Medicine. Third American edition. 1872.
[2] Murchison: The Continued Fevers of Great Britain. Second edition. 1873.

An eruption of herpes about the lips and nostrils is often observed at the close of the attack. Convalescence is rapid.

Murchison describes four varieties of this form of fever, as follows:

I. Abrupt seizure with chills or rigors; the febrile action high; quick, full pulse; hot skin; white, furred tongue; great thirst, and no appetite; constipation; scanty, high-colored urine; intense headache, with sometimes restlessness, sometimes drowsiness; pains, as from bruises, in the limbs. The attack comes to an end in twelve, twenty-four, or thirty-six hours, and is properly called *Ephemera*.

II. The pyrexia is occasionally prolonged over several days—rarely, however, exceeding ten. The pulse is frequent, full, hard, and bounding; thirst and the heat of skin are intense; headache is sharp and distressing; delirium sometimes occurs. Termination abrupt, with copious perspiration. This is the Synocha, or Inflammatory Fever of English writers of the last century. It is separable from ephemera only by the difference in duration.

III. The Ardent Continued Fever of the tropics, as observed by Dr. Murchison among the European troops at Calcutta in 1853, and in Burmah in 1854, appeared to be merely an exaggerated form of the now rare synocha of Britain. Young, plethoric persons not yet acclimated were chiefly attacked. The fever prevailed during the hot, dry months, when the mercury usually ranged from 33.3° C. (92° F.) to 41° C. (106° F.) and never fell below 29° C. (84° F.). The symptoms in many cases commence after incautious exposure to the sun. A chill, or occasionally nausea and vomiting, ushered in the attack. To these speedily succeeded the frequent, full pulse, burning skin, flushed face, giddiness, intense headache, ringing in the ears, intolerance of light, restlessness, and sleeplessness, which mark a difference from synocha in degree rather than any difference in kind. About the fourth or fifth day active delirium set in, followed by more or less unconsciousness, with contracted pupils and sometimes complete coma. Between the sixth and ninth days death took place, the patient remaining comatose to the end, or a copious perspiration occurred, followed by an increased flow of urine depositing copious urates, and convalescence. The subsidence of the fever was in some instances followed by sudden, or even fatal collapse.

IV. The term Asthenic Simple Fever is suggested by Dr. Murchison for a variety of the form of continued fever under consideration, in which the febrile action is less intense and the duration more prolonged than in the varieties above mentioned. The patient loses appetite and strength; the pulse is frequent, but rather feeble than tense; the tongue is slightly furred; the bowels are confined; some headache is present, and sleep is disturbed. These symptoms may extend over a period of two or three weeks without change, except as regards the patient's strength, which gradually fails. Such attacks have been known to follow great bodily or

SIMPLE CONTINUED FEVER.

mental fatigue. It is to be borne in mind that these **cases are never fatal, and that enteric fever** often presents the collection **of symptoms just described.**

The following case, observed and narrated by Prof. Flint, is a typical example of febricula **as it is usually** seen in **childhood in the United States:**

A **child, six** or seven years of age, while playing out of doors, apparently **in perfect health,** at noontime complained of illness and was taken home. Soon afterward **the** axillary temperature was 104° F. There was no evidence of any local affection; **no** remedies were prescribed. At midnight the fever was diminished, after seven **hours** it was slight, and at noon the thermometer showed absence of fever. There was no **return** of the febrile condition, although no preventive treatment was employed, and **the** usual health was at once regained.

ANALYSIS OF SYMPTOMS.

The temperature.—The suddenness of the rise and of the **rapidity with** which the maximum, 39.4° C. (103° F.) or 40.5° C. (105° F.), is **reached, have** been pointed out. In cases terminating by well-marked crisis **with critical discharges,** either in the common form of copious perspiration or from the bowels, the decline of the temperature is never so rapid **as its oncoming.** In cases **of longer duration the fall usually takes the form of a prolonged and gradual defervescence.**

The circulation.—The pulse is in almost all cases frequent and full. In the severer forms it is usually tense and binding.

The digestive system.—The tongue is white, furred; thirst is constant, often distressing; loss of appetite is usually complete; the bowels are, **as a rule from which there is little variation, constipated till the termination of the attack;** vomiting, save in cases brought about by excesses in eating or drinking, and occasionally at the onset of the severe variety of the fever met with in the tropics, is uncommon; it occurs in some cases as the febrile action subsides.

The urine.—It is diminished in quantity, dark in color, and of very high specific gravity, 1030–1035, with increase of solids, and particularly of urea. It presents the very type of febrile urine.[1] With the decline of the temperature the volume of urine speedily augments, and copious deposits of urates occur. Albuminuria does not occur. In six cases, examined by Parkes, throughout the **whole course of the disease the urine was** never albuminous.

The skin.—The **face is flushed, the surface hot and dry. There is no** characteristic eruption. Occasionally an erythematous blush is to be ob-

---

[1] Parkes on the Urine. 1860.

served upon the loins and thighs; it disappears with the fever. The eruption of herpes upon the lips and nose is so common at the close of simple continued fever that this disease has by some persons been called *Herpetic Fever*.

The nervous system.—Chills or rigors are rarely absent at the onset, except in young children. Headache is a constant symptom. It is acute

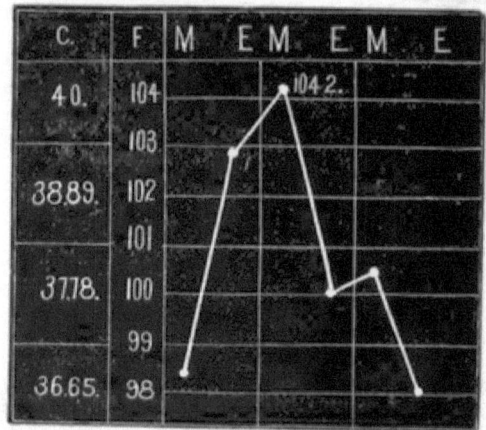

FIG. 1.—Temperature in Simple Continued Fever. (Wunderlich.)

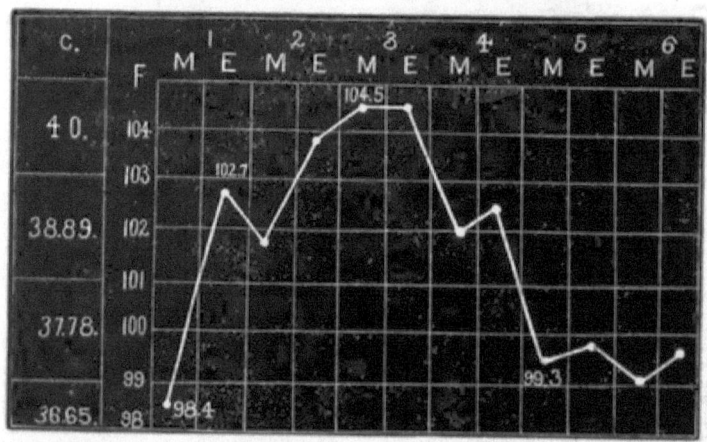

FIG. 2.—Temperature in Simple Continued Fever; more gradual defervescence. (Wunderlich.)

in character, and is sometimes described as throbbing or darting. It is in severe cases intense. Delirium may follow it. Restlessness and sleeplessness are common; on the other hand, the patient is in some instances dull and drowsy. In the variety above described as the ardent fever of hot climates, giddiness, intense headache, ringing in the ears, intolerance of light, muscæ volitantes, restlessness and inability to sleep, pass into delirium, to be followed by stupor with contracted pupils, and this condition may deepen into coma, in which the patient dies.

Duration.

The whole sickness, as in the case of the child seen by Prof. Flint, above narrated, frequently does not last more than a few hours. Its duration, in the mild form of simple continued fever usually encountered in the temperate climates, is from three or four to six days, rarely longer than ten. Several cases seen by the writer in the summer of 1876, in Philadelphia, came to an end with free perspiration about the sixth day. The variety characterized by less active fever, and described as asthenic, may continue two or three weeks.

Diagnosis.

It is obvious, from what has been stated, that the diagnosis of simple continued fever cannot in all instances be positively established. This statement is not only true of the disease when seen early, but it is also true in some cases after the fever has come to its close and the patient has regained his health. A doubt, arising from the absence of sufficient evidence, must under some circumstances remain in the mind of the candid practitioner as to whether the case has been in fact a simple non-specific fever, or a mild, abortive, not well-characterized instance of one of the specific fevers. In order to arrive at a satisfactory diagnosis of simple continued fever, the following considerations are to be taken into account:

The occurrence of the fever after events that are thought to be adequate to cause it, as exhausting over-exertion, exposure to heat, excesses at table, and the like.

The absence of any discoverable local inflammation, or of the history of any recent injury.

The abrupt beginning, without prodromes; the rapid temperature-rise. The early severity of the febrile symptoms, commonly greater at the commencement than in either enteric fever or typhus, may sometimes aid in diagnosis.

The duration, commonly short.

The absence of eruption.

Constipation and the absence of the abdominal symptoms of enteric fever.

The absence of joint-pains, of jaundice, of the enlargement of the liver and spleen, which are early present in relapsing fever.

Its sporadic occurrence and the absence of epidemic diseases.

Prognosis and Mortality.

The prognosis of simple continued fever in temperate climates is in the highest degree favorable as regards a complete recovery. Death rarely if ever occurs. There are no sequels, and perfect convalescence is rapid. Deaths reported as due to this cause are probably the result of enteric fever with latent abdominal symptoms. In the tropics simple fever becomes a formidable and frequently fatal disease.

The *post-mortem* examinations conducted by Dr. Murchison in India revealed "great congestion of all the internal organs, particularly of the lungs, liver, and spleen. The right side of the heart was full of firmly coagulated blood. The sinuses of the brain, and the pia mater were also very vascular, and occasionally there was an increased amount of intra-cranial fluid."

Martin[1] speaks of "congestive states of the cerebro-spinal organs." No characteristic lesions are met with.

Treatment.

The diagnosis of simple continued fever being established, no special treatment is required in temperate climates. The disease tends to recovery. Neither complications nor sequels are apt to occur. The sufferings of the patient may be, however, greatly mitigated by judicious symptomatic treatment.

A purge, to be followed by saline diaphoretics and diuretics, may be ordered. Sponging the surface with cold water, or vinegar mingled with water, is grateful. If the arterial excitement be great, with a full, bounding pulse and throbbing head, aconite may be given in the form of the tincture of the root, gtt. j.—iij., q. s. h., the effect upon the pulse being closely watched. Restlessness and vigil may be relieved by the bromides or by chloral hydrate in gramme doses (gr. xv.), p. r. n. Thirst calls for the unstinted use of the alkaline aërated waters, Apollinaris, seltzer, Vichy, carbonic acid water; weak iced tea with lemon-juice is an acceptable draught. By reason of the short duration of the fever in most cases, the supporting diet is not called for; custards, blanc-mange, jellies, and light broths, are all that the patient requires till with the defervescence appetite returns.

The happiest results have seemed to follow, in my own practice, the treatment of ephemera and febricula in children by purgation, followed by the frequently repeated administration of small doses of chloral hydrate 0.06—0.20 gramme (gr. j.—iij.) quôque horâ vel q. s. h., with cool

---

[1] James Ranald Martin, F.R.S.: The Influence of Tropical Climates. New edition. 1856.

drink as craved. But it is to be borne in mind that the inherent tendency of the sickness is to a speedy, and—as compared with other fevers—an abrupt termination, and that its course is in childhood almost always of brief duration, so that the most guarded deductions are to be drawn as to the apparent success attending measures of treatment aimed at cutting short the duration of the attack. There exists no doubt of the value of treatment in alleviating the urgency of some of the more distressing symptoms.

Quinine, the mineral acids, a nutritious, readily digested diet, and wine are indicated in the so-called asthenic variety of simple continued fever.

The ardent fever of the tropics demands, from its intense pyrexia and the urgency of the danger to life, energetic anti-pyretic measures. Cold affusion—the effect upon the temperature being carefully watched, ice-water enemata, quinine, digitalis, jaborandi, are remedies that would appear to be most likely to do good in a disease often fatal by the very intensity and persistence of the fever-process, and the effect of the high temperature upon the tissues of the body, and more particularly upon the blood and the nervous system. This variety of simple fever is scarcely less closely allied to sunstroke in its pathology than in its causation, and demands, in fact, analogous therapeutic measures.

II.

INFLUENZA.

DEFINITION.—A continued fever, usually of mild intensity, occurring only in widely extended epidemics, and due to a specific cause; it is essentially characterized by early catarrh of the mucous membrane of the respiratory tract, and in many cases also of the digestive tract; by quickly oncoming debility out of proportion to the intensity of the fever and the catarrhal processes; and by serious nervous symptoms. There is a strong tendency to inflammatory complications, especially of the lungs; uncomplicated cases are rarely fatal except in feeble and aged persons. The attack does not confer immunity from the disease in future epidemics.

SYNONYMS.—Febris catarrhalis; Defluxio catarrhalis epidemicus; Catarrhus à contagio; Rheuma epidemicum; Cephalalgia contagiosa; Epidemic catarrhal fever; Tac; Horion; Quinte; Coqueluche; Ladendo, also written La Dando; Baraquette; Générale; Coquette; Cocote; Allure; Follette; Petite poste; Petit courier; Grenade; La grippe; Ziep; Schaffhusten and Schaffkrankheit; Huhner Weh; Blitz-Katarrh; Modefieber; Mal del Castrone. There are also several names indicating its supposed origin; thus, it has been called in Russia "Chinese catarrh;" in Germany and Italy, "the Russian disease;" in France, "Italian fever," "Spanish catarrh," and so forth.

OF these names several are scientific, but the most are popular. The latter seem to be in many instances the more expressive and important. It is indeed a remarkable fact that in two instances at least the popular name for the disease under consideration has found its way widely into medicine and medical literature, almost to the exclusion of the studied terms by which science has sought to designate it; these are "influenza" and "La grippe."

I have omitted from the list of synonyms such obsolete and now meaningless terms as Peripneumonia notha (Sydenham, Boerhaave), Peripneumonia catarrhalis (Huxham), Pleuritis humida (Stoll), as being of interest rather to the student of medical history than to the student of medicine.

Febris catarrhalis, defluxio catarrhalis, catarrhus epidemicus, rheuma epidemicum, are terms which retain with difficulty the place given them in the literature of influenza by the medical authorities of a past century.

Catarrhus à contagio (Cullen) and cephalalgia contagiosa are derived from a view of the nature of the disease which has been the cause of much controversy, and which must, as will be shown farther on, be now regarded as settled by a compromise.

Epidemic catarrhal fever is, with its Latin equivalent, perhaps the most satisfactory of the so-called scientific names for the disease.

In the popular names for the affection there is to be noted an indication of the natural character of some, at least of the peoples who have suffered from its frequent visitation. Among the English it is known as cold, or epidemic cold, or, in deference to medical authority, as catarrh, or epidemic catarrh, and at present, both among the folk and with doctors, as influenza. Englishmen are not then either quick to see in the disease a resemblance to some common circumstance or thing, nor are they disposed to make a joke about it.

The Germans find obvious resemblances. In the labored respiration and the character of the cough they find a suggestion of a common epizoötic affecting the sheep; hence Schaffhusten (sheep-cough) and Schaffkrankheit (sheep-sickness); or, because the cough is like the crowing of a cock and the disturbance of respiration and the rapid prostration suggest some resemblance to a common disease of the domestic fowl, it has been called Huhner Weh (chicken-disease, whooping-cough), and Ziep, which is about equivalent to "pip." They call it also, from its rapid invasion, Blitz-Katarrh (lightning catarrh) and Modefieber (the fever in vogue).

But the French make a jest of everything, and the more serious the subject the better the joke. Hence, they have found a new name for almost every great epidemic of influenza, and each more trivial than the last. Hence, tac (rot); horion (in jest, a blow); quinte, because the spells recur at intervals of five hours (sic); coqueluche (a hood, or cowl), from the cap worn by those suffering from the malady; and so on through the long list given above.

La grippe is from the Polish chrypka (Raucedo); it is thought, however, by some writers, to be derived from agripper (to seize).

Influenza is of Italian derivation. It is said that the disease received this name because it was attributed to the "influence" of the stars, or from a secondary signification of the word indicating something fluid, transient, or fashionable.

Historical Sketch.

Epidemics of influenza have been clearly recorded only since the beginning of the sixteenth century. There are numerous accounts of earlier epidemic diseases resembling it, but they are neither sufficiently particular nor distinctive to warrant us in inferring its undoubted existence from them. It is supposed to be referred to by Hippocrates, who yet gives no exact description.[1] An outbreak in the Athenian army in Sicily (415 B.C.), recorded by Diodorus Siculus, *has been supposed to be influenza*. In spite of these statements, and those of others to the effect that it is a disease known from a remote antiquity, it may be said that no accounts can be confidently established, as referring to the disease now known as influenza, in the writings of classical antiquity.[2]

As early as the ninth century several epidemics of catarrhal fever, Italian fever and the like, which were probably influenza, were made matter of history. In the year A.D. 827, a cough, which spread like the plague, was recorded. In 876 there appeared in Italy a similar epidemic, which spread rapidly over all Europe. It is related that dogs and birds suffered with symptoms not unlike those characterizing the affection in man. In 976, Germany and all France suffered from a fever of which the chief symptom was cough. No epidemic is noted until two centuries later, when, in 1173, a widespread malady, of which the symptoms were chiefly catarrhal, raged in Europe; while less important epidemics of a like character are recorded as having occurred during the following century (1239-1299).

Parkes states that in the fourteenth century there are to be found records of six epidemics, and in the fifteenth seven great visitations of influenza are described.

Aitken[3] speaks of a very fatal prevalence of influenza throughout France in 1311, and of an epidemic in 1403, in which the mortality was so great that the courts of law in Paris were closed in consequence of the deaths.

Influenza is mentioned in the "Annals of the Four Masters" as having prevailed in Ireland in the fourteenth century, and a disease, expressed by similar symptoms, is alluded to in early Gaelic manuscripts under the name of Creatan (creat, the chest). The disease is described also in an Irish manuscript of the fifteenth century, under the terms Fuacht and Slaodan.[4]

The first epidemic that prevailed in the British Isles, of which any ac-

---

[1] Parkes: Reynolds' System of Medicine, vol. i. 1868.
[2] Zuelzer: Ziemssen's Cyclopædia of Medicine, vol. ii. 1875.
[3] Aitken's Practice of Medicine, vol. i. 1872.
[4] Theophilus Thompson: Annals of Influenza. 1852.

curate description remains, is that of the year **1510.** The disease came from Malta and invaded first Sicily, then Italy and Spain and Portugal, whence it crossed the Alps into Hungary and Germany as far as the Baltic Sea, extending westward into France and Britain. Its track widened over the whole of Europe from the southeast to the extreme northwest, and it is said that not a single family and scarce a person escaped it. It was attended by "a grievous pain in the **head**, heaviness, difficulty **of breathing,** hoarseness, loss of strength and appetite, **restlessness,** retchings from a terrible tearing cough. Presently succeeded a chilliness, **and so violent** a cough, that many were in danger of suffocation. **The first day it** was without spitting ; but about the seventh or eighth **day** much viscid phlegm was spit up. Others (though fewer) spat only water and froth. When they began to spit, cough and shortness of **breath** were easier. None died except some children. In some it went off with a looseness ; in others by sweating. Bleeding and purging did hurt."[1] Blisters were commonly employed ; two each upon the arms and legs, and one to the back of the **head.**

The description of influenza is sufficiently clear to place the nature of this epidemic beyond all doubt.

The epidemic of 1557, starting westward **from Asia, spread over** Europe and then **crossed** the Atlantic **to** America. **It circumnavigated** the globe. The malady broke out in England after a season of unusual rain and a period of great scarcity of corn, in the month **of September.** " Presently after **were** many catarrhs, quickly followed **by** a most severe cough, pain of **the** side, difficulty of breathing, and a fever. The pain was neither violent nor pricking, but mild. The third day they expectorated **freely.** The **sixth,** seventh, or at the farthest the eighth day, all who had **that pain of the** side, died ; but such as were blooded on the first **or** second day, recovered on the fourth **or fifth** ; but bleeding on the last two days did no service." "Some, but very few, had continual fevers along with it ; many had double tertians; others simple slight intermittent. All were worse by night than by day; such as recovered were long valetudinary, had a weak stomach and hypped." Gravid women either aborted or died. This epidemic spread with frightful rapidity. Thousands **were** attacked at the same moment. The entire population of Nismes, with scarcely an exception, fell ill of **it upon the same day. It was** extremely fatal. In Mantua Carpentaria, a small town near Madrid, it broke **out** in August, **and so** fatal was the bloodletting **and** purging which **constituted the treatment at first, that of** the two thousand persons who were bled, all died.[2] The disease raged in some **parts** till the middle

---

[1] **Thomas Short:** A General Chronological History of the Air, **Weather, Seasons, Meteors,** etc. London, 1749. Quoted in the Annals of Influenza.

[2] Dr. Short : loc. cit.

of the following year (1558), and carried off, in Delph alone, five thousand of the poor. In all cases mild treatment was called for, with warm broths and speedy immersals "to recall the appetite, and keep the vessels of the throat open."

In 1580 a great epidemic of influenza spread from the **southeast toward the northwest over Asia, Africa, and Europe.** From Constantinople and Venice it overran Hungary and Germany, and reached the farthest regions of Norway, Sweden, and Russia. It spread into England, and has been described by Dr. Short. In Italy it prevailed during August and September, in England from the middle of August to the end of September, and in Spain during the whole summer. In most places its duration was about six weeks. As a rule the termination was favorable, although the disease ran a somewhat protracted course. In the account of Dr. Short it is stated that "few died except those that were let blood of, or had unsound viscera." In some places, on the contrary, the course of the disease was very severe. In Rome two thousand died of it, according to the author just cited, but Zuelzer informs us that the victims of this epidemic in the Eternal City were not less than nine thousand, and adds that Madrid must have been almost depopulated by it. This high mortality has been attributed to the bloodletting practised in the treatment of the disease. The symptoms were similar to those of the previous epidemics, with a greater shortness of breath, which continued in many cases for some time after the disappearance of the catarrhal trouble. There was great sweating at the end of the attack.

The plague, measles, and small-pox prevailed also, and with considerable violence during the year 1580.[1]

The disease, unfelt for several years, reappeared in Germany in 1591; an epidemic, extending from Holland through France and into Italy, occurred in 1593; in 1610, catarrh is said to have prevailed throughout Europe. In 1626-27, epidemic catarrhal fever made its appearance in Italy and France; in 1642-43, in Holland; in 1647, in Spain and in the colonies of the Western World, and again in 1655, in North America.

According to Webster,[2] this epidemic of 1647 was the first catarrh mentioned in American annals.

In 1658 and 1675, it again visited Austria, Germany, England, etc. The first of these two epidemics is described by Willis,[3] and the second by Sydenham,[4] as they occurred in England, and the accounts are to be

---

[1] Theophilus Thompson: **Annals of Influenza.**
[2] **Noah** Webster: A Brief **History of** Epidemic and Pestilential Diseases. London, 1800.
[3] Dr. Willis: The Description of a Catarrhal Fever Epidemical in the Middle of the Spring in the Year 1658. Practice of Physic. 1684.
[4] The Epidemic Coughs of the Year 1675, with the Pleurisy and Peripneumony that supervened. From the Works of Thomas Sydenham. M.D.

found in the already oft-quoted "Annals of Influenza." It is about this period that the disease began to be known as influenza, and it is not without interest to observe that the "influence" of the stars suggested itself, in connection with its sudden appearance and wide prevalence, to the minds of the physicians of this date. Willis writes that "about the end of April (1658), suddenly a distemper arose, *as if sent by some blast of the stars*, which laid hold on very many together; that in some towns, in the space of a week, above a thousand people fell sick together."

Epidemics are recorded as having occurred in Great Britain and Europe in 1688, 1693, and in 1709. The disease raged in 1712 widely over Europe from Denmark to Italy.

In 1729-30, a widespread epidemic swept over Europe. In five months it extended over Russia, Poland, Germany, Sweden, and Denmark. In Vienna sixty thousand persons fell ill of it. In the autumn it spread to England, and reached France and Switzerland; from there it extended to Italy, and by February it had reached Rome and Naples. Spain did not escape its ravages, and it is said to have found its way to Mexico. The symptoms did not differ in any important respect from those already described as characterizing previous epidemics. Pains in the limbs and fever marked the onset of the attack; catarrh, oppression, hoarseness, cough, followed. In some cases, delirium, drowsiness, and faintings occurred. A petechial eruption was observed, in some instances, between the fourth and seventh days. Zuelzer suggests that spotted fever may have prevailed at the same time. Turbid urine, copious sweats, bilious stools, and nose-bleeding were often noted. In Switzerland, only children and old persons died. The disease was not very fatal.

Two years later (1732-33) an epidemic, starting from Saxony and Poland, overran Germany, Switzerland, and Holland, and invaded Great Britain in the month of December. Toward the end of January it spread in a southeasterly direction to France, Italy, Spain, and westward to North America, thence southward to the islands of the West Indies, and on to South America. The course of the disease in this epidemic was favorable. It terminated in from three to fourteen days, with sweating, bleeding from the nose, or an abundant discharge from the nasal passages. The aged and those suffering from chronic pulmonary diseases mostly perished. In Scotland three forms of the affection were described, namely: the cephalic, the thoracic, and the abdominal. The epidemic slowly spread over Eastern Europe and in a southeasterly direction, and may be said to have lasted till 1737.

Concerning this epidemic, John Huxham, of Plymouth, wrote as follows:[1] "About this time a disease invaded these parts, which was the

---

[1] Observations on the Air and Epidemical Diseases. Translated from the Latin. London, 1758.

most completely *epidemic* of any I remember to have met with; not a house was free from it; the beggar's hut and the nobleman's palace were alike subject to its attacks, scarce a person escaping either in town or country; old and young, strong and infirm, shared the same fate." The malady had raged in Cornwall and the western parts of Devonshire from the beginning of February; it reached Plymouth on the 10th, which was on a Saturday, and that day numbers were suddenly seized. The next day multitudes were taken ill, and by the 18th or 20th of March scarcely any one had escaped it.

"The disorder began at first with a slight shivering; this was presently followed by a transient erratic heat and headache, and a violent and troublesome sneezing; then the back and lungs were seized with flying pains, which sometimes attacked the heart likewise, and though they did not long remain there, yet were very troublesome, being greatly irritated by the violent cough which accompanied the disorder, in the fits of which a great quantity of a thin, sharp mucus was thrown out from the nose and mouth. These complaints were like those arising from what is called *catching* cold, but presently a slight fever came on, which afterward grew more violent; the pulse was now very quick, but not in the least hard and tense like that in a pleurisy; nor was the urine remarkably red, but very thick, and inclining to a whitish color; the tongue, instead of being dry, was thickly covered with a whitish mucus or slime; there was an universal complaint of want of rest and a great giddiness. Several likewise were seized with a most racking pain in the head, often accompanied by a slight delirium. Many were troubled with a *tinnitus aurium*, or singing in the ears; and numbers suffered from violent earaches, or pains in the *meatus auditorius*, which in some turned to an abscess. Exulcerations and swellings of the fauces were likewise very common. The sick were in general very much given to sweat, which, when it broke out of its own accord, was very plentiful, and continued without striking in again, and did often in the space of two or three days wholly carry off the fever. You have here a description of this epidemic disease such as it prevailed hereabouts, attacking every one more or less; but still, considering the great multitude that were seized by it, it was fatal to but few, and that chiefly infants and consumptive old people. It generally went off about the fourth day, leaving behind a troublesome cough, which was very often of long duration, and such a dejection of strength as one would hardly have suspected from the shortness of the time.

"On the whole, this disorder was rarely mortal, unless by some very great error arising in the treatment of it; however, this very circumstance proved fatal to some, who, making too slight of it, either on account of its being so common, or not thinking it very dangerous, often found asthmas, *hectics*, or even consumptions themselves, the forfeitures of their inconsiderate rashness."

Arbuthnot also described this visitation of the disease.[1] He regarded the uniformity of the symptoms in every place as most remarkable, and tells us that, during the whole season in which it prevailed, there was "a great run of hysterical, hypochondriacal, and nervous distempers; in short, all the symptoms of relaxation." Most observers looked upon the continued changes of temperature as active in producing this widespread and long-continued epidemic.

During the years 1737-38, influenza again swept over England, North America, the islands of the West Indies, and France; in 1742-43, it prevailed in Western Europe and the British Isles; in 1757-58, in North America, the West Indies, France, and Scotland. In 1761, it overran the North American Colonies and the West Indies.

The epidemic of 1762 extended very generally over Europe and Great Britain. In Germany nine-tenths of the population were attacked by the disease.

Widely extended epidemics prevailed in Europe and America in 1767 and 1775; in 1772 it raged in North America; in 1778-80 in France, Germany, and Russia. Noah Webster found influenza prevalent in North America in 1781; the next year, one of the most remarkable epidemics of this disease (described as the epidemic of 1782) appeared in Europe. It came from the East, from Asia into Russia. From St. Petersburg it spread during the winter and spring over Sweden, Germany, Holland, and France. In the autumn it was in Italy, Spain, and Portugal. The crews of Dutch and English ships were taken ill with the disease upon the high seas.

In Vienna three-fourths of the population fell ill of it with such suddenness that it got here for the first time its name of "Blitz-katarrh" (lightning catarrh). It was characterized by great pain in the back, breast, and throat, and by extraordinary enfeeblement. Relapses occurred, and inflammation of the lungs and bowels were common. Children remained relatively exempt from its seizure. This epidemic broke out in England about the end of April, and raged until the end of June. "The duration of the malady in some was not above a day or two; but it usually lasted near a week or longer. In a few the symptoms seemed to abate in two or three days, but some returned and raged with more violence than at first."[2] The disease was not regarded as in itself fatal, and few could be said to have died of it, "but those who were old, asthmatic, or who had been debilitated by some previous indisposition." Yet its influence upon the weekly bills of mortality in London, where it made its

---

[1] An Essay Concerning the Effects of Air on Human Bodies. London, 1751.
[2] An Account of the Epidemic Disease called the Influenza of the Year 1782. Collected from the Observations of several Physicians in London and in the Country, by a Committee of the Fellows of the Royal Colleges of Physicians in London. Read at the College, June 25, 1783.

appearance between May 12th and 18th, was so great that it seems worth while to transfer the record from the report of the College of Physicians to these pages.

The total weekly returns stand thus:

Tuesday, May 7th............ 299	Tuesday, June 11th.......... 560
" " 14th............ 307	" " 18th.......... 437
" " 21st............ 336	" " 25th.......... 434
" " 28th............ 390	" July 2d............ 296
" June 4th............ 385	

Numerous recurring outbreaks took place in Europe and America during the years 1788–90. One of these, as it occurred in America, is well described by Dr. John Warren,[1] of Boston, in a letter to Lettsom. This letter is dated May 30, 1790, and among other matters of great interest respecting the disease, it is stated that "Our beloved President Washington is but now on the recovery from a very severe and dangerous attack of it in that city" (New York).

Webster mentions an epidemic in America in 1790, one in Europe in 1795, and another in Europe in 1797; but there seems to have been no general epidemic of sufficient importance to attract the attention of other writers upon the subject until 1798, when the malady again broke out in Russia and spread over the greater part of Europe, continuing to prevail in various regions till 1803, when it again appeared in England, and is described by several writers of that country.

From 1805 to 1827, influenza prevailed (according to Zuelzer, who tells us that few years during this interval were free from it) in frequently recurring epidemics in Europe and America. Thompson mentions no visitation in England between 1803 and 1831.

In the year 1830, began a series of epidemics remarkable for their wide diffusion and the rapid succession with which they followed one upon another. The disease began in China; in September it reached the Indian Archipelago; it swept into Russia, and invaded Moscow in November; in January, 1831, it was raging in St. Petersburg; March found it in Warsaw; April in Eastern Prussia and Silesia; in May it prevailed in Denmark, Finland, and a great part of Germany, and in the same month it fell upon Paris; in June it affected England and Sweden; it still was creeping about Middle Europe, and lingering in Great Britain at the end of July; in the early winter it swept southward into Italy, and westward across the Atlantic to North America, and was still harassing the inhabitants of certain regions of the United States in January and February, 1832. Meanwhile it continued in the East, spreading to Java, Farther India, and the

---

[1] Thomas Joseph Pettigrew: Memoirs of the Life and Writings of J. Coakley Lettsom. 1817.

Indian Archipelago. It continued in Hindostan after it had died out in Europe. But in January, 1833, it again visited Russia and rolled thence southward and eastward over the most of Europe. It is recorded that by February it had reached Galicia and Eastern Prussia; in March it was in Prussia, Bohemia, and Warsaw, and had extended to Syria and Egypt; in April to many parts of Germany and Austria, and to France and Great Britain. Midsummer found the disease yet prevailing in some districts of Germany and Northern Italy, and in the early autumn it was in Switzerland and Eastern France; in November it visited Naples. Epidemics so frequent, so widespread, and so unsparing of individuals wherever the disease appeared, could not fail to excite a deep and general interest. From this period the literature of the subject has been voluminous.

A brief period of repose ensued. For three years no epidemic occurred which was of sufficient importance to attract the attention of medical historians.

In December, 1837, influenza reappeared, and first, as so often before, in Russia; Sweden and Denmark were almost simultaneously affected; in January, 1837, it broke out in London, and rapidly swept over all England, and into France and Germany. In January it appeared in Berlin and shortly afterward in Dresden, Munich, and Vienna. The disease spread by February into Switzerland, and into Spain as far as Madrid by the end of March. In London almost the whole population was attacked, and the mortality was enormous. Dr. Bryson[1] states that the deaths were quadrupled during the prevalence of the disease. Large populations suffered most.

This epidemic spread into the southern hemisphere and prevailed at the same time, and consequently at exactly the opposite season that it prevailed north of the equator, in Sydney and at the Cape of Good Hope.

From 1837 till 1850–51, numerous epidemics of influenza occurred. Few years were exempt from them. The epidemic of 1847–48 has been described by many writers, and more particularly, as it occurred in London, by Peacock[2] with great exactitude. It is estimated that one-fourth of the entire population of that city were more or less affected by the disease. The epidemic prevailed in London for six months, and although the deaths registered for the entire period, as from influenza, amounted to only one thousand seven hundred and thirty-nine, it is stated in the report of the Registrar-General that during the six weeks the epidemic was at its height, not less than five thousand persons died, in the metropolitan districts, in excess of the average mortality of the period, the excess showing itself in nearly every class of disease; the local maladies

---

[1] Annals of Influenza.
[2] Thomas Berill Peacock, M.D. : On the Influenza, or Epidemic Catarrhal Fever of 1847–48. 1848.

which had been the predominant affections being doubtless, in many cases, assigned as the cause of death.

This epidemic affected between one-fourth and one-half of the population of Paris, and in Geneva the proportion of those attacked was not less than one-third of the entire population.

More or less widespread epidemics of influenza are recorded as having occurred in 1857–58 and 1860; in 1864 in Switzerland; in 1867 in Paris in the spring; and at various times in the United States and Canada.

A mild epidemic occurred in 1874, in Berin.

Influenza prevailed over a wide area in the United States during the early months of 1879. The characteristics of this visitation have been well described by Da Costa.[1]

For the most part the disease, since the great epidemic of 1847–48, has affected a smaller proportion of the inhabitants of the localities visited, and has run a less dangerous course than in the epidemics previously described. It has for this reason occupied a less conspicuous place in the medical literature of recent years. It is nevertheless true that even in the mildest epidemics, when a relatively small number of persons are seized, and the symptoms are in most cases almost insignificant, cases do here and there occur which are of a serious or even fatal character, and that the death-rate from other diseases is for the time considerably increased.

Affections of a catarrhal kind have frequently prevailed among the domestic animals at the same time that influenza has been epidemic. Horses, dogs, and cats are subject to these disorders; neat cattle, goats and sheep have been more rarely affected; chickens and pheasants have suffered, and it is stated by some of the older writers that birds, and particularly the sparrow, have deserted localities in which influenza was prevailing, while migratory birds have taken flight earlier than usual.

These epizoötics have sometimes preceded the appearance of influenza among men by a period of some weeks or days, in other instances they have appeared contemporaneously; and in a widespread outbreak among horses in the United States in 1872, in which the symptoms and morbid anatomy, accurately observed, were undoubtedly those of influenza,[2] the disease did not affect man except to a very limited extent. A want of fulness of description, and the inaccuracy of diagnosis common in the consideration of general diseases of the lower animals, leave the precise nature of most of the epizoötics described by the earlier writers in great uncertainty.

An extensive but mild influenza has prevailed as an epizoötic, chiefly

---

[1] The Prevailing Epidemic of Influenza—Its Characteristic Phenomena—Pulmonary, Gastro-intestinal—Cerebral and Nervous—Its wide Distribution, Mortality, and Treatment. Medical and Surgical Reporter. Philadelphia, March 8, 1879.

[2] F. Woodbury, M.D.: Morbid Anatomy of the Epizoötic. Philadelphia Medical Times, December 14, 1872.

affecting horses, during the latter part of the summer and the autumn of 1880, in Canada and the United States east of the Mississippi River. Dogs were also affected, but less generally, and human beings to a still slighter extent. In several localities where this invasion of the disease was observed by the writer, the horses were first affected, the dogs next, and after the lapse of some weeks, as the animals were recovering, it became epidemic; but those persons who took care of horses, and were much in contact with them, neither suffered earlier nor more severely than others not so exposed.

Etiology.

1. *Predisposing Causes.*

Large as has been the place in medical literature occupied by the histories of epidemics of influenza, the nature of the "epidemic influence" which gives rise to the disease is still unknown.

There are no well-established facts upon which to base the existence of individual peculiarities that may be regarded as *predisposing causes*. When the disease appears, a large proportion of the population is attacked without distinction of age, sex, social condition, or occupation. Previous illness, whether acute or chronic, local or constitutional, affords no protection.

Aged and infirm persons, and those of nervous temperament, are thought to be especially liable to attack; but the robust possess no immunity. All races and dwellers in every climate are the victims of influenza. In a community invaded by the disease, females are apt to be the first attacked, the adult males next, and the children last. It has been observed that in some epidemics children are but little liable to be attacked.

An attack in one epidemic confers no exemption from the disease in another epidemic, and, independently of relapses, which are not infrequent, persons have been known to experience a second attack during the prevalence of the same epidemic.

Persons dwelling in overcrowded and ill-ventilated habitations, and in low, damp, and unhealthy situations, have in certain epidemics especially suffered, and, according to the report of the Registrar-General, the increase of deaths by influenza during the epidemic of 1847, in England, was much greater in the districts in which there is ordinarily a high mortality than in healthier places. This, as Dr. Parkes observes, must indicate greater prevalence or greater severity of the disease.

Influenza appears at all seasons of the year, and affects every latitude, though it is somewhat more common in cold climates. It has no connection with known atmospheric conditions. Many of the earlier writers sought to establish a causative relation between low temperatures and sudden variations of temperature and influenza, and, by reason of the confusion in the minds of the people between the disease and common "colds," there has always existed an opinion that such a relation obtains. There is no evidence to sustain this view, and all the later writers upon this subject concur in the statement that neither low temperature nor abrupt changes give rise to the affection. It has prevailed in hot and dry seasons, in the West Indies, on the sea-coast of Java, in India, in Egypt, at the Cape of Good Hope, in the Riviera in summer.

The condition of the air, as regards moisture or dryness, does not influence the spread of the disease. It has occurred at sea, on low sea-coasts, and in the dry air of Upper Egypt.

Prolonged east and northeast winds have often prevailed at the time of influenza. This fact is in accord with the observation that many epidemics have extended from east to west and southwest, as, for example, from Russia over Europe. The spread of the disease is, however, not influenced greatly by local winds. It does not move with the same velocity, and even sometimes moves against them. In several well-authenticated instances a dense and foul fog has preceded or attended the outbreak of epidemics. The much greater number of epidemics that have occurred altogether without such manifestations make it in the highest degree probable that this has been a coincidence. Ozone in large quantities, artificially produced, may give rise to the symptoms of ordinary catarrh, but it is not a cause of influenza. The disease is not in any way connected with the condition of the soil, elevation, volcanic eruptions, or any other local cause. The history of every epidemic goes to prove this statement. Moreover, without this assumption, its diffusion over whole countries and continents—indeed, over several quarters of the globe—would be beyond our comprehension.

Before taking up the consideration of the exciting causes of influenza, it is necessary to state the known facts concerning the march of epidemics and the spread of the disease in affected localities. It has prevailed with greater or less frequency in most parts of the world. Epidemics have recurred at irregular periods. It was at one time thought that the course of the disease was cyclical, with a return at intervals of about one hundred years. This view was long ago proved to be unfounded. About every twenty-five or thirty-five years great epidemics have swept over vast areas of the globe, and influenza may be said to be, at such times, pandemic. Less widely extended epidemics have taken place with greater or less frequency in the intervals of the great outbreaks. But it is not possible to establish anything like a cycle by which the returns of the disease are governed.

It has been supposed in some instances to prevail within restricted localities, as, for example, in a single city, but it is probable that such local epidemics are due to local causes, and that they are of the nature of simple ordinary catarrhal fever, rather than true influenza.

The epidemics extend in great areas, usually in a direction from the east or northeast toward the west and south. At other times they take the opposite direction, and in some years they have appeared to radiate in various directions from several centres. It is in consequence of these facts in regard to the spread of influenza, that two views have arisen in the minds of scientific men concerning the origin of the affection. The first of these is that each epidemic starts out from some single unknown source, and spreads thence from point to point, invading more distant localities successively as it advances, until at length it dies out in regions most remote from the starting-point. This opinion is in accord with the popular belief. Thus, the Italians have called it the German disease; the Germans, the Russian pest; the Russians, the Chinese catarrh; and the geographical relation of these nations indicate the usual track of the great epidemics, as shown in the foregoing historical sketch.

The other opinion is that it arises not from a single place, but may start anywhere, and that a widespread epidemic may be due to the successive outbreak at many distinct points of origin.

The evidence that the great epidemics of influenza are due to some general and pandemic influence, is to my mind conclusive. The point of origin of the great epidemics has not yet been indicated with precision, and must remain beyond conjecture until further facts bearing upon the question of their source are brought to light.

When it has prevailed over a large portion of the earth's surface, its progress from place to place has usually been rapid. In this respect, however, the epidemics show a great diversity. It sometimes travels exceedingly slowly. It is said to have overrun Europe in six weeks, and it has again taken six months to do so. It sometimes attacks places widely remote from each other within short intervals of time, and it has appeared at the same time in different quarters of the globe. It does not follow the great lines of travel and commercial intercourse.

When the influenza enters a city, it continues to prevail, as a rule, from four weeks to two months, but exceptionally it remains a longer time; for example, the epidemic of 1831 was prevalent in Paris for the greater part of a year. It in all instances finally disappears, and sporadic cases do not occur in the intervals of the epidemics.

In rare instances, however, the epidemics are heralded by sporadic cases. But commonly they seize simultaneously upon numbers of the inhabitants of affected districts, so that, when the epidemic is severe, the sick are in a short time to be counted by thousands, and business is paralyzed as by a blow. They rapidly reach their height and subside almost as suddenly as they began. In a large city the disease frequently, perhaps always, makes its appearance nearly at the same time in several different localities, affecting certain streets and quarters solely or more generally than others for a time, and spreading thus from several centres through the entire community. Large towns and cities are generally affected earlier than the villages around them, and the latter, though closely adjacent, sometimes escape for weeks.

The crews of ships upon the high seas, not sailing from an infected port, are said to have suffered from the seizure, and epidemics have crossed the Atlantic from the Old World to the New, and in some instances in the opposite direction.

2. *The Exciting Cause.*

The question of the contagiousness of influenza is one of grave interest, and has been the subject of much controversy. The great rapidity of the spread of epidemics, the vast areas they overrun, the fact that they do not follow the lines of human intercourse, the suddenness with which great numbers of the inhabitants of an invaded district or city are seized, the fact that the most complete seclusion from intercourse with affected persons, or even the shutting up of houses, affords in most instances no protection whatever, all go to show that the disease spreads, in the main, independently of indirect contact, and this opinion has been almost universally entertained. On the other hand, there is evidence to show that the disease is to some extent contagious; and so convincing have the facts bearing upon this point appeared to some, that they have believed it to be propagated entirely by human intercourse. Haygarth[1] declares, as the result of his observations during the epidemics of 1775 and 1782, that the influenza spreads "by the contagion of patients in the distemper;" and Falconer,[2] writing of the epidemic of 1803, says, "I have no doubt that it is contagious in the strictest sense of the word." Watson[3] regards the instances in which the complaint has first broken out in those particular houses of a town at which travellers have arrived from infected places, as too numerous to be attributed to mere chance. Very often those dwelling near the invalids are attacked next in the order of time, and when the disease affects a household all do not usually manifest the symptoms at the same time, but one member after another is stricken down with it.

In some rare cases the isolation or seclusion of a community has appeared to give protection, as in cloisters, prisons, garrisons, and the like; at all events, there are instances on record where segregated communities of this kind have escaped attack. This is, however, merely negative evidence, and cannot carry conviction.

---

[1] John Haygarth, M.D., F.R.S.: On the Manner in which the Influenza of 1775 and 1782 Spread by Contagion in Chester and its Neighborhood.

[2] William Falconer, M.D., F.R.S.: An Account of the Epidemic Catarrhal Fever, Commonly called the Influenza, as it appeared at Bath in the Winter and Spring of the Year 1803. Bath, 1803.

[3] Principles and Practice of Medicine.

A recent, carefully conducted observation, under somewhat unusual circumstances, shows that influenza may be brought from an infected city in such a way as to give rise to a localized outbreak in a remote community, in which, however, the disease, in the instance under consideration, did not become epidemic.

Drs. Guitéras and White [1] narrate that, influenza prevailing in Europe, and particularly in Paris and London, an American gentleman in bad health contracted the disease in London, improved, suffered a relapse shortly afterward in Paris, and died there at the end of December, 1879. His body was embalmed and sent home. Following the exposure of the remains of this person to the view of his family in Philadelphia, there was an outbreak of influenza with characteristic symptoms, which affected, in the first place, members of that family; afterward, friends living in close intercourse with them, next the medical attendants of some of them, and finally the housekeeper, and a patient or two of one of the physicians who wrote the paper, the whole number affected in Philadelphia being eighteen, at the time of the publication of the account. Subsequently two or three other cases were developed, but the disease did not extend beyond the immediate circle of those in direct communication with the invalids.

Between those holding the opinion that influenza is not contagious, and those imbued with the opposite view, there must be, it seems to me—regard being had to the foregoing facts—a compromise of the question which is to be based upon the degree of contagiousness. This will be conceded by all modern authorities to be but slight.

Influenza has been supposed to develop at once without a period of incubation, persons in perfect health being struck down with it as by lightning-stroke. It is now ascertained that a period of incubation, varying from a few hours to several days, and usually without subjective symptoms, exists. Numerous instances are recorded in which persons coming into an infected city have remained well for one, two, or three days, but have eventually shared the sufferings of those into whose midst they had come. There are cases also in which the period of incubation could not have been less than two or three weeks. There is no sufficient evidence of a genetic or causal relation between influenza and any other epidemic disease. The statement that other prevalent diseases abate in frequency and intensity upon its outbreak, is not borne out by well-observed facts. Graves [2] holds that those suffering with acute diseases are less liable during the febrile stage, but that they are attacked as convalescence sets in.

---

[1] John Guitéras, M.D., and J. W. White, M.D.: A Contribution to the History of Influenza; being a Study of a Series of Cases. Philadelphia Medical Times, April 10, 1880.

[2] Clinical Medicine.

Some writers have thought that an attack of influenza may degenerate into intermittent fever. It is more probable, that the instances observed were endemics of intermittents, making their appearance upon the subsidence of epidemics of influenza.

The facts in reference to the spread of epidemics of influenza and the course of the disease in infected localities, are comprehensible upon no other theory than that of a specific principle of disease as its exciting cause. What this principle may be, is not yet known to us; where it originates is equally unknown, and our knowledge of the influences that from time to time call it into activity, and send it forth in definite directions over the earth, is no less negative.

So general a disease can only be disseminated by the most general medium, the atmosphere, and its exciting cause must be capable of reproducing itself in that medium; otherwise it would be lost by dispersion in traversing distances measured by the boundaries of continents and oceans. The rapid diffusion of influenza, sweeping over continents in a few weeks at one time; its slow migration, creeping about a city and its environs for months at another, are, as Biermer[1] suggests, to be most easily explained upon the theory of a living miasm, capable of being transmitted by the air, and possessing at the same time an independent existence. Such an entity would find certain localities more favorable to its growth, reproduction, and prolonged existence, than others. From this point of view influenza is a miasmatic disease.

From a fair consideration of what has been written concerning its local dissemination, it must be admitted that its causes are, to a slight extent, capable of being reproduced in or about the human body, and transmitted by personal intercourse, as well as conveyed from place to place by the persons or clothing of those affected, or travelling from localities in which the disease prevails.

We are thus led to the conclusion that it is also contagious, though feebly so.

Influenza, in view of these theories of its exciting causes, may be described as a miasmatic-contagious fever.

CLINICAL HISTORY.

The course of the disease, in individual cases, presents the greatest variations as regards intensity, from the most trifling indisposition to an illness of the gravest kind, terminating in death.

These variations are dependent upon: 1st, the previous health of the individual, his age, and the power of resisting depressing influences which he

---

[1] Biermer: Virchow's Handbuch der speciel. Pathologie u. Therapie. Band V. 1te Abth. 4te Lieferung. Erlangen, 1865.

possesses; 2d, the energy and the amount of the specific cause of the disease to which he has been exposed—in other words, the dose of the fever-producing poison; and 3d, the character of the prevailing epidemic.

It is, however, important to observe that, as has already been stated, cases of very great severity are occasionally encountered during the prevalence of mild epidemics.

In every epidemic, on the contrary, a considerable part of the community suffers from influenza in the mildest, or what has been called the "rudimentary" form. This is characterized by general "malaise," an easily oncoming weariness of bodily and mental effort, a disinclination for business, some inability to fix the attention, and slight mental confusion; to these nervous disturbances are added slight catarrhal symptoms, as coryza, sore throat, a tickling cough, and the like; but the indisposition is subfebrile—it does not amount to a fully developed fever.

Another portion of the cases in most epidemics present the symptoms of an ordinary attack of acute coryza, laryngitis, bronchitis, pharyngitis, with great increase in the constitutional disturbances, distressing headache, and pains in the back and limbs. The fever in this class of cases does not range high, yet the patients are ill enough to betake themselves to bed.

The onset of the attack in severe cases is usually abrupt. It begins with shivering or a chill, or with fits of chilliness alternating with heat. Fever is rapidly established. It is usually moderate, though it sometimes reaches a high grade. It shows a tendency to morning remissions. Sensations of chilliness are apt to occur; they are called forth by even slight changes in the external temperature. These chilly sensations are apt to be followed during the course of the fever by the sensation of flushes of heat, and are, in many cases, attended by annoying sweats.

The febrile outbreak is sometimes preceded for a little time by intense frontal headache, with pain in the orbits and at the root of the nose. In other cases these pains quickly follow the chill. Sneezing, redness of the eyes and edges of the nostrils, a more or less abundant thin discharge from the nose, and lachrymation, now occur. In some instances there is bleeding from the nose. The throat becomes sore, there is a tickling sensation in the upper air-passages, a dry cough sets in, attended by more or less hoarseness and shortness of breath. The cough is paroxysmal, hard, distressing. It sometimes causes vomiting, like that which occurs in the paroxysms of whooping-cough. Chest-pains, stitches in the side (not pleuritic), frequent sneezing, loss of the sense of smell and of taste, attend the development of the general catarrhal manifestations.

The fever is attended by great depression, pains in the limbs, loss of appetite, thirst, constipation, and diminished secretion of urine. The pulse is full, but, as a rule, only moderately increased in frequency. There is in many cases slight, or even decided blueness of the lips and fingertips. The patient is distressed by restlessness and want of sleep. At the end of four or five days the febrile symptoms decline, at times gradually, oftener rapidly, with copious sweats or spontaneous flux from the bowels. The fever continues, however, when severe complications have taken place ten or twelve days. The defervescence is marked by an increased flow of sedimentary urine, and considerable amelioration of the subjective symptoms. The catarrhal symptoms outlast the fever two or three days, but cough and expectoration may not disappear for some time.

Attendant upon these symptoms and proportionate to the severity of the fever or the catarrh, or both, that is to say, in proportion to the gravity of the attack in general, are the evidences of functional disturbance of the nervous system. There is remarkable nervous depression, loss of strength and lowness of spirits, combined with mental weakness, and even stupor and delirium. In some cases slight convulsions take place. Cutaneous hyperæsthesia occasionally occurs, and Da Costa states that areas of burning pain in the skin are to be met with. Neuralgia, muscle-pain, and aching referred to the bones, are very common and often severe.

In other cases abdominal symptoms are prominent, while those referable to the head and chest are less urgent. The disease assumes the guise of a more or less severe catarrh of the gastro-enteric mucous membrane, with hepatic disturbance. The fever and the peculiar nervous depression, spoken of in the foregoing account of the course of the affection, are the same. Cases likewise present themselves, but less commonly, in which but little of the usual tendency to localization of the catarrhal processes is to be observed; there is fever of varying intensity, with great depression, and simultaneous and equal implication of the head and the organs of the chest and abdomen.

Many writers have sought to arrange the foregoing different forms of influenza in definite categories. It would be a useless task to reproduce their views upon the subject, or even to enumerate the varieties that they have described. In truth, it is open to doubt whether it would serve any useful didactic purpose to do so, while in practice the various described types merge so gradually into each other, and are so modified by the individual peculiarities of the sick, and by the complications which arise in the course of the attack in consequence of such peculiarities or of previously existing diseases or tendencies to special forms of disease, that particular cases cannot in most instances be referred to theoretical categories. In illustration of this remark, it is to be stated that hysterical persons and those of what we may term a nervous constitution, are prone to suffer especially from the peculiar nervous symptoms of influenza. So also the disease is modified by the age of the subject of the attack, and children manifest, in a high degree, the signs of cerebral congestion, while old persons are subject in a peculiar manner to dangerous pulmonary complications, and those of a gouty or rheumatic constitution suffer from muscular pains more than others.

The duration of the mildest form of influenza is from two to three days; in well-developed cases without complications, convalescence sets in between the fourth and tenth days, and severe cases with complications may last much longer—several weeks elapsing before recovery is complete.

Analysis of the Symptoms.

For the purpose of separate consideration, it will be found convenient **to take** up the symptoms belonging to the fever first, then those of the special catarrh, and finally those more particularly referable to the nervous system; but we encounter, in the present state of our knowledge of the pathology of influenza—or it would be perhaps better to say, our ignorance of its pathology—no little difficulty in deciding under which of these headings particular symptoms are **properly to be** classed, by reason of the close interdependence **of the chief processes of the disease and the anomalies of its phenomena viewed as a whole.**

The Fever.

Temperature.—The older observers **concluded, from the diminished** frequency of the pulse by day as compared with that of the night, the less urgent subjective symptoms and the relatively cooler skin, that the type of the fever was remittent or subcontinuous. This is doubtless the case, although accurate thermometric observations, by which alone the type of any fever can be with certainty established, are as yet, even in the most recent epidemics, wanting in sufficient numbers to enable us to formulate any law.

The intensity of the fever-process **is variable.** As a rule it is moderate or slight; occasionally it is intense. I observed, in several cases during the epidemic of 1879 in Philadelphia, an evening temperature of only 39° C. (102.2° F.). Da Costa, in the same outbreak, found the febrile movement not high; the highest temperature he observed was 40° C. (104° F.). Biermer found a temperature of over 39° C. in moderate cases of catarrhal fever, and does not doubt that under certain transient conditions the temperature **may reach** the height of that of pneumonia or typhus. In weakly persons and the aged, the fever is adynamic.

The circulation.—The pulse is variable. Its frequency is moderately increased; it is sometimes full, sometimes weak. **It has no constant** character. Some observers have noted a frequent irregularity. Graves informs us that in many cases the condition of the pulse was very changeable, and that it often became quite different in character in the course of a few hours.

The secretions.—The urine is usually diminished, sometimes its secretion is temporarily suppressed. It often shows but little change, but is more commonly, as in other fevers, concentrated and high-colored. It deposits, on cooling, a sediment of urates, which toward the close of the fever is often very abundant. The defervescence is in **many instances attended by a copious secretion of urine.** Exact observations as to the composition of the urine in twenty-four hours are wanting.

At first the skin is hot and dry; sometimes frequent sweats occur, **free sweating generally marks the** febrile remissions, and the defervescence not rarely sets in with **copious, acid, ill-smelling sweats. In some cases the tendency to sweat shows itself early** and continues throughout the attack. **Sudamina occur in great numbers.** An outbreak of herpes about the lips is occasionally seen.

The digestive system.—**Disturbances of the digestive tract are more or less prominent in almost all cases.** Only in the rudimentary and subfebrile forms, and even then most rarely, are they absent. In many cases they are such as are usually seen in febrile disorders, namely, **loss of appetite, thirst, impaired taste, pasty, coated tongue, tenderness in the epigastrium, and constipation.** Nausea and vomiting sometimes usher in the attack. In other cases (the so-called abdominal form) all the above symptoms are **more severe, and diarrhœa, colicky pains and vomiting are superadded.**

The countenance is changed, in part by the appearance, characterizing an ordinary **attack of coryza, of considerable or great** severity, and in part by an expression of anxiety and depression. It is pale. Where the pulmonary catarrh is excessive and the dyspnœa great, the lips become bluish. The facies sometimes suggests **that of** typhoid **fever.**

THE CATARRH.

A more or less extensive hyperæmia of the mucous membrane of the **respiratory** tract is invariably present, and may be said to characterize the disease. The symptoms are essentially of a catarrhal nature.

There is **cold in the head, more severe in most cases than ordinary simple coryza.** The eyelids are swollen and reddened, there is abundant **lachrymation, sneezing is frequent, and** the discharge from the nose is abundant. Epistaxis is **not** rare. Sore throat, with tickling sensations **and difficulty in swallowing,** are due to inflammation of the pharynx and neighboring parts. **In many** instances the catarrhal symptoms are referable to a pharyngitis and tonsillitis only, the lower air-passages escaping. Hoarseness is common.

The cough.—This has been in most epidemics a prominent symptom. It is apt to be frequent and distressing—sometimes paroxysmal from the beginning of the sickness, almost always so during its course. Its spasmodic character in some of the older epidemics led to the confounding of epidemic catarrhal fever with whooping-cough. It is apt to be worse toward evening and at night, but the sick are often tormented day and night by the racking, loud cough. It often leads to vomiting, and, by its violence and **persistence,** gives rise to pain and soreness **in the muscles of respiration** (myalgia), and occasionally to hernia. **It is at first dry, or attended** with a scanty muco-serous expectoration; later on, the sputa **become** opaque and muco-purulent, and in consumptive or full-blooded persons, or those having mitral disease, they are sometimes streaked or mingled with blood. Toward the close of the attack the cough becomes less urgent and loses its spasmodic character. In some epidemics cough is not a prominent symptom, and a few cases are encountered in most epidemics in which well-developed influenza runs its course without unusual, peculiar, or excessive cough. **If the cough be** due to bronchitis, we find on auscultation the physical **signs of that affection. They are** of course wanting when it is due simply to laryngo-tracheal irritation. Hence, we frequently detect sonorous and sibillant, or mucous and subcrepitant râles upon both sides of the chest in the course of the attack, as in non-epidemic **acute bronchitis ;** and, on **the other hand, cases occur** where the auscultatory signs are but little, or not **at all, altered from those** of health. It is scarcely necessary to add there are no special characterizing signs that can be regarded as diagnostic of influenza.

Dyspnœa.—Many patients suffer from this symptom. It is due, in some instances, to complications; but it also occurs with remarkable frequency in those in whom none of the objective signs of any lesions can be discovered in the lungs. It is here of nervous origin. Graves assumes a direct disturbance **in the function of the** vagus as its cause. **This view is sustained by the observation that** the dyspnœa is now and then intermittent, or shows rhythmically recurring remissions, which are unattended by alteration of the physical signs. To Biermer it appears more probable that the congestions so common in influenza, not attended by marked **physical** signs until they lead to œdema, are to be regarded as the cause of the dyspnœa.

It varies greatly in intensity. In many patients it goes on to marked oppression, great shortness of breath, precordial pain, and the like. In certain epidemics, orthopnœa and suffocative attacks were very common.

Stitches in the side, **and pain under the** sternum, are observed without appreciable physical signs.

Symptoms Referable to the Nervous System.

Debility.—Great prostration of muscular strength is a **very early** symptom, and constitutes, in most epidemics, one of the **remarkable features of the disease**. Patients from the onset feel extremely weak, and are exhausted by the slightest bodily effort. The ordinary strength is not regained until convalescence is far advanced.

Headache.—**Severe** frontal pains are scarcely ever absent. They extend **across the brow**, and deeply about the orbits and at the root of the nose, having their seat in the Schneiderian mucous membrane and its prolongations lining the frontal sinuses and the nasal ducts. Sometimes the pain is referred also to the region of the antrum of Highmore, and to the Eustachian tube and the middle ear. It occasionally extends over the whole head. Cutaneous hyperæsthesia of the head and neck, and stiffness of the neck-muscles, are also met with. The headache is often most intense; it lasts commonly till the end of the attack, and may even outlast it. It increases in severity toward evening, with the fever and mental agitation. The occurrence of epistaxis affords some relief.

Pain.—Among the more constant symptoms of influenza are very severe pains in the limbs. Patients experience sensations of soreness and bruising, such as follow **the most severe and unaccustomed muscular effort**. Dull, **tearing,** and burning pains are felt sometimes in particular muscles or tendons; sometimes **they are diffused over the** whole body. Distressing pains of a dragging or boring character, in the loins and the calves of the legs, are complained of. These pains are neither relieved nor aggravated by gentle movement or by moderate pressure. A sense of contraction of the chest and precordial distress also occurs, and stitches in the side (pleurodynia), substernal pain, and pains in the throat and nape of the neck, are common.

General nervous symptoms.—Patients, when the case is severe, are usually restless, sleepless, and anxious. **Dizziness and a tendency** to faint occur on rising, particularly in women. Mild delirium is not **uncommon;** but the more intense forms are **also** observed. Active delirium was thought to be a mortal symptom in some of the older epidemics.

The inability to sleep bears no direct relation to the intensity of the fever. It is seen in some cases where fever is slight or **even absent.**

Somnolent states also occur. Great hebetude and torpor have marked some epidemics. That of 1712 was called the sleepy sickness, by reason of the prevalence of these symptoms.

In the gravest cases, painful muscle-cramps, subsultus tendinum, twitchings of particular muscles, and tremblings of the hands, are observed.

The mental power is enfeebled, and the acuteness of the special senses **is diminished.**

COMPLICATIONS AND SEQUELS.

The most important complications of influenza are inflammatory diseases of the lungs. The intense hyperæmia and bronchitis, already described as occurring in the severer cases, cannot properly be looked upon as complications. They constitute rather essential processes of particular forms of the disease. But capillary bronchitis, catarrhal pneumonia, and, less frequently, croupous pneumonia, arise as complications in the course of the disease. Pleurisy, except as associated with lobar pneumonia, is rarely met with. Satisfactory statistics are wanting, but Biermer estimates that from five to ten per cent. of the whole number of patients suffer from inflammatory lung-complications, and holds that the bloodletting so frequently practised by the older physicians was due to a desire to combat inflammation. The comparative frequency of chest-complications in different epidemics varies greatly, but the estimate of Biermer may be accepted as an approximate average.

Owing to the masking of the physical signs in the early stages, and the pre-existing pulmonary œdema, it is not always easy to recognize at once the occurrence of capillary bronchitis. It is attended with increasing dyspnœa, decided lividity of the face and extremities, and greater prostration. Crepitant and sub-crepitant râles at the lower portions of the posterior dorsal regions, rapidly spreading over all parts of the chest, without dulness at first and with increased resonance later, instead of the signs of condensation which are met with in pneumonia, are the signs which attend its appearance.

Catarrhal pneumonia occurs insidiously, with gradual intensification of the bronchitic symptoms about the fourth or fifth day; but it may set in as early as the second day, or much later, during convalescence. It is developed without chill, as a rule, or great increase in the fever. Old persons and those of feeble constitutions are more liable to the foregoing complications. Lobar pneumonia is less common. It is a late complication, occurring toward the close of the attack, or even when the patient is beginning to get about. It is easily recognized, and differs in no wise from acute lobar pneumonia occurring under other circumstances.

In October, 1880, influenza being in Philadelphia, both epizoötic and epidemic, but very mild among both horses and men, I attended a medical student, who, having had what he regarded as a "cold" for about a week, had kept at his work without treatment, until, upon the occurrence of a chill followed by grave thoracic symptoms, he was obliged to betake himself to bed. I first saw him the following day, in the Hospital of the Jefferson College. There were the symptoms of acute lobar pneumonia, with the signs of extensive consolidation of the left lung and pleurisy of the right side. Moreover delirium and jaundice were present. The urine was non-albuminous. The next evening he died. At the same time many members of the class suffered from unquestionable influenza, and a careful inquiry into the history of the case of this young gentleman satisfied me that the pneumonia had arisen as a complication of a neglected and moderately severe catarrhal fever. Until the eighth day before his death he was in excellent health. No examination of the body was permitted.

Graves[1] thought that a kind of paralysis of the lungs, with great œdema, takes place in some cases, and attributed it to an affection of the vagus. It was his conviction "that the poison which produced influenza acted on the nervous system in general, and on the pulmonary nerves in particular, in such a way as to produce symptoms of bronchial irritation and dyspnœa, to which bronchial congestion and inflammation were often superadded."

It is certain that localized collapse of the lung often occurs. Drs. White and Guitéras attributed the consolidations of the lung to congestive collapse due to enlargement of the tracheal and bronchial glands, and "disturbance of the great nervous tract about the root of the lung." They were enabled to satisfy themselves of the existence of the glandular enlargement—*adénopathie bronchique*—in nine of their eighteen cases, by percussion practised in the method of M. Géneau de Mussy,[2] who was, as they believe, the first to call attention to the information that may be gained by percussion of the spinous processes of the vertebræ over the course of the trachea. Following this line in the healthy subject, a distinct tubular (high-pitched and slightly tympanitic) sound is elicited by percussion, down to the point of bifurcation of the trachea, on the level of the fourth dorsal vertebra. Opposite the fifth, and downward, we get the lower-pitched pulmonary resonance. When the tracheal and bronchial glands are enlarged, the tubular sound over the upper dorsal vertebræ is replaced by dulness, which may contrast sharply, above with the tracheal, and below with the vesicular resonance.

They point out some well-recognized peculiarities of the so-called pneumonias of influenza, as giving weight to their view that the consolidations are not, in the beginning, pneumonia at all. Thus, we have at first weakness of the vesicular murmur, then its absence; the respiration soon becomes bronchial without being preceded by dulness or the crepitant râle; the extension of these consolidations from one part of the lung to another is very irregular; the process is more apt to involve both sides than one; the disappearance of the consolidation is frequently very rapid.

The physical signs in one of their patients were very interesting, as supporting the theory of collapse from the nerve-disturbance consequent upon enlargement of the lymphatic glands. The case presented, one day, pectoriloquy and bronchial breathing at the root of the left lung; the next day there was dulness of a large portion of the left lower lobe, with bron-

---

[1] Annals of Influenza.   [2] Clinique Médicale. Paris, 1874.

chophony and bronchial breathing over an area extending from above the angle of the scapula to the base, and out to the axillary line. That is to say, there was, first, engorgement of the left **bronchial glands, and the next day the congestive collapse of portions of the lung.** On the day after this, no traces could be found of the consolidation. This is certainly not the history of catarrhal pneumonia.

The relations of cause and effect between collapse and catarrhal pneumonia are so **close,** that it is not difficult to see how the condition spoken of may lead to secondary lobular or catarrhal pneumonia. In truth, this is a frequent result of collapse from any cause.

They do not adduce any post-mortem facts in support of their theory. Peacock,[1] however, observed in the epidemic of 1847, softening and enlargement of the bronchial glands in several cases, and in one instance where there was no antecedent disease of the lungs, and where the physical signs corresponded to some extent with those of the cases upon which Drs. White and Guitéras base **their views.** These complications are the chief cause of the danger of influenza in the aged, the debilitated, and those suffering **from previous** disease of the thoracic organs.

Pleurisy is rare, except where there is coexisting inflammation of the lungs. It may be associated with pericarditis. In old persons, serous **effusions into the** pleural sac are now and then encountered.

Troublesome laryngitis and chronic bronchitis may **follow the attack.** In consequence of the extension of the catarrhal processes along the Eustachian tube, an actual inflammation of the middle ear is, **in rare instances,** set up. Parotitis with salivation sometimes occurs, **likewise** aphthous inflammations of the mouth.

Herpes labialis may occur as a favorable indication.

Latent phthisis may be developed by an attack of influenza, and if phthisis be already established it may run a more rapid course. Emphysematous affections are aggravated; diseases of the heart are unfavorably influenced; nervous affections may be made worse, and, in particular, neuralgias are aggravated. Old neuralgias, that have long ceased to give trouble, have been known to reappear during the convalescence from influenza.

Many of the older observers have recorded the intermittent character of influenza in certain epidemics, and its tendency to run into intermittents, particularly of a tertian type, during convalescence. This has not been observed in the outbreaks of later years, **and it is probable that in** such instances an endemic malaria has modified the epidemic **catarrhal** fever, or the former has broken out as the latter passed **away.**

Pregnant women are in danger of aborting.

Women who have **suffered from** amenorrhœa have had the menses

---

[1] Loc. cit.

re-established after an attack of influenza. This statement has been verified repeatedly in several epidemics.

PATHOLOGY.

From the absence of sufficient pathological investigation, our knowledge of influenza is as yet very incomplete. Biermer has described it as the sum of a series of catarrhal manifestations which have developed under common epidemic influences. The close association of the various local affections arises from their almost simultaneous occurrence in considerable numbers, and the identity of the primary pathological processes which underlie them. As regards mucous surfaces, these lesions consist principally in hyperæmia. The nature of the lesion underlying the nervous phenomena is altogether unknown. Nevertheless, it may be assumed that the causation of both the catarrhal and the nervous symptoms is essentially the same; while the relation of both to the fever, forces us to the conclusion that it is also a separate result of the same cause, since no constant relation either as to time or intensity exists among the three groups of symptoms in question, although they appear in a general way nearly simultaneously. That is to say, headache, weakness, malaise, may precede the fever, with coryza and the like, or the coryza may precede the fever, or even occur without marked fever, in which case more or less fully developed nervous phenomena are never absent; or, as is more frequently the case, the chill or chilliness which forms part of the fever and ushers it in, is the first sudden manifestation of the disease. At the same time the nervous symptoms are in some cases marked, with very moderate coryza, bronchitis and so on, or a considerable febrile movement may be attended with comparatively mild catarrhal symptoms, and a degree of nervous perturbation that is relatively slight. In severe cases, the catarrhal and the nervous symptoms are apt to be severe, while the febrile movement may be in correspondence with them, or may be of only moderate energy.

These facts point to a common cause for the varied phenomena of influenza, and it may be confidently asserted that each of the three groups of symptoms constitutes a distinct factor of the disease. This view is at variance with the opinion—based upon the fact that acute common catarrh, bronchitis, tonsillitis, and other acute affections, sometimes run their course in a similar way to influenza, with fever, nervous depression, and a serious sense of illness—that influenza is in essence simply an epidemic catarrh. Moreover, the sudden onset of influenza, its not infrequent abrupt termination—which suggests crisis—its unsparing seizure of great numbers of the population, the severity of the nervous symptoms, and the amount of laryngo-bronchial irritation—often out of measure with the lesions of the mucous membrane,—all point to **the action of a morbid agent** affecting the body at large. The severity of the symptoms also, in many cases, is much greater **than** in acute non-specific local affections, while the complications, and in **particular the recrudescence of fading** neuralgias and the tendency to abortion, and the sequels, **as cough,** weakness, headaches, flying pains, which **often** remain long **after** convalescence, are evidences of its belonging to the group of infectious diseases **rather than** to that of simple acute inflammatory diseases.

In conclusion, it must be urged that the similarity of the symptoms in **many** epidemics, occurring during the course of several centuries and under different social conditions and even different degrees of civilization, forcibly demonstrates the specific and definite character of the causes which give rise to influenza.

Morbid anatomy.—Very little light is thrown upon the pathology of the disease by the anatomical changes found after death. Uncomplicated influenza is rarely fatal. **As a rule,** it is found that the unfavorable termination is **in** consequence of lung complications. The essential lesions are congestion and catarrhal swelling of the mucous membrane of **the upper** air-passages **and** the bronchial tubes. **These changes may be restricted,** in the lungs, to the trachea and larger bronchi, or they may extend to the finest twigs. They may amount to **great thickening and** deep capillary injections **of the mucous lining of the tubes, which contain** at the same time **clear, frothy mucus, or thick, viscid masses of muco-purulent** secretion unmixed with air.

More or less congestion of the gastric mucous membrane, and more rarely of that of the intestine, is also met with. The solitary and agminate glands of the intestine are not affected save as the **result of special complications.** A few observations relate to the finding of enlarged and softened bronchial glands. More extended researches are needed, not only upon this point, but also in the whole domain of the pathological anatomy of the disease.

Hyperæmia, œdema, hypostatic congestions, splenization, catarrhal pneumonia, and hepatization, affect the lung-tissue in cases fatal by the complications which are associated with such changes. The tissue-changes of diseases existing prior to the attack of influenza, such as old catarrhal consolidations, tubercle, brown induration, emphysema, and so forth, are of course frequently discovered at the necroscopy.

Diagnosis.

The discrimination of influenza from other affections **having some points of resemblance to it, is,** under ordinary circumstances, unattended **with difficulty.** The march of the epidemic, the number of persons attacked, the prominence of the nervous symptoms, the rapidly developed debility, and the character of the cough, usually severe out of proportion to the physical signs, distinguish it from all other epidemic diseases.

It is only to be differentiated from non-specific catarrhal affections attended by fever, considerable malaise, weakness, severe headache, and pain in the extremities, by a due regard to the causative relations of the two affections. Simple catarrhs not rarely present the group of symptoms which characterize epidemic catarrhal fever, but they occur almost constantly as the result of great and sudden changes in the weather, and are therefore met with in greatest frequency in bad seasons, and are particularly common at the end of winter and in the spring.

Influenza is not in any way dependent upon the vicissitudes of the seasons, and may occur, as has been shown, at all times of the year, in wet or dry, mild or cold seasons, equally, and in every variety of climate. It is of course diagnosticated without difficulty from the sporadic catarrhal fevers, which lack the characteristic depression, neuralgic and rheumatoid pains, the irritative cough, dyspnœa, and so on.

Cases of influenza are met with that bear a strong resemblance to beginning enteric fever. Malaise, headache, obtunded hearing, mental depression, high fever, coated tongue, tender belly, diarrhœa, are symptoms to be observed in both affections. But influenza lacks the temperature-curve, the usually rapid pulse, the splenic enlargement, and the eruption of enteric fever, and the progress of the disease will in a few days clear up the most doubtful cases.

Prognosis and Mortality.

A fatal issue is rare in uncomplicated cases. The very young bear influenza badly; the old bear it more badly still. Nevertheless, children have enjoyed a considerable proportionate immunity in some epidemics. Healthy persons in the middle periods of life bear it well. Certain pre-existing diseases modify its course unfavorably. Among these are chronic bronchitis, emphysema, and fatty heart. The debility of advanced phthisis and other exhausting diseases render influenza dangerous. Death takes place, in by far the greatest number of cases, as the result of the complication of the attack, either with some pre-existing affection, or with an acute disease arising in its course. The commonest of the latter are inflammations of the parenchyma of the lungs.

Patients presenting very severe symptoms generally recover if they be not the subjects of complicating maladies, or very young or very old.

Relapses are not uncommon; second attacks have been known to occur during the continuance of an epidemic; it is often the case that an individual in the course of his life passes through several epidemics of influenza and is the subject of the disease in each of them.

The prognosis is greatly modified by the character of the prevailing epidemic. In some epidemics the deaths are few, and the mortality from other diseases does not appear to be greatly augmented. In others, many die of the epidemic disease, and the fatality of certain endemic affections is markedly increased. In some of the older epidemics the high mortality was doubtless due to injudicious measures of treatment, among which bloodletting and other depressing agencies were conspicuous. Some of the older accounts also warrant the suspicion that a coexisting typhus had to do with the high death-rate. It is estimated that in the epidemic of 1837, which was a very severe one, two per cent. of those attacked, died. The proportion of fatal cases in particular epidemics, varies in different countries and even in different quarters of the same city.

Treatment.

No efficient means of prophylaxis are known. Unfavorable hygienic surroundings, overcrowding, a damp, unhealthy locality, appear to increase the prevalence and severity of influenza. The opposite conditions of living do not, however, secure immunity from the attack. During an epidemic, young infants, aged persons, those enfeebled by chronic diseases, and in particular those subject to chronic bronchitis, consumption, emphysema, and fatty heart, should be cared for with unusual diligence and solicitude, since they constitute the classes most prone to the graver complications of the disease, and from which its fatal cases are almost wholly derived. Such individuals should be warmly clad; they should shun, as far as possible, the vicissitudes of the weather, even if practicable keeping within warmed and well-ventilated apartments; they should exercise unusual prudence in diet and lead a carefully regulated life, with long hours of sleep. It is true that these measures are not preventive of the attack. Families not quitting the house, living in the greatest seclusion, even the bedridden, do not always, or even as a rule, escape. Yet it has frequently been observed that those whose occupations are carried on in the open air are attacked earliest and in greatest numbers. On the other hand, in some instances, persons isolated from the community with strictness—in prisons, cloisters, hospitals—have remained free from the disease prevailing around them. It therefore appears probable that, under certain favorable circumstances, not as yet perfectly understood, the avoidance of the open air, and of the direct influences of the weather, may confer some degree of immunity from the attack, and it is desirable that the class of persons most liable to the graver consequences of the disease should avail themselves of even the most uncertain precautions.

The treatment of influenza is expectant and supporting. Not only is the epidemic self-limiting, tending to exhaust the susceptibility of the infected community, as we have seen in most instances, in the space of a few weeks, but the attack is also of definite duration, and the perturbations set up by the action of the influenza-poison upon the individual subside spontaneously in three or four, or at most, ten or twelve days. The susceptibility of the individual is also, for the time being, exhausted; for second attacks in the same epidemic are not very common. In cases where the duration of the attack is prolonged beyond the period indicated, it is kept up by complications, and we have to do not so much with the pathological processes of influenza, as with secondary diseases that the influenza has excited, either by the intensity of its action or by reason of some peculiarity of the subject of the attack.

By far the greatest number of cases are light and unattended by real danger. The treatment is, therefore, for the most part, an extremely simple one.

These lighter cases rarely require medical measures. The patients are uncomfortable and anxious, easily fatigued and unfitted for business. It is best that they keep the house, and, if willing, the bed or sofa, for the space of two or three days. The diet should be restricted to a few simple and easily digested dishes. Meat should be avoided. Parkes regards the common custom of taking hot beef-tea as an extremely bad one. He thinks that it invariably increases the headache and languor, and agrees with Pearson that warm foods, which force sweating, are not only useless, but that they also do harm. Moderate quantities of cold drinks may be taken. The fruit-syrups, lemonade, raspberry vinegar, a weak solution of citrate of potash, or of cream of tartar, and barley-water with lemon, are useful. Very weak wine-whey is often liked. The effervescing mineral waters will be preferred by some. The best of such drinks is the mixture of equal parts of iced seltzer-water and milk. If the stomach be irritable, koumiss may be tried as a combined beverage and food. In the mild cases stimulants are not always needed. Some of the older writers think them positively injurious in the early stages of the disease. Sound claret, with or without seltzer-water, is not contraindicated.

Quinine in moderate doses should be taken from the onset. The head-pains are not increased by it. Dover's powder, if well borne, should be taken at night. Some form of opiate may be required, even in mild cases, to counteract wakefulness. A compressed pill, containing extract of opium 0.030 gramme (gr. $\frac{1}{2}$), camphor 0.135 gramme (gr. ij.), and ammonium carbonate 0.165 gramme (gr. ijss.), will be found useful when Dover's powder cannot be employed. During convalescence, iron and barks will often be requisite.

The coryza, tonsillitis, laryngitis, bronchitis, are to be treated according to general principles, if they require treatment at all. In most mild cases the catarrhal symptoms call for no special measures of treatment.

Free inunctions of fatty substances about the brow and over the bridge of the nose may be of use as regards the coryza.

Morphine dissolved in cherry-laurel water, one part in fifty or sixty, is recommended by Zuelzer for the relief of the head-pains associated with the coryza. A few drops may be snuffed up from time to time. These pains are of a surety mitigated to some degree by wearing a flannel cap, or wrapping the head in a silk handkerchief. Warm applications sometimes give comfort, while cold almost invariably adds to the distress.

Distress in the upper air-passages, and the tickling cough, call for steam inhalations, and the air of the apartment may be rendered moist by the evaporation of water kept boiling in a broad, shallow vessel.

Gargles of potassium chlorate, or potassium chlorate combined with sumac, exert a soothing influence upon the congested tonsils.

Severe cases call for more energetic measures of treatment.

The more prominent indications are the control of the fever; the diminution of the hyperæmic fluxion to the mucous tracts; the arrest of increasing debility ; **measures of support;** the mitigation of pain and the induction of sleep ; and finally, **the prevention of the pulmonary congestion,** to which the depression leads, by enfeeblement **of the circulation.** The last indication is especially urgent in infants, the very old, **and those** previously debilitated from any **cause.**

Inflammatory complications require special treatment or modifications of treatment.

The febrile movement is not, as a rule, high; grave nervous symptoms and serious catarrh may be associated with moderate fever. The management of the fever must be in accordance with views expressed in the Introduction, regard being had to the tendency to depression which is so prominent an element in almost every case of influenza.

An antifebrile regimen is to be observed. The moderate duration of this fever, as compared with enteric fever, renders it less important that large amounts of fever-food should be given, while the tendency to depression makes it of the utmost importance that the administration of food be systematic and carefully looked after by the medical attendant. The disinclination to take food is so great that it is often with difficulty that a sufficient quantity can be given in the early days of the attack, and it is to be doubted whether benefit follows anything in excess of the most moderate amount. It is necessary to observe regular hours, as in the management of all the low fevers. As soon as convalescence begins the patient should be urged to eat; the quantity of food taken at once is to be augmented, and the intervals between the periods of its being offered are to be lengthened.

A favorable action upon the excretory function of the skin and kidneys will result from the free drinking of water, or of the beverages spoken of already. In all cases, at least enough fluid should be taken to relieve thirst.

Diaphoretics have been much used, upon the theory that by determination to the skin they correspondingly diminish the tendency to hyperæmia of the affected mucous tracts. Dover's powder, solution of the acetate of ammonia, and other mild diaphoretics, are to be selected. Jaborandi should be employed with great caution. The wet pack and other hydrotherapeutic measures have been employed in Germany to act upon the skin and to effect a direct reduction of temperature in influenza. For old and feeble persons warm packs are to be employed. A profuse sweating at the onset of the attack is said to occasionally cut it short. Biermer states that early diaphoresis often brings about a rapid and lasting amelioration of the symptoms. It is to be borne in mind that the fever is rarely excessive, and that sweating is not infrequently a troublesome symptom. In some epidemics it has been a very troublesome one.

General bloodletting is not to be resorted to in influenza. Its danger was apparent to some of the early writers. As has been pointed out, the high mortality of some of the older epidemics is to be explained by the venesections practised at the beginning, and even during the course of the attack. It has no favorable effect upon the catarrhal processes, and but little upon the subjective symptoms. Parkes states that the fever is not relieved by it; the nervous depression is increased and the risk of lung-congestion is augmented. Bleeding is not likely to be practised in epidemic catarrhal fever while the present views of its place in therapeutics continue to influence practice. Cautious local bloodletting, for the relief of local inflammatory trouble, is spoken of in most of the modern books. It seems to me that the occasion for its employment must be so rare in the treatment of this disease, that the statement may be henceforth omitted. In influenza, as it is known to medical men of the present, from the descriptions of the old and personal experience of the few recent and milder epidemics, bloodletting, either general or local, is clearly uncalled for.

Emetics hold a high historical place. It was of old customary to begin the treatment with a vomit. As late as the epidemic of 1837, Lombard, of Geneva, believed that they shortened the attack and lessened the intensity of the symptoms, when administered at the beginning. In cases attended by a good deal of early gastric disturbance and nausea, they are said to be of vast use. They sometimes set up great irritability of the stomach, with vomiting that it is difficult to control. On the whole, the cases in which an emetic would do good must be extremely rare. Purgatives were formerly regarded as important in the treatment. This view no longer prevails. In case of constipation, gentle purgation, *ex indicatione symptomatica*, is a necessary part of the proper management of the case. For this purpose the laxative mineral waters, as Friederichshalle,

Hunyadi, Pullna, are excellent. Castor-oil may be given, and calomel is in some cases, and particularly in childhood, of great service. Simple enemata of warm water, or soap and water, will often suffice. The tendency in some cases to exhausting and troublesome diarrhœa, and the fact that diarrhœa occurs sometimes in the course of the largest **number of cases,** should inspire caution in the use of purgatives. Repeated purgation during the progress of the attack is not only useless—it is positively in**jurious.**

Quinine is to be given in full doses. It exerts at **the same time a** powerful influence upon the temperature, upon the tendency **to local hy**perœmias and upon the nervous symptoms, and in particular the headache. Da Costa states that its use is of "primary importance," and Rawlins,[1] as early as 1833, states that excellent results followed its administration, the effect being the better the earlier it was given. It has even been lauded as a specific for influenza.

The mineral acids may be **given, with a view to realizing** their antifebrile and tonic effects.

For the most **part, the foregoing measures directed against the fever,** will exert a favorable **influence upon the catarrhal processes. Expecto**rants are recommended ; ipecac is preferred. The preparations of antimony are inadmissible by reason of their tendency **to depress. Ammo**nium chloride is indicated in the earlier stages of the bronchitis. **Among** recent drugs, yerba santa (eryodiction glutinosum) and the oil **of euca**lyptus, are likely to prove of use in mitigating the symptoms in epidemic catarrh, as they do in certain forms of simple, sporadic **catarrh.**

It is of great importance that the peculiar dry, racking cough, so often present in the early days of the attack, should be relieved. It is not useful in removing bronchial accumulations, being, as has been shown, in most instances out of proportion to the lesions of the bronchial mucous membrane; on the other hand, it tends to increase the hyperœmia of the upper air-passages by the mechanical violence of the cough-paroxysms. Further, it is distressing and exhausting, and, if uncontrolled, contributes to the muscular and nervous prostration. Benefit will be derived from keeping the air of the apartment moist, and from the occasional inhalation of the steam from hot water, either used alone or poured upon the compound tincture of benzoin, a pint to the teaspoonful, or upon paregoric, a pint to the tablespoonful, in a proper vessel or inhaler.

No drugs are more potent to this end **than opium and its derivatives,** and in particular morphia and codeia. The hypodermic use of **the mor**phia salts, judiciously resorted to, constitutes our most **valuable thera**peutic resource in **fulfilling the threefold** indication of **relieving cough,** alleviating both the head-pain and the pains in the extremities, and in

---

[1] London Medical Gazette. May, 1833.

procuring sleep. The former dread of opium in bad cases was not well founded. Its moderate administration is attended with advantages that far outweigh any danger of increasing the tightness across the chest and retarding expectoration. It is necessary to observe the same or even greater caution, in giving it to infants and aged persons in influenza, that is essential under other circumstances. The favorable influence of carbolic acid in restraining cough makes it probable that it would be of great service in this disease. It may be combined with codeia as follows:

℞. Acidi carbolici liq.................... 0. 530 c.c.   ℥ viij.
   Codeiæ sulph....................... 0. 530 gram. gr. viij.
   Aquæ lauro-cerasi,
   Aquæ............................ āā 32 c.c.   āā f. ℥ j.
M. Signa.—A teaspoonful q. s. h.

The chest-pains and substernal pains may be combated with sinapisms, turpentine stupes, repeated inunctions of fatty substances containing extract of belladonna and the like. Pleurodynic stitches call for similar measures; a long strip of machine-spread belladonna-plaster about 5 ctm. (two inches) in width, applied very firmly to the side of the chest from the spine in a direction downward and forward parallel with the ribs, and reaching to the median line in front, affords great relief to the lateral chest-pains.

The control of the debility must be regarded as the most important indication in old and feeble persons. Wine, spirits, milk-punch, ammonia, spirits of chloroform, are to be used, not in accordance with fixed rules, but as occasion may require. In many cases wine or whiskey will be indicated from the beginning, the quantity being determined rather by the effect upon the circulation and the general condition of the case, than by rule. Women and others unaccustomed to the use of alcoholic drinks, often, as Da Costa states, take wine and brandy in considerable quantities, with striking benefit and without flushing or other evidences of its disagreeing.

Chloral is inadmissible as a hypnotic by reason of its depressing effect upon the heart. The bromides may be used in connection with opium, if the latter alone be not well borne.

Diarrhœa must be managed in accordance with general principles. If slight, it does not require special treatment. It is apt to occur at one period or another in the course of most cases, and not infrequently marks the beginning of convalescence. Colic may be treated with warm fomentations and carminatives; if it be due to constipation, mild laxatives are to be combined with them.

Severe cases of influenza demand the careful attention of the physician, who must be on the alert to detect the inflammatory lung-complications which so often lead up to the fatal issue, as early as possible. Their treatment must be regulated by the circumstances of the case, the nature of the particular complication, the age of the patient, and so on, in accordance with general therapeutical indications.

In conclusion, it is to be urged that all measures, of whatever kind, that tend to depress the general nervous system, or the functional activity of the respiration, and especially the heart-power, are to be sedulously avoided in the management of influenza.

During the convalescence good nursing and abundant nourishment are needed to build up the strength. Unfavorable influences of the weather are to be guarded against. It is important to warn the patient that a severe attack of influenza renders him liable for some time afterward to pulmonary disorders. The sequels, and in particular those implicating the respiratory tract, are to be appropriately treated. A course of tonics, iron and quinine, and the like, is nearly always useful, and a brief change of climate is often of advantage.

III.

CEREBRO-SPINAL FEVER.

DEFINITION.—A malignant continued fever, occurring in general or limited epidemics and caused by some unknown specific external influence. It is of sudden onset, mostly of rapid course, and very fatal. The symptoms point to profound disturbance of the functions of the brain and spinal cord; associated headache, vomiting, and painful contraction of the muscles of the back of the neck, are characteristic. Delirium, stupor, coma, cutaneous hyperæsthesia, and motor palsies occur. In many instances eruptions, chiefly herpetic and petechial, attend the disease. There is uniformly great nervous depression. The lesions found after death are constant, varying only in the degree of their development. They are the results of an acute diffuse inflammation of the pia mater of the brain and spinal cord, and consist of intense hyperæmia with dense cell-infiltration, and fibrino-purulent exudation.

SYNONYMS.—Epidemic cerebro-spinal meningitis; Epidemic meningitis; Fever with cerebro-spinal meningitis; Cerebro-spinal arachnitis; Typhus syncopalis; Cerebro-spinal typhus; Cerebral typhus; Typhus cerebralis apoplecticus; Petechial fever; Malignant purpuric fever; Malignant purpura; Pestilential purpura; Febris nigra; Fièvre cérébrale; Phrénésie; Céphalalgie épidémique; Méningite cérébro-rhachidienne; Méningite purulente épidémique (France); Febbre soporoso-convulsiva; tifo apoplectico tetanico (Italy); Nacksjucka; Dragsjucka (Sweden); Geinck Krampf; Genickstarre; Hirnseuche (Germany); Spotted fever; Congestive fever; Cold plague.

As our knowledge of a disease becomes more exact, and we are enabled to discriminate it as a substantive affection from maladies which resemble it, the names by which it is designated become fewer in number and more precise. The long list of synonyms for cerebro-spinal fever, given above, is rather of historical than of present practical interest. Many of them have been employed in ignorance of the real nature of the disease, and their multiplicity has arisen in part from that ignorance, by which observers were led to confound it with other diseases, and particularly with typhus, and in part from the great variety of forms which the disease assumes in different epidemics. It is only within two decades that careful attention has been generally directed to the study of the anatomical lesions which almost constantly occur, and which on the one hand explain to a great extent many of the symptoms that are characteristic, and on the other separate the affection from all other diseases which it in any way resembles. It is only, therefore, within comparatively recent years that distinctive terms embodying the idea of the localized lesions have been applied to it. With the introduction of these names, due to such a knowledge of the nature of the disease as enables the profession to diagnosticate it from other affections with precision, other names for it, not embodying the idea of the spinal lesions, nor that of its being a distinct affection, as well as those derived from symptoms neither constant nor characteristic, have become obsolete.

As will appear from the following account of the disease, it is a general affection of febrile character, and is therefore properly described among the fevers. The fever is of continued type; it is for that reason classed with the continued fevers. It is not symptomatic of the local inflammatory processes, which are, however, constant. The term cerebro-spinal fever, or epidemic cerebro-spinal fever, appears to be, for the reasons just given, preferable to that of epidemic meningitis, or cerebro-spinal meningitis, and finds its analogue in enteric fever, the only other of the continued fevers attended by a constant anatomical lesion. The affection is designated cerebro-spinal fever in "The Nomenclature of Disease" of the Royal College of Physicians, and this term, indicating as it does the infectious character of the malady and the anatomical seat of the constant lesion, has, in the present state of nosology, the warrant of sound logic and is worthy of general adoption.

Historical Sketch.

Cerebro-spinal fever was first recognized as a distinct affection about the beginning of the present century. It is to be conceded to those writers who have sought to show that it has existed from a remote period, that their opinion is probably correct, for the first recognition of an infectious disease as an independent affection is more likely to mark a period of advance in medical science than the period of origin of the disease. Doubtless many accounts of epidemics of unusual forms of fever with cerebral symptoms, recorded by the older writers, included cases of this disease, but the fact cannot be established by adequate proof. In truth, it is only in the light of knowledge obtained later in the century by systematic *post-mortem* investigations, that the earlier epidemics, described by numerous observers both in America and Europe, under many different names, are now to be recognized as instances of the disease under consideration.

The history of **cerebro-spinal** fever must then begin with the circumscribed epidemics which occurred nearly simultaneously in Middle Europe and in the United States, shortly after the setting-in of the present century.

Of these, the first of which we have any record arose in Geneva and its environs, in February, 1805. The disease appeared at nearly the same time in different parts of the city, and prevailed until April. This outbreak was described by Vieusseux,[1] whose account remains, according to Hirsch,[2] the only contribution to the knowledge of this affection which has come to us out of Switzerland.

The researches of Stillé[3] have brought to light the fact that the disease prevailed in Prussia, Holland, the Rhine Provinces, Bavaria, and the east of France, but not elsewhere in Europe, in limited epidemics during every one of the following years, until 1816. An epidemic occurred at Grenoble, in 1814, which attracted some attention. In the same year an outbreak of the disease occurred in Paris, and in the following year (1815) at Metz. These three visitations were almost exclusively confined to the garrisons.

Meanwhile, it arose in the United States at Medfield, in Massachusetts, in the year 1806, and prevailed at various points in New England, Canada, the State of New York, Pennsylvania, and elsewhere in the Western and Southern States, from that year until 1816.

Hirsch is unwilling to regard these early American outbreaks, which were described by those who observed them under such names as "Sinking Typhus," "Typhus Syncopalis," and "Spotted Fever," as instances of the disease under consideration, and goes so far as to discuss them under the heading "Typhus Syncopalis" as a form of typhus.[4] This opinion he reiterates with emphasis in his work[5] on Cerebro-spinal Meningitis, published later. Nevertheless, the evidence in the various accounts published at the time has been sufficient to convince most American writers upon the subject that these epidemics were, in fact, outbreaks of cerebro-spinal fever.

This view is also held by Radcliffe in his scholarly article upon the

---

[1] Journ. génér. de méd., xxiv., p. 163.
[2] A. Hirsch: Handbuch der historisch-geographischen Pathologie. Erlangen, 1864.
[3] Alfred Stillé: Epidemic Meningitis, or Cerebro-spinal Meningitis. Philadelphia, 1867.
[4] Band I., p. 165.
[5] A. Hirsch: Die Meningitis Cerebro-spinalis Epidemica. Berlin, 1866. P 11, footnote.

subject, in Reynolds's System of Medicine,[1] and may be accepted as correct. No record of the occurrence of the disease from the year 1816 exists, until 1822, when, during the spring, numerous cases appeared in Vesoul, in France. This epidemic chiefly affected the civil population. According to Ziemssen,[2] the symptoms of a disease which prevailed in Dorsten during the winter of 1822–23, and was described as a "myelitis, sometimes complicated with encephalitis," correspond with those of cerebro-spinal meningitis.

At each of these places the appearance of the disease was limited both as to the locality and the duration of the epidemic. Stillé informs us that it prevailed in a like local and temporary measure in 1823, at Middletown, Conn.; in 1828 in Trumbull County, O.; in 1830 at Sunderland, England, and in 1833 at Naples.

A period of quiet ensued. During four years nothing further was heard of the disease. But in the early part of the year 1837 it again made its appearance in France. This outbreak was the beginning of a widely extended and long-continued prevalence of cerebro-spinal fever. Its ravages were no longer confined to a restricted locality or to a season.

From the Pyrenean frontier in the southwest, where it was first felt in Bayonne, and from the south, where it made its appearance a little later in the same year (1837), at Foix and Narbonne, it spread in a northerly direction over the greater part of France, the middle regions and the high table-lands alone escaping.

Almost at the same time it broke out in Dax, Auch, Perpignan, and in Bordeaux. It visited La Rochelle in the same year, and early in 1838 showed itself in Rochefort, being at first confined to a regiment that had come from the department of Landes in the south, in which the disease was prevailing. It prevailed from 1838 to 1842 in the southeast of France, and particularly along the valley of the Rhone. The garrisons of Toulon, Marseilles, Nismes, Avignon, and other cities, suffered. During four years following 1838 the disease showed itself successively among the troops stationed at Metz, where it prevailed during the winter of 1839–40; Strasbourg, where it appeared in the autumn of 1840 and continued till the following summer, affecting the civil community also; Nancy, early in 1841, where, with the single exception of a lady dwelling in the city, its ravages were restricted to the soldiery, and Colmar, where, in the spring of 1842, scattered cases occurred in the garrison.

Early in 1839 cerebro-spinal fever broke out in Versailles, and it is worthy of remark that the first cases occurred in the very same regiment from the department of Landes in which it first showed itself the previous year at Rochefort, and which had come at the end of the year 1838 from

---

[1] A System of Medicine, edited by I. **Russell** Reynolds. Vol. ii. Philadelphia, 1868.
[2] Cyclopædia of the Practice of Medicine. Vol. ii. Amer. ed. New York, 1875.

Rochefort to Versailles. This outbreak was confined to the garrison. From this time until 1842 it prevailed among the military forces in and around Paris. A little later the disease appeared at various points in the valley of the Loire. We first encounter it here in the spring of 1840 at Laval, where it prevailed until the following year, being confined to the soldiery until the close of the epidemic, when a few scattered cases broke out among the civil population. In the winter of 1840–41, many cases were observed at Le Mans, and at Château-Gonthier; at the last-named place among troops that had come from Laval. The same and the following winter, cerebro-spinal fever appeared either as an epidemic, or with notable frequency, at Poitiers, Tours, Blois, Ancenis, and in the early part of the year 1842 it showed itself at Nantes, not far from the mouth of the Loire, where it spread indifferently among the soldiers and citizens. In 1840–41, it visited the northwest coast and prevailed in Brest, Cherbourg, and Caen. In 1842 it appeared in Lyons. With the close of that year the virulence of the disease in France seems to have passed away, for, although it continued to occur in that country until the winter of 1848–49, the outbreaks were limited, and in most instances took place in localities previously visited by the disease in its epidemic form.

At the period when the disease began to spread itself over wide areas in France, namely, during the winter of 1839–40, it entered Italy, where it continued to prevail until 1845. It first appeared at Naples and at various points in the Papal States, but the following winter it spread over the greater part of the kingdom.

The same year cerebro-spinal fever made its appearance in Algiers, where it prevailed in an epidemic form at various points, and in particular in the central and eastern districts, until 1847, selecting its victims among the civil population, both native and European, as well as among the troops.

In the spring of 1844 a transient outbreak took place at Gibraltar. The civil population chiefly suffered, only a few scattered cases occurring among the British troops of the station.

Denmark felt the scourge in 1845. During the spring it prevailed as an epidemic in parts of Jutland, while elsewhere, and especially in Zealand and in the city of Copenhagen, many scattered cases occurred. In the winters of 1846–47 and 1847–48, it again appeared, but in an endemic rather than an epidemic form.

The year 1846 brought the malady into the British Islands. Here, however, its ravages were relatively limited. It broke out in several workhouses of Dublin and Belfast, and a number of cases occurred among the citizens of Dublin. In the spring of the same year many cases were observed in Liverpool, though the disease cannot be said to have assumed the proportions of an epidemic.

While the disease was thus spreading over Europe, it again visited the

United States. It made its appearance in this country in the beginning of the year 1842, at points widely distant from each other and remote from intercourse with the Old World. Outbreaks occurred at about the same time at Louisville, Ky., in Tennessee, and at Montgomery, Ala. During the following years until 1850, a series of epidemics took place in Arkansas, Mississippi, Illinois, Pennsylvania, Massachusetts, New York, and North Carolina. In 1848, Montgomery, Ala., was a second time visited. In January and February, 1850, New Orleans suffered, and with this, the last of the local epidemics, the long prevalence of cerebro-spinal fever in the Western hemisphere came, for the time, to an end, almost simultaneously with its temporary extinction in Europe.

For four years cerebro-spinal fever ceased its ravages in the Old World. So far as can be learned, the New World enjoyed a longer period of immunity; but the disease had ceased to rage, not to exist.

Suddenly, in the early part of the winter of 1854, it made its appearance in Sweden, a country which had, up to this date, escaped. In no land that this terrible disease had scourged in previous epidemics had its prevalence been so general and so destructive. It continued until 1861. Hirsch has pointed out this peculiarity of the Swedish epidemic, that, starting from the province of Göthenburg, in the Skagerrack, in the southwest, it crept steadily toward the north, cases occurring in every season of the year, and districts affected one year almost wholly escaping the next; while the southern boundary of the new area visited by the disease, corresponded very nearly with the northern boundary of that in which it had existed the year before. Strange to say, Norway wholly escaped until 1859, when the fury of the disease was beginning to abate in Sweden, and then experienced it only in circumscribed outbreaks.

Scarcely had the epidemic in Sweden ceased when the disease appeared in the Netherlands—winter of 1860–61, and the next winter witnessed a very general outbreak in Portugal.

A little later, Germany, which, with the exception of slight epidemics early in the century, had almost entirely escaped the ravages of cerebro-spinal fever, became the seat of a severe and long-continued epidemic. The disease appeared in the north in the summer of 1863; it spread rapidly, but at no place assumed the form of a severe epidemic. Middle Germany felt it in 1864, from April to September, and again after an interval of several months—in February, 1865, when an extensive outbreak took place at Eisenach and the neighboring districts. The prevalence of the disease at this time became much more alarming in Southern Germany. It was first observed by Ziemssen at Erlangen, in July, 1864, but it was not until the following winter that it became distinctly epidemic. This author regards it as probable that its earliest appearance dates farther back, as five cases of fatal suppurative cerebral meningitis were examined *post-mortem* at the Polyclinic in Erlangen, in the winter of 1862–63. The violence of the disease in Germany seems to have abated toward the close of the year 1865. Traces of it were, however, met with till 1872. Ziemssen regards it as naturalized in Germany. Small and circumscribed epidemics appeared at various points in the Austrian Empire during the period from 1865 to 1867. It again prevailed to some extent in Vienna in 1872.

The most extended, and at the same time the most destructive outbreak that has visited the British Isles, showed itself in Ireland in March, 1866, and reached its greatest development in the following winter. Dublin suffered chiefly; many of the smaller towns had each a few cases; the soldiery supplied proportionately more cases than the civil population, as in the French epidemics; the disease was almost wholly restricted to Ireland. The continued freedom from this disease which Scotland has enjoyed is most remarkable.

It existed in Russia from 1864 to 1868, and was encountered in the beginning of 1868 in the Crimea.

Small epidemics were observed in Turkey, Greece, Asia Minor, Smyrna, and Jerusalem, between 1868 and 1872. These visitations were not characterized by great severity. On this side the Atlantic, from the time of the outbreak in New Orleans in 1850, no epidemic arose until 1856. During this year it appeared in North Carolina, and continued to prevail in that State till 1857; in the latter year it broke out in the central and western parts of the State of New York, where it raged in its most malignant form. The same year Massachusetts became the seat of a somewhat extensive epidemic.

In 1861-62, it appeared in Missouri, both among the troops and the civil population, and almost at the same time in the army encamped in the vicinity of Washington, D. C. The same year we read of it in Indiana, Kentucky, and Connecticut. In 1862-63, it attacked the troops in Newbern, and this and the following winter those at Memphis, Tenn. Ohio felt its ravages from 1860 to 1864, and within this period the inhabitants of Illinois, Rhode Island, New Jersey, Pennsylvania, and most of the Southern States suffered from it. In 1872, it prevailed extensively in the city of New York and its environs.

It made its appearance in Philadelphia in 1863, and prevailed annually in an epidemic form until toward the close of the last decade, while occasional cases were observed up to 1873, in which year a small but fatal epidemic prevailed in every district of the city, even those most widely separated. Since 1873 it has not occurred here as an epidemic.

Within the last few years cases of cerebro-spinal fever have been observed from time to time in localities in which it has prevailed earlier as an epidemic. These cases have occurred sometimes singly, oftener in small groups. The remark of Ziemssen, that the disease appears to have become naturalized in Germany, seems also to hold good of our own and of other countries that it has scourged. Cases have recently been observed in Ireland, France, Austria, and at Warsaw, and on this side the Atlantic in St. Louis and Atlanta ; but the disease fortunately is at this time nowhere epidemic.

But, in studying the history of this strange disease, our attention is attracted to the fact that it is not alone in those regions that have known it in its epidemic form that cases and groups of cases afterward occur; but that such extremely restricted outbreaks also take place in countries where it has never been, so far as we know, epidemic. England is a notable and authentic instance of such a country. Cerebro-spinal fever has never prevailed in that island as an epidemic. With the exception of a number of cases that broke out in Liverpool in the year 1846, it has been rarely met with within her borders, and then only in single cases or small groups of cases. Thus, in 1807, four cases were observed by Dr. Gervis, of Ashburton. A single case was seen by Dr. Wilks in each of the three years, 1856, 1858, 1859, in London ; in 1859, Dr. Day saw two cases at Stafford, and another at the same place in 1865 ; in the same year three cases were seen by Dr. Wilks in London, three by Dr. Ogle, and one by Dr. Martin.

The foregoing historical notes indicate in a general way the march of the disease and the geographical limits within which it has thus far shown itself. They are the record of outbreaks that have fallen under the notice of medical men who have published their observations. Without doubt they imperfectly outline its actual distribution and prevalence ; yet we are warranted in assuming that no extensive outbreak has escaped notice, and that the appearance of so formidable an affection elsewhere on the civilized globe would have been known and recorded.

Stillé suggests that the history of cerebro-spinal fever, during the present century, may be divided into three periods, each of which comprises the account of a widespread prevalence of the disease upon both sides of the Atlantic, lasting for a series of years. The first of these outbreaks lasted eleven years, from 1805 to 1816 ; the second, thirteen years, from 1837 to 1850 ; and the third, which began in Europe in 1854, and in North America in 1856, was still in force at the time of the issue of his book (1867). It came to a close, however, in 1873, after a continuance of from seventeen to nineteen years, with, however, a period of repose in the United States from 1857 to 1861. If this view be correct and the outbreaks described are in fact distinct epidemics, separated by intervals in which the affection does not occur, it is to be looked upon as a true pandemic—seeing that it appears in places as widely separated as Europe and America at nearly the same time, and overruns the greater part of a continent. In this respect cerebro-spinal fever resembles influenza; but it differs from it in recurring from season to season during a long term of years.

It is to be remembered, however, as a matter of fact, that in the periods of non-activity, cases and even small epidemics of the disease have been observed, as at Vesoul and Dorsten, at Naples, and in this country in Connecticut and Ohio, as recorded by Dr. Stillé himself, and that in recent years occasional limited outbreaks have led so careful an observer as Ziemssen to believe that the disease has been an abiding one in Europe.

There is thus some warrant for regarding cerebro-spinal fever, contrary to the suggestion of Stillé, as having had during the century a continued existence within certain geographical limits, as having had its periods of epidemic outbreak and its periods of quiescence, but never as having wholly ceased to exist within those limits, or as having disappeared in the same sense that cholera and influenza disappear from the same countries in the intervals of their epidemic visitation.

If we call to mind the fact that the disease under consideration was not at all known as a substantive affection until early in this century, and that for many years it was regarded as a variety of typhus fever by most practitioners and by many learned writers upon epidemic diseases, it will not appear unreasonable to assume as probable that many isolated cases and small groups of cases may have occurred during the periods of quiescence between 1816 and 1822, between 1822-23 and 1833, between 1833 and 1837, and between 1850 and 1854-56, and yet have failed to find their way into print and to the general knowledge of the medical profession. Surely, in view of our knowledge of such cases during recent years, it is entirely within the range of possibility that instances of cerebro-spinal fever have escaped recognition as such, at periods when its pathological anatomy was little understood, and the great majority of medical men were ignorant of its existence as a distinct affection.

It has been stated that the lower animals are subject to this disease, that it is at times epizoötic as well as epidemic. It is well known that the epidemic in New York, in 1872, was preceded for some months by the prevalence of a disease among the horses, which presented the same symptoms and *post-mortem* appearances. In no case was it observed to spread directly to the stablemen, veterinary surgeons, or others having charge of them; and upon the appearance of the affection as an epidemic at a later period, those persons were not found to be more liable to its attack than others.

Cerebro-spinal fever presents certain peculiarities in its mode of attack, its extension, its course and duration as an epidemic disease, that separate it widely from other epidemic diseases, and which it seems proper to consider at this point in the discussion.

First, of its mode of invasion. The disease has more than once broken out with activity almost at the same time in the New and the Old World; in many instances it has appeared simultaneously at points as far distant from each as the diameter of a kingdom, while the intermediate regions have remained free from it, not only while it raged in the points attacked, but afterward. This was the case in many of the epidemics of France and Algiers, in those which occurred in Ireland, and in those which visited the United States, especially in the earlier history of the disease. It is also true that in general epidemics, such as have prevailed in Sweden, some portions of Germany, and in districts of our own country, certain localities in the midst of the infected regions have wholly, or almost wholly, escaped its ravages.

Secondly, this fever differs from other epidemic diseases in its mode of extension. In general epidemics it has much more frequently been observed to spread by a series of isolated outbreaks of irregular distribution than by a direct advance from place to place, or by radiating lines from an infected centre. This is not, however, an invariable rule, as is seen, for example, in some of the French epidemics, where the advance of the disease went hand in hand with the movements of troops, or where it corresponded with the course of a river, as in the epidemics which traversed the valley of the Loire; or in that great series of epidemics which passed over Sweden from the southwest toward the north. When, however, we trace the march of the disease more closely in these and similar epidemics, we are struck with the fact that its progress is still by a series of isolated outbreaks—not, in these cases, of irregular distribution, but in the general direction of the line of advance.

Not less remarkable is the course of the disease in an infected population. The wide geographical distribution sketched in the above account by no means represents a general diffusion among the inhabitants of the cities, districts, and countries in which it has prevailed. Scattered cases and groups of cases may occur over a wide area, without any great tendency to a concentration of the violence of the epidemic, as in the Swedish and most of the German epidemics; while, on the other hand, the whole number of cases may occur within restricted limits, the latter being the more common rule. Many epidemics have attacked a single class in the community. This was the case in France, in 1837 and the following years, when the disease chiefly affected the soldiery, often being confined to a garrison, or a section of a garrison, sometimes even to a single regiment, without extending to the surrounding populations, and in 1844, at Gibraltar, where the civil population bore the brunt of the attack. The same limitation of the cases to a class among the people was observed during the epidemic in Italy, where in 1840 an outbreak at Procida was almost exclusively confined to the convicts in the galleys; in Ireland, where in 1846 the inhabitants of the workhouses principally suffered; and in the late American war, at Newbern, Memphis, and in the neighborhood of Washington, where the troops alone suffered.

Finally, this disease presents remarkable differences from other epidemic diseases, in regard to its duration as an epidemic. In this respect it has, at different places and in different outbreaks, shown the most extreme variations. The greatest number of epidemics have lasted from three to six months; others have been of shorter duration, coming to an end in a few weeks, while it has frequently happened that new cases have appeared throughout an entire year, or from the spring of one year till the end of the following winter. The duration of the epidemic depends upon causes not yet known. It cannot be said to be influenced by the size of the population, for on one hand we read of comparatively brief outbreaks in populous cities like Berlin, Vienna, and on the other of lingering epidemics in such relatively sparsely inhabited countries as Algiers and Sweden.

The epidemics are often, in spite of a duration of several months or even of a year or more, limited to a relatively small number of cases in the infected community—a few individuals here and there being attacked, and the mortality being moderate. In other instances, on the contrary, considerable numbers suffer and the death-rate is high, and, as Hirsch points out, the proportionate number of persons attacked, and of fatal cases, are not seldom in inverse ratio to the duration of the epidemic, a relatively great number of cases occurring, with a high mortality, in epidemics that came to an end in a few (six to eight) weeks.

Sometimes the outbreaks do not, as is the case with most epidemic diseases, rise steadily to an acme and then gradually decline, but seem to run an irregularly intermittent course, a number of persons being attacked, then the disease to all appearance vanishing, only, however, to return after a time to seize new victims, and this disappearance and return being repeated till the close of the epidemic, after many weeks or months.

ETIOLOGY.

The cause of cerebro-spinal fever is as yet unknown. Much less is known of the laws which control its origin, its distribution, its action in communities and upon individuals, than is known of the active causes of most of the other infectious diseases. The unaccountable appearance of the disease at the same time in widely separated localities, its diffusion by isolated attacks rather than by direct advance, its variable and often long-continued prevalence in epidemics, its sporadic occurrence between the epidemics, the extraordinary diversity of the symptoms in different epidemics and in different cases, baffle the comprehension and render futile every effort to formulate even a satisfactory hypothesis of its cause and origin.

There appears to be no longer any question as to the infectious nature of this disease. The constant local lesions suggest the idea of its being essentially an inflammatory process—to this suggestion the prominence and constancy of the symptoms due to meningeal inflammation bear support; but a closer examination of the clinical history of the affection, and a wider study of its pathological anatomy show that this view is untenable.

The onset of cerebro-spinal fever, the initial chill, and the febrile phenomena, are analogous to those of the infectious diseases, especially the eruptive fevers. The rapid course of the attack to its fatal termination—which often takes place within a few hours at the beginning and at the height of a malignant epidemic, and for which no satisfactory explanation is found in the lesions—is without parallel in simple inflammations, but is not infrequent in severe infectious diseases. The appearance of various eruptions, and especially of herpes, and the rapidity of the discoloration of the body after death, also point to the infectious character of this disease. But when we come to consider the tissue-changes outside of the region of the cerebro-spinal axis, we are still more fully impressed with the resemblance which exists between this and the other infectious diseases. We find blood-changes, and degenerations of the heart and the voluntary muscles, to be almost always present. The change in the blood is as constant as the meningeal inflammation and exudation.

Although the specific cause of this, as of other infectious diseases, is not known to us, the observations of the past quarter of a century furnish data upon which to base knowledge of a very positive kind relative to its predisposing causes.

It may be safely asserted, in spite of the fact that the limits within which cerebro-spinal fever has been known to prevail are limited, as compared with the inhabited surface of the globe, that climate has no direct influence in producing this disease. Epidemics have, as we have seen, occurred in the Eastern hemisphere, in Central and Western Europe and Algeria, between the thirty-fifth and sixty-third degrees of north latitude—therefore in all kinds of climates, from the subtropical of the Mediterranean coast to the rigorous of mid-Sweden. In the Western Hemisphere, the range has scarcely been less extensive, for we meet with records of the disease in the eastern portions of North America, from the Gulf Coast to Canada, a territory whose southern boundary is the thirtieth and whose northern boundary is about the forty-eighth degree of north latitude, and which embraces the greatest variety of temperate climate. The disease is not known to have prevailed within the tropics.

The season of the year, and the weather, appear to exert an important influence. Winter and the cold months are unquestionable and powerful predisposing causes of epidemic cerebro-spinal fever. Of 226 local outbreaks in France, 166 occurred between December 1st and May 31st, 60 during the other six months of the year; while in Sweden, of 397 local outbreaks, 311 took place in the former period, 86 in the latter.[1] In the

---

[1] See Hirsch: Die Meningitis Cerebro-spinalis Epidemica, p. 113. The figures are incorrectly given by Stillé, and by Radcliffe, who follows him.

United States the outbreaks have taken place with great uniformity in the winter and spring, the exceptions to this rule being very few.

The epidemics which have occurred in the winter and spring have usually been widely extended, the outbreaks of the summer and autumn being milder and more limited.

We possess no accurate detailed investigations as to the effect of temperature, moisture, and the direction of the wind.

The development of this disease is clearly favored by cold weather, yet it is impossible to say in what way this agent acts. The specific cause of the disease, the fever-poison, may find in a low temperature conditions favorable to its existence, or the influence of cold upon the bodily constitution may call forth an especial predisposition for the affection; or, finally, in the modifications of the mode of life peculiar to the season of cold weather, this disease, as some others, may find the conditions most favorable to its development.

Locality does not act in any way as a predisposing cause of cerebrospinal fever. Low, marshy regions, high plateaus and the inhabited districts of mountains, have alike suffered from its ravages. The condition of the soil exerts no influence upon its development. It has prevailed in swampy bottom-lands, and upon dry, sandy soils, if not with equal frequency, at least with no preference for one or the other that the records collected make manifest. Densely populated cities and the scattered populations of agricultural regions, may alike stand in terror of its return, for they have alike known its horrors in the past.

In contradistinction to locality in the broad sense with reference to communities, the place of abode and the mode of life of the individual have much to do with the development and prevalence of this fever. Damp, overcrowded, and unclean habitations favor it. Those living wholly upon the ground floor are apt to suffer. Ziemssen regards the overcrowding of dwelling- and sleeping-rooms, and the consequent loading of the air with animal emanations, and perhaps the saturation of the soil with garbage and the products of its decomposition, as being agents as powerful in the germination of the contagion of this disease as they are in that of cholera.

The common occurrence of limited outbreaks in detached dwellings, in a row of houses or a street, in schools, in prisons, in workhouses, as in almost every one of the Irish epidemics, and in garrisons, as in most of the epidemics in France in 1837 and the following years—and this, in most instances, without the spreading of the disease to the civil population, force upon us the conclusion that whatever may be the nature of the specific cause, the unfavorable dwelling-place of the individual acts as a powerful collateral cause of the disease. That such conditions are, however, merely collateral causes of the disease and altogether inoperative in its production in the absence of the specific cause, is made clearly manifest by their permanence within the limits of its merely occasional occurrence, and in other localities where it is unknown.

To similar unfavorable modes of living is to be attributed the severity of the epidemics that have so often prevailed among the negroes in the South. It is impossible to trace any direct influence of race as a predisposing cause of cerebro-spinal fever. The negro is not more susceptible to it than others living under the same unfavorable hygienic conditions as did the slaves formerly, and do the freedmen of to-day, upon the cotton- and rice-plantations.

While it has been observed in by far the greatest number of outbreaks, that those subjected to privations and consequent foulness of person and dwelling, by reason of poverty, suffered most, it is by no means to be inferred that the opposite conditions of life secure entire immunity. On the contrary, those living in affluence and under hygienic conditions of the most favorable kind, have in many epidemics fallen victims to the disease. In the United States, the households of respectable, well-to-do farmers have often contributed a large percentage of the cases in infected regions.

It may be stated, as a general rule, that in adult life the proportion of males attacked is much greater than of females. In some outbreaks, as where the disease has been confined to a garrison, males only have suffered; but the rule holds good, where civil populations alone are visited, more males than females suffer. The explanation of this fact lies undoubtedly in the difference of the mode of life of men and women in civilized communities. It is to be observed, however, that in a few epidemics the females attacked have outnumbered the males (Hirsch). Among children the number of males and females is nearly equal.

Age is of the greatest importance among the predisposing causes. Immunity is reached at no period of life. Ziemssen has examined, after death, persons as old as seventy and seventy-seven years; beyond forty, cases are, however, rare; more common between twenty and forty years, but by far most common in the first twenty years of life. In many epidemics only children under fifteen years have been attacked. Of 1,267 fatal cases occurring in Sweden from 1855 to 1860 inclusive, where the age was stated, 889 were under fifteen years, 328 between sixteen and forty, and 50 over forty years old. Of 779 fatal cases of which statistics were collected by Hirsch in the districts of Carthaus and Brebant, in Dantzig, 208 occurred within the first year of life, 337 between the first and fifth years, 151 between the fifth and tenth years, 41 between the tenth and fifteenth, 16 between the fifteenth and twentieth, and 26 at all ages over twenty.

Again, Dr. Sanderson[1] found that, in the Lower Vistula, 218 fatal cases

---

[1] A Report of the Results of an Inquiry into the Epidemics of Cerebro-spinal Meningitis Prevailing about the Lower Vistula, etc. Official paper. **Eighth Report of the Medical Office of the Privy Council.** London, 1865.

were under the age of fourteen years, and only seventeen above that age.

Children are not only more susceptible to the disease, but the death-rate is higher during childhood than at any other period of life. The figures given above, being derived from mortuary statistics, are misleading, unless due allowance be made for the fact just stated. According to the official report of the cases occurring in Central Franconia, from June, 1854, to the date of the report in 1865, as quoted by Ziemssen, of 456 persons attacked—

    257 were from 0– 9 years of age.
    126   "   10–19 "   "
    41   "   20–29 "   "
    32 were over   30   "   "

Of 42 cases observed by Ziemssen himself—

    14 were from 0– 9 years of age.
    13   "   10–19 "   "
    9   "   20–29 "   "
    6 were over   30   "   "

The following are the statistics of the Board of Health of the city of New York, relative to the age of the persons attacked:

Of 975 cases—

    125 were under 1 year of age.
    336 " from 1– 5 years of age
    204 "   " 5–10 "   "
    106 "   " 10–15 "   "
    54 "   " 15–20 "   "
    79 "   " 20–30 "   "
    71 " over   30 "   "

Finally, out of 81 cases coming under the personal knowledge of Dr. Smith,[1] in the city of New York—

    8 were under 1 year of age.
    18 " from 1– 3 years of age.
    20 "   " 3– 5 "   "
    17 "   " 5–10 "   "
    7 "   " 10–15 "   "
    11 " over 14 "   "

All occupations and professions are liable to this disease. Military life appears to be attended with, however, an especial liability. Most of

---

[1] J. Lewis Smith, M.D. Cerebro-spinal Fever, with Facts and Statistics of the Recent Epidemic in New York City. American Journal Medical Sciences. October, 1873.

the epidemics of France following the outbreak of 1837 occurred almost exclusively among the soldiery. During the American Civil War several outbreaks occurred among the troops, without extending to the neighboring civil populations.

Military observers have very generally regarded unusual **exertion, hardship and fatigue, especially** in connection **with exposure to cold,** as powerful predisposing causes. These influences play but little part in the production of the disease in the civil population. Nevertheless, many observers are of the opinion that exciting and perturbating influences of some sort not infrequently precede the attack, and stand in a causal relation to it, in so far as they render the individual prone to the epidemic influence.

Dr. J. Lewis Smith mentions over-work, fatigue, mental excitement, prolonged abstinence from food, followed by over-eating and the use of improper and indigestible food, as rendering the subject liable to the attack. He found that children who had been **subjected to the** severe discipline of the public schools, returning home tired and hungry, and eating heartily at a late hour, were especially **liable to the** attack. **The condition of the individual, as regards previous health, was not seen to exert any causal** influence of importance. **Persons attacked are generally strong and in** robust health, and, as has been seen, in the earlier periods of life. Chronic invalids **do not, however, escape, nor do those suffering from other** acute affections.

But these influences, however potent as predisposing causes they may be, **are** utterly inadequate to the production of cerebro-spinal fever in the absence of the single, specific, exciting cause.

The nature of this cause is unknown.

It can no longer be maintained that **cerebro-spinal fever is a form of** typhus, or that it is allied **to it. On the** contrary, it is now universally admitted that **it is a disease** *sui generis.* The essential point of difference between these two diseases will **be** considered under the head of diagnosis.

Still less can this disease be regarded by thoughtful observers as a variety of malarial fever. It can be readily **shown** that between the cause or infecting principle of cerebro-spinal fever **and that of** malarial diseases there exists no identity whatever.

Cerebro-spinal fever not infrequently, as will be seen **when we come** to consider its clinical **history, presents a distinctly intermittent course.** For this reason it **has been thought that some relationship between the** two affections exists.

A brief consideration of the **following** points shows that **this opinion cannot be maintained.** Cerebro-spinal **fever** shows no preference for ma**larious localities;** it prevails commonly during the **cold** weather of the **winter and** spring months, and is relatively infrequent in the seasons in which malarious diseases are rife ; thirdly, when it prevails in malarious regions, it shows neither a greater malignancy, nor a more marked tendency to run an intermittent or remittent course than it does in high, sandy regions, where native malarial fever is unknown ; and in conclusion, there is, beyond the tendency of the fever to remit and intermit, no further resemblance between the two diseases. Marked enlargement of the spleen is not common in cerebro-spinal fever, enlargement of the liver is not one of its sequels, and quinine has no power to check it, even when its course most closely resembles that of a malarial attack.

Cerebro-spinal fever has been almost unanimously pronounced to be either absolutely non-contagious, or contagious to a very slight degree, by those whose observations have rendered them competent to judge of the matter. The vast majority of cases break out without the most remote possibility of personal communication. The cases occurring at the beginning of an epidemic appear here and there in the community, in distant quarters of a city, or miles apart in rural districts. As the disease spreads, new cases and groups of cases arise in distant localities whose inhabitants have held little or no communication with the sick or their attendants. In many instances only one, or perhaps two members of a family, are attacked throughout the course of an epidemic. The members of a household occupying the same apartments, or passing freely in and out by the bedside of the patient, are apparently not more liable to contract the disease than others having no contact with the sick, unless the dwelling be foul and ill-ventilated. When several cases occur in the same house, it is often at such irregular intervals as to preclude the idea of infection at the same time, or from one another, as the periods of incubation of the contagious disease is within certain limits, uniform. Thus, Dr. Sewall,[1] in the epidemic of 1872 in New York City, met with an instance where six cases occurred in a single family, but at intervals of five, seven, eleven, twenty-five, and forty-five days respectively. In the same epidemic Dr. J. Lewis Smith encountered in each of 39 families a single case, in 16 families 2 cases each, and in 3 instances, 3 cases in a family.

It is a notable fact that physicians, nurses, and other attendants upon the sick have contracted the disease only in the rarest instances, and patients in hospital wards into which cases of cerebro-spinal fever have been brought have remained, as a rule, unaffected by the cause of the disease.

In view of the foregoing facts, it is to be concluded that the opinion so universally entertained by medical men who have made this disease the subject of personal study, namely, that it is non-contagious, is correct; that is to say, that it is not contagious in the sense in which we are in the habit of using the term in speaking of small-pox, scarlet fever, typhus, etc. But that it is capable of being communicated from the sick to the

---

[1] New York Medical Record, July 1, 1872.

well, under certain favorable circumstances which are as yet unknown to us—in other words, that it is in a modified sense contagious—seems probable. The difficulty in discussing the subject of the contagiousness of any disease arises in part from our habit of using the word in its broadest and most positive meaning, as when we speak of the diseases just named. As long as there is room for doubt as to the possibility of a malignant disease spreading by contagion, it is, from a practical standpoint, important to regard it as to a degree contagious; so that, if errors occur in respect of this point, they may be committed on the right, that is to say, on the safe side.

Among the facts which suggest the possibility of the specific exciting cause of this fever being communicable from one person to another, or at least portable on the person or among the effects of those who have been exposed to it, are the following: the outbreaks in that regiment of French soldiers already mentioned, which, having suffered from the disease in Bayonne in 1837, again suffered from it in 1838, after having been transferred to Rochefort, where the disease had not previously existed, and in which, in the following year, it having been transferred to Versailles, the first cases of a lingering but limited epidemic again showed themselves; in 1840 the disease broke out at Laval, and accompanied the troops marched from that place to Mans and Château-Gonthier; the appearance of the disease among the French soldiers in Algiers, in 1840, at a time when it prevailed extensively in the army in France. Hirsch points out the importance that would attach to this extension of the disease across the Mediterranean, if it were known whether or not the malady appeared first among troops fresh from France. The same authority narrates the following reliable observations:

"According to Fraentzel, the first case of the disease among the troops in Berlin was in one of the reserve of the Alexander regiment, who a few days before had come from Leignitz, where the meningitis was then prevailing as an epidemic. In the second company of this regiment, to which he belonged, five more cases afterward occurred; in the two companies lying adjacent on each side, there were three cases in one, two in the other; in the four companies beyond them, only one and that the lightest case, while there was no illness at all in the second battalion, which occupied the same barracks and was separated from the first only by a small court, the two battalions as is generally the case, associating but little with each other, but sharing in all respects the same labors and mode of life, and eventually the evils of overcrowding."

Still more striking is the following observation cited by Hirsch as occurring in the epidemic of West Prussia, in 1865:

"On February 8th, K., aged twenty, fell ill in the township of Sczakau. He was nursed by the girl W., who had hastened to him out of the village of Sullenczyn. After the death of K. his nurse returned home and there died on February 26th, of epidemic meningitis. This was, with the exception of one on January 15th, the first fatal case of meningitis in Sullenczyn. To the burial of this maid came the family of the farm-steward K., to Sullenczyn from the township of Podgass, accompanied by the servant D. and the **four-year-old daughter** O. of the teacher R. in Podgass. After their return from the funeral, a little child of K.'s fell **ill and** died in twenty-four hours, then **the servant D.**, **who died on** March 4th, and **finally** the girl R., **on** March 7th."

The same authority cites also the following example:

"At another village, two children of **one** family, aged respectively one and a half and three and a half years, died of the epidemic, one on January 27th, and the other **on February 7th.** The **clothes of** the deceased were taken to a neighboring village, **and** came into the possession of a girl aged five years. She soon sickened of the epidemic, **and died** on February 14th."

The following cases are less striking. The first is cited by Stokes, the other by J. Lewis Smith:

"A child was seized with epidemic **cerebro-spinal meningitis and died.** A second child of the same family was attacked with the malady a few days later. The day following the attack of this child, the mother, who slept in the same bed with it, sickened of the disease."

"A boy, twelve years of age, died of **cerebro-spinal fever,** and was buried **on Saturday** or Sunday. On the following Monday the mother washed the linen of the **boy,** which had accumulated, and within two days was **herself** affected with the **disease.** She and her infant, who was also seized with it, **died.**"

Such instances are exceedingly uncommon. Were they less so, the argument in favor of the contagious nature of the disease would be stronger.

All that our present knowledge warrants us in saying upon this point is that the usual mode of epidemic spread of cerebro-spinal fever renders it highly probable that the fever-producing poison is of the nature of a miasm which becomes active at places remote from each other at the same time, and spreads independently of human intercourse; while in rare instances it appears to have been carried upon the persons or among the personal belongings of those who have been in contact with the sick, and in the rarest cases it would seem to be capable of direct transference from the sick to the well.

This fever may be classed among miasmatic diseases.

CLINICAL HISTORY.

Cerebro-spinal fever presents a great diversity of symptoms in different cases. Like other epidemic diseases, its course is attended by the greatest variations in intensity, duration, and the prominence of particular phenomena, not only in different epidemics, but in the same epidemic.

In this respect, however, it not only resembles other epidemic diseases, but it also far surpasses them. No acute disease whatever appears in such various arrays of symptoms. Stillé has well called it a "chameleon-like disorder." It is this that has rendered it more difficult to describe satisfactorily than to recognize at the bedside. It is this also that has led to the great diversity of opinions concerning it that have been entertained by different observers.

It is to this extraordinary difference in the symptoms attendant upon different cases and predominating in different epidemics, that are due the efforts of systematic writers upon the subject to simplify their descriptions of cerebro-spinal fever by the arrangement of the cases under separate groups or headings—an effort which in most instances has failed signally of its object.

Thus, Forget[1] classifies the cases which came under his observation at Strasbourg as follows:

A. Cerebro-spinal:
   1. Fulminant.
   2. Comatose-convulsive.
   3. Inflammatory.
   4. Typhoid.
   5. Neuralgic.
   6. Hectic.
   7. Paralytic.

B. Cerebral:
   1. Cephalalgic.
   2. Cephalalgic-delirious.
   3. Delirious.
   4. Comatose.

Ames[2] and others separate the cases into two groups:
   1. The Congestive.
   2. The Erethetic or Inflammatory.

Wunderlich[3] classifies the cases into groups according to their relative degree of severity, and describes—
   1. The Gravest.
   2. The Less Grave, and
   3. The Lightest.

---

[1] Gazette médicale de Paris, 1842. 15–20.
[2] New Orleans Medical and Surgical Journal. November, 1848.
[3] Archiv der Heilkunde. 1864, 1865.

Radcliffe arranges them as—

  A. Simple.
  B. Fulminant.
  C. Purpuric.

Stillé as—

  A. The Abortive.
  B. The Malignant.
  C. The Nervous.
    a. The Ataxic:
      1. The Delirious.
      2. The Cephalalgic.
      3. The Neuralgic.
      4. The Convulsive.
      5. The Paralytic.
  and *b.* The Adynamic:
      1. The Comatose.
      2. The Typhoid.
  D. The Inflammatory.
  E. The Intermittent.

Bartholow[1] as—

  1. The Ordinary or Common.
  2. The Fulminant.
  3. The Petechial.
  4. The Abortive.

And finally it has been suggested that the cases may be classified with reference to type into—

  A. Continued.
  B. Remittent.
  C. Intermittent.

It is, I think, simpler, more convenient, and more in accordance with modern methods of describing diseases, to adopt the plan followed by German writers (Hirsch, Ziemssen), and, shunning all artificial classifications, to give a general sketch of the course of the disease as it appears in the greater number of cases, presenting the symptoms as they arise, group themselves and succeed each other, and regarding those groups of cases that differ broadly from this general picture, as varieties requiring separate attention rather than as distinct forms of the disease.

In this way will we more readily keep in mind the essential unity of the

---

[1] Practice of Medicine. New York, 1880.

malady as shown by its infectious character and constant lesions, under the widest diversity of symptoms in different cases.

The onset of the disease is, in most cases, abrupt. If prodromes occur, they are of variable duration, and consist of symptoms referable to disturbance of the nervous system, such as headache, dragging muscular pains in the neck and extremities, vertigo, and a sense of fatigue. Githens'[1] statement that the disease came on gradually, with usually "about a week of prodromata," is not borne out by other observers. Ziemssen, on careful inquiry, found prodromes present in only five out of forty-three cases. When present, they last from a few hours to several days. They are apt to disappear shortly before the outbreak of the disease. In some instances slight repeated shivering has preceded the attack.

In by far the greatest number of cases the attack is ushered in by symptoms of the most formidable character. The patient is seized with a violent chill; agonizing headache, nausea, vomiting—which is repeated and provoked by movement or any attempt to rise—supervene. He is restless, tossing about the bed and oppressed with an overwhelming sense of illness. His countenance betokens his profound distress. His face is seldom flushed, usually pale or cyanotic, sometimes wearing the expression of those under the influence of narcotic poisons. In a short time dragging pains in the neck come on, which spread to a greater or less extent along the spine and into the extremities, and are soon followed by that tetanic stiffness of the muscles of the spinal region that is one of the characteristic features of the disease. Pain is now experienced in attempting to bend the head forward, or to turn it from side to side. The muscular stiffness extends to the extremities, and movements are made with awkwardness and difficulty. In a few hours or days this symptom deepens into complete opisthotonos. The head is drawn back, the spine curved, the forearms flexed upon the arms, the legs upon the thighs. Cramps in the muscles of the legs and elsewhere, and spasmodic twitchings of the lips, eyelids, etc., come and go. General convulsions may occur, especially in children. With these symptoms of irritation of the roots of motor nerves are associated those of a not less profound disturbance of sensation. Hyperæsthesia of the entire surface is present, but the sensitiveness of the face, forehead, and neck is most marked. A slight pinch, or an attempt to separate the eyelids for the purpose of examining the eye, will often call forth an expression of pain, even when insensibility is profound. The greatest suffering is, however, from the headache, which often causes restlessness and expressions of suffering during insensibility. It is described as sharp, lancinating or boring, and may be either in the forehead

---

[1] Dr. W. H. H. Githens: Notes of Ninety-eight Cases of Epidemic Cerebro-spinal Meningitis, treated in the Philadelphia Hospital. American Journal Medical Sciences, July, 1867.

or occiput, or may shoot about in all directions. Sometimes it is felt as a constricting band; sometimes it cannot be located, but is spoken of as an unutterable anguish. Pain of a like nature is felt in the lumbar, epigastric and umbilical regions. The abdominal pain is usually an early symptom, and sometimes precedes the vomiting. Vertigo persists. It recurs upon every attempt to rise, and is often distressing when the patient lies quiet, compelling him to seize hold of the bed. The vomiting continues. At first the contents of the stomach, afterward bilious matters and gastric mucus, are thrown up.

The high mental excitement which marks the onset of the attack passes into delirium—which may be active, even maniacal, so that restraint is required, or of a busy, wandering character. In a short time it passes into somnolence or stupor, which is, however, still attended with more or less restlessness and continual movement upon the bed. Various disturbances of the special senses occur. Intolerance of light is constant, double vision and temporary or even permanent blindness, sometimes supervene. Intolerance of sounds, ringing in the ears, and dizziness usher in deafness, which is more or less pronounced and not infrequently persistent. Taste is lost— the patient refuses food; nevertheless, the vomiting persists. Constipation is present at first, often throughout the sickness. Toward the end of the attack, diarrhœa and involuntary discharges may take place. In this respect epidemics differ greatly, and in truth it cannot be said that the symptoms referable to the alimentary canal are at all characteristic. The tongue is, as a rule, lightly covered with a whitish fur; where there is great depression it becomes dry and brown, and sordes collect; but this again gives place to a moist and whitish fur.

The fever is generally moderate, very irregular, and does not observe a typical course. The pulse is about normal in frequency or moderately quickened. It presents, however, the most remarkable variations in respect of its frequency, being at one time unaccountably quickened, at another unaccountably slowed, and these conditions succeed each other with great rapidity and no less irregularity; the changes in tension are not less notable. The respirations are likewise irregular. They are at first quickened, later they become shallow and irregular in rhythm. Sometimes this irregularity is of that form known as Cheyne-Stokes respiration.

Cutaneous eruptions appear after the first few days. Herpes is common; erythema, roseola and urticaria occur. Petechiæ also, the sign of blood-disintegration, are common. In many epidemics no eruption has been observed. The herpetic eruption is most frequent on the face; it occurs elsewhere on the body, and is sometimes symmetrically distributed. The others are of irregular distribution.

The disease develops rapidly to its height; from the third to the sixth day, the symptoms have reached their full intensity. If the attack progress to an unfavorable termination, the symptoms of motor and sensory excitation yield to those of depression; the rigidity passes away and is replaced by palsies; the stupor deepens into coma; the fixed expression of pain fades into blankness; the eyes are sunken, the pupils dilated; no noise disturbs the patient. The temperature rises—40.5° C. (105° F.), or even 42.2° C. (108° F.), being attained; the pulse becomes rapid, small, and scarcely perceptible, the breathing weaker and more shallow, and, with convulsive muscular movements ending in the most profound coma, death puts an end to the scene of horror.

If the case run a favorable course, the symptoms of depression is less marked, and they continue for a much shorter time. Intense headache, backache, and muscular pains in the extremities, are complained of; the patient generally lies quiet in one position. There is intolerance of light and of sound. The vomiting after a few days, or early in the progress of the case, comes to an end. The headache and other pains slowly subside, and with them the muscular rigidity; the strength is slowly regained, and after a period varying from two or three to several weeks the patient regains his health. The convalescence is often protracted for a long time by the back- and head-pains, and by disturbances of sight and hearing.

In certain instances, after the fever has lasted some time, the patient passes into that condition to which the term "typhoid state" has unfortunately been applied, and lies in a condition of semi-stupor, with muttering delirium, a dry, cracked tongue, sordes upon the teeth and gums, a feeble, rapid pulse, involuntary discharges, and like symptoms of vital depression. These are cases of prolonged meningeal inflammation where the infection has been severe and the blood-alterations profound. They have been described as the "typhoid form" of cerebro-spinal fever by many writers. To classify these cases together as a group, seems to add to the obscurity of the description of the disease rather than to simplify it. Certain cases of other infectious diseases tend to run on into a similar state. It is profitable to regard them as simply grave and protracted examples of such affections, rather than to describe them as separate forms. It is above all unfortunate and unprofitable to designate them by the term "typhoid," in view of the confusion that has existed as to the relation of cerebro-spinal fever to typhus.

Cases occur in every epidemic that differ so much from the general description of the ordinarily severe and mild forms of the disease, that they require separate consideration as varieties. These are the fulminant, the abortive, and the intermittent varieties.

The fulminant variety—the meningitis cerebro-spinalis siderans of Hirsch, and the méningite foudroyante of French writers. The poison seems to fall upon the patient like a thunderbolt. He is struck down without warning, in the midst of health, and speedily falls into a state of collapse. In a few hours death may ensue. There is usually a violent chill;

the patient becomes cyanosed; the skin is cold—it may be clammy to the touch, or bathed in a profuse perspiration. The face is shrunk and livid, **the eyes deep-sunk in the orbits.** There is shivering at intervals; intense headache alternates **with drowsiness, and, after** brief delirium, unconsciousness supervenes. There is contraction of the neck, and general convulsions may usher in profound coma, which is in most cases the forerunner of death. The respiration is slow and labored; the pulse, weak from the onset, rapidly grows more rapid and more faint. The urine is scanty **and loaded with albumen.** Purpuric blotches appear on the surface of the body and pass quickly on to vesication and sloughing. These cases occur in **nearly all epidemics, and with** greatest frequency at the beginning. Some observers think that they have been less frequent in the later epidemics. They do not occur, as far as is known, sporadically. They are, with **the rarest exceptions,** fatal, death usually taking place within a **few hours (five to twelve),** though life may be prolonged till the third day.

Tourdes,[1] speaking of the suddenness of **the attack, states** that soldiers **full of youth and strength were stricken in the street, at drill, in the** barracks, **whilst at meals, and succumbed in a few hours. Ziemssen in 43 cases encountered four, which proved fatal in 12, 24, 28, and 30 hours respectively.**

The following case, abridged from the article by the same author, will serve **as an example of this** variety of the disease:

A child, **aged eight years, was suddenly taken ill,** April 22d, with severe headache whilst at play, and **came home complaining and in tears.** After being put to bed she suffered from nausea, **active vomiting, and vertigo.** The headache increased, the eyes **became distorted, the fingers of both hands firmly clutched. This condition is** said to have lasted about **two hours, during which** the patient remained apparently conscious and often screamed loudly.

In the evening she laid **quietly in** bed, perfectly conscious, complaining of headache and intense thirst; her neck was not stiff; temperature 39.5° C. (103.3° F.), **taken** in the rectum; pulse 100. Late in the evening there were several attacks of **vomiting;** she slept badly.

On the following morning the headache was entirely gone; the skin, especially that of the face, was very pale, moderately warm; temperature 38.3° C. (100.7° F.) in the rectum; pulse 100.

Toward noon the patient felt better and got up. She even fetched beer from a neighboring public house.

About two o'clock, after having amused herself for some time with her sisters, she suddenly became quiet, laid down upon the floor, and complained of severe headache. In a little while the eyes became drawn, marked contractions of the hands and feet ensued, which were soon followed by violent general convulsions, with constant groaning and screaming. Consciousness was said to have been lost only for a short time. After four hours the convulsions gradually ceased. **The child** asked for a drink, sank into a stupor, and died in half an hour.

The *sectio cadaveris* revealed a small amount of sero-purulent infiltration of the

---

[1] *Histoire* de l'épidémie de méningite à Strasbourg en 1840 et 1841. **Paris,** 1843.

arachnoid and subarachnoid spaces in the brain and spinal cord, anæmia and œdema of the brain and spinal cord, bronchial catarrh, and partial collapse of the lungs, swelling of the solitary follicles of the small and large intestines.

The abortive variety occurs at the height and during the **decline of all epidemics**. Many cases do not require confinement in bed. **The patients complain of headache, stiffness in the neck and spine, and malaise.** Vomiting occurs, but fever is, as a rule, absent. Such cases are instances of the incomplete action of the epidemic influence. The diagnosis rests chiefly upon the presence of associated headache, spinal stiffness, and **vomiting** during the prevalence of the epidemic. All observers record such cases. Stillé narrates the case of a girl who was fully convalescent on the fifth day, and states that he observed many slight but distinctly **marked** cases that were fully convalescent within a week.

The following examples of this variety of **the disease came** under the observation of D. J. Lewis Smith:

"A boy of eight years, previously well, **was** taken with headache, vomiting, and moderate febrile movement, on April 20, 1872. The evacuations were regular, and no local cause of the attack could be **discovered.** On the following day the symptoms continued, except **the vomiting, but** he seemed somewhat better. On April 4th the febrile movement **was more pronounced, and in the afternoon he was drowsy and had** a slight convulsion. **The forward movement of the head was apparently somewhat re**strained. **On the 6th the symptoms had** begun to abate, **and in about one week from the commencement of the attack his health** was fully **restored."**

"A boy, six years old, was well till the second week in May, 1872, **when he became** feverish and complained of headache. **On** May 14th he still had **headache, with a pulse of** 112. The pupils were sensitive to light, but the right **was wider than the left. The bromide** and iodide of potassium **were** prescribed, with moderate counter-irritation be**hind the** ears. The headache and febrile movement in a few days abated, the equality of the pupils was **restored, and within a little** more than a week from **the first symptoms** he fully recovered."

During the prevalence of epidemics of cerebro-spinal fever, some **of the symptoms characteristic of that disease** have occasionally been observed **as incidental** complications of inflammatory affections of the lungs, pleuræ, and tonsils. There was chill, followed by headache, nausea, and vomiting, and stiffness of the spinal **muscles;** vertigo, unconsciousness, and **slight** elevation of temperature. In such cases the symptoms are usually mild. It is open to doubt whether **or not there be not, in** fact, mild cases of the epidemic disease complicated with **local inflammations.** Two cases cited by Ziemssen would warrant the **latter conclusion.**

The intermittent variety has been observed in many epidemics. The intermissions show themselves **either** at the beginning of **the attack, both** in short and **protracted cases, and at** its close, during the **period of conva**lescence. **They are** attended by **complete** or almost **complete** subsidence of the fever **and all the** other manifestations of the disease. The recurrence of the exacerbations assumes **the** quotidian or tertian type, but careful investigation has shown that it lacks the regularity of true malarial diseases, to which the resemblance of these cases is more apparent than real. It has already been shown that this variety resembles intermittent fever only in the one feature of alternations of periods of activity of the symptoms and periods of repose. The intermittent course of the disease often lasts for several weeks and finally suddenly ends in the death or recovery of the patient. At other times the prodromal stage consists of several brief attacks, and again the convalescence may be broken by a series of severe paroxysmal seizures, led in by chills and attended by fever of a high grade—40° C. (104° F.).

The following case serves at the same time as an illustration of the mildest form of cerebro-spinal fever and of its intermittent variety. It occurred under Dr. J. Lewis Smith's care.

"A girl of thirteen was seized in the last week of December, 1872, with vomiting, followed by headache. During a period of from six to eight weeks, or till nearly the first of March, she presented the following symptoms: daily paroxysmal headache, often most severe in the forenoon, neuralgic pain in the left hypochondrium and sometimes in the epigastric region; pulse and temperature sometimes nearly normal, at others accelerated and elevated, both with daily vomiting; inequality of the pupils, the right being larger than the left during a portion of the sickness. This patient was never so ill as to keep the bed, usually sitting quietly during the day in a chair, or reclining on a lounge, and she never fully lost her appetite. Quinine had no appreciable effect on the paroxysms of pain or fever."

Analysis of the Symptoms.

SYMPTOMS PERTAINING TO THE NERVOUS SYSTEM.

A *chill* of more or less decided character, often very severe and sometimes lasting from one to two hours—less frequently a shivering, which speedily passes over—marks the onset of the attack. This phenomenon is occasionally repeated several times in the course of the first day.

The attack begins most frequently in the evening, during the night or on rising, less commonly during the early hours of the day. The patient in some instances, having gone to bed in perfect health, awakes with symptoms of the most alarming kind. During the subsidence of epidemics, the beginning of the sickness is less apt to be abrupt. It is also, as a rule, of more gradual approach in children.

Headache is among the earlier, the most distressing, constant and more persistent symptoms of the disease. It is absent only in those cases of the fulminant variety in which the patient is overcome as by a lightning-stroke, and falls directly into collapse. The seat of the pain, as has already been pointed out, is variable. Its agonizing character is almost constant. The patient is thrown by it into the greatest restlessness. He throws his arms about frantically, or, pressing his hands against his head, groans and cries in agony. Even in deep stupor or in coma he gives expression to this pain by moving his hands to his head and by frequent groaning. Headache usually continues throughout the attack, with slight daily variations; in many instances it shows marked remissions and sometimes distinct intermissions. Its cessation during the attack is a most favorable indication. In many cases the liability to headache of a severe kind continues for years after the attack of cerebro-spinal fever, being excited by unusual bodily or mental exertion and like depressing causes.

The seat of the headache does not necessarily indicate the position of the inflammatory products found after death, nor the points of most intense meningeal inflammation; nor does its violence correspond to the intensity or danger of the disease. It is not, therefore, a symptom of great prognostic value, seeing that it is most intense in many favorable cases, and absent in some that lead to a fatal issue in the course of a few hours.

Vertigo is also an early symptom, sometimes appearing as a prodrome. Associated with the headache, it adds not a little to the distress of the patient. Suddenly seized by it, patients have staggered like drunken men; others have fallen, unable to rise again. It is occasionally, even in the recumbent posture, troublesome.

The recurrence of vertigo and headache during the convalescence, after they have entirely ceased, is to be regarded with anxiety. It often, especially when associated with vomiting and convulsions, betokens the development of hydrocephalus.

Vomiting is to be regarded as a symptom referable to the disturbance of the nervous system. It is an initial symptom, often occurring without any previous nausea, and lasting one or two days. It is provoked by movements, and in particular by rising. It recurs, in some cases, with frequency throughout the sickness, and adds to the depression by interfering with the assimilation of food.

Hallucinations and delirium, though much less constant and persistent than headache, are present in almost all the severe cases. They occasionally occur early in the attack, more commonly not until the second or third day. The delirium varies greatly in degree and kind. It appears in some cases to be the direct result of the violence of the headache. It is often transient and seldom continuous throughout the course of the attack, longer or shorter periods of lucidity commonly intervening between the outbreaks, or else the delirium alternates with periods of somnolence or stupor. It is not seldom of a furious kind, the patients falling into a state of frantic excitement, and being restrained with difficulty. Such violent manifestations may alternate with periods of placidity. Sometimes the sick lie in a quiet, wandering state, from which they may be aroused so as to make intelligible replies. Again it may resemble intoxication or hysteria. I conducted the *post-mortem* examination of the body of a young married woman, who died during the epidemic of 1873 in Philadelphia, of cerebro-spinal fever, with obscure symptoms and a delirium resembling hysteria. The mental state of many patients may be expressed by such terms as apathy or indifference. In fatal cases the delirium passes into the coma which precedes death. In mild cases there may be slight, transient delirium, occurring often only at night, or hallucinations upon particular subjects—a form of monomania.

Restlessness, though less frequent than the foregoing conditions, is not uncommonly present, especially early in the progress of the case. The patient tosses about in the bed and keeps his arms and legs in constant movement, or he seeks to spring from the bed, and has to be restrained by force.

Sleeplessness is present in a considerable number of cases, and is often met with in the history of the stage of the prodromes. Hirsch regards this condition as so constant at the height of the disease, that "if one find a patient who has suffered from it apparently sleeping, he may be almost sure that he has fallen into a stupor." In such cases the occurrence of a long, quiet sleep is to be regarded as an exceedingly favorable change in the course of the disease.

Coma, as has been incidentally pointed out, occurs in by far the greatest number of fatal cases, and is generally the forerunner of death. It may occur in the graver and in rapidly fatal cases without the intervention of delirium.

Stiffness or contraction of the neck is an almost constant and very characteristic symptom. It is rarely seen on the first day, and is then only slightly manifest. It becomes marked between the second and fifth days, and may continue, in cases terminating favorably, from three to five weeks, or even far into the convalescence, relaxing gradually. It varies greatly in degree from a slight stiffness, not easily perceived, but becoming apparent upon attempting to bend the head forward, to a contraction so great that the back of the head is held firmly between the shoulders, at almost a right angle with the spine. In the latter case swallowing is performed with difficulty, and active or passive attempts to bend the head forward are alike futile, partly by reason of the rigid, tetanoid character of the contractions, partly because of the great pain to which they give rise. The degree of contraction is by no means always proportionate to the neck-pain. On the contrary, this pain is sometimes absent as the patient lies at rest, and is called forth only upon efforts to overcome the contraction.

Contraction of the other erector muscles of the spine is also very often present. This leads to a straightening of the spine (orthotonos), with stiffness and the prominence of the contracted muscular masses and the disappearance of the spinous processes between them, or more rarely, in its highest grade, to complete tetanoid opisthotonos. It adds not a little to the discomfort of the sufferer, who is debarred from lying upon his back, and turns from one side to the other with the utmost pain and difficulty.

The duration of this, as of most of the symptoms of cerebro-spinal fever, is very variable. Sometimes it lasts only a few days, at others weeks, and patients have been confined to bed by it till the fourth and even the sixth week from the beginning of the attack. Ziemssen states that he has seen convalescents going about with rigid spines.

Pleurosthotonos, or *contraction of the spinal muscles of one side*, has been encountered by some observers. Levy saw it twice in fifty-seven cases. It is extremely rare.

The stiffness of the neck is sometimes absent. This observation has been made in some instances where the ordinary anatomical changes, and particularly the inflammation and exudation, have been present in the spinal meninges to an extent as great as that usually met with. Its absence cannot be explained.

Trismus has been observed only in patients who were extremely ill and comatose. It is highly ominous. Of five cases encountered by Hirsch, four speedily perished.

Stiffness and *contraction of the muscles of the extremities* occur in a considerable proportion of the cases. Active movements are executed in such instances with awkwardness and pain, and passive movements are

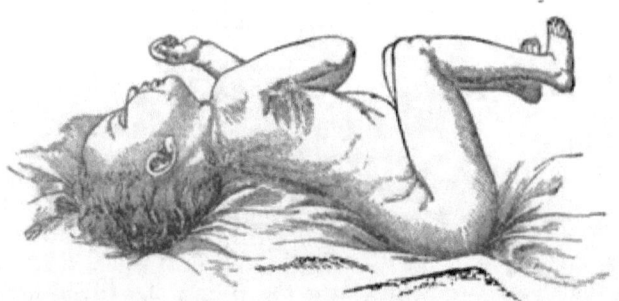

FIG. 3.—Attitude of Child in severe Cerebro-spinal Fever. (J. Lewis Smith.)

resisted. The usual position of such patients, in bed, is with the head drawn back, the forearms flexed upon the arms, and the knees drawn up upon the abdomen, with or without forward arching of the spine (Fig. 3).

Clonic spasms or convulsions occur with less frequency than muscular rigidities. They are met with, however, in a considerable proportion of the cases, particularly in children, with whom they sometimes occur as an early symptom, replacing the initial chill. General convulsions have been observed to usher in the attack in adults also, but much more rarely. They vary in degree from twitchings of single muscles or groups of muscles, to violent epileptiform seizures attended with loss of consciousness; the latter constitutes in adults a symptom of great gravity.

Tremors and *subsultus tendinum* are *less* frequently observed.

Paralysis is of much less common occurrence than the character of the lesions would lead us to expect. It occurs in a small proportion of the cases, and affects one or both extremities, upper or lower, and may be more or less complete. Hemiplegia may also occur. Palsies of certain associated groups of muscles, as those of deglutition, articulation, and others, are relatively more common. Paralysis is very rare as an early symptom; it appears toward the close of the disease. If the patient recover, it usually passes away in the course of a few days or weeks; exceptionally it is of long persistence, or even permanent. More or less complete general paralysis is encountered as one of the phenomena of approaching dissolution.

The facial expression is indicative of the severe pain which attends the disease. The features wear a fixed and rigid look, which passes with the exhaustion into an expression of apathy, without relaxing into the flushed dulness of typhus, nor the languid expression of enteric fever. The face is usually pale.

Pain in the spine (rachialgia), and especially in the neck, is a frequent symptom. It varies greatly in extent and intensity, as well as in duration. It appears sometimes coincidently with the headache, and has been observed as an occasional symptom late in convalescence. A dragging pain in the neck has already been mentioned as one of the prodromes. This pain, like the headache, is subject to remissions and exacerbations.

Severe pains in the extremities, especially in the legs, also frequently occur. They are often evoked or intensified by movements of the spine. Lightning-like pains invade other parts of the body, and an intense sickening neuralgic pain in the abdomen, particularly in the epigastric and umbilical regions, is very common. This pain is sometimes associated with uncontrollable vomiting. It was so common a symptom in the epidemic observed by Dr. Sanderson in the Lower Vistula, that, as he informs us, the disease acquired among the people the trivial designation of "The Belly-Ache."

Pain of a similar nature is sometimes referred to the chest, and at times is associated with difficulty in breathing. Asthmatic attacks are spoken of as occurring in some cases. It is probable, from their rarity, that they are incidental symptoms.

Hyperæsthesia of the skin, the joints, and other soft parts, though far from being a constant symptom, is to be regarded as characteristic when it does occur, and as sharply drawing the boundary line between this and any other disease with which it can possibly be confounded. It is absent altogether in many cases and in some epidemics; in others it is a very common symptom, and was frequently observed in the later epidemics in the United States, and in that of the Lower Vistula in 1865. It occurs early in the course of the attack, often on the second or third day, and is often so extreme as to cause great additional suffering, the patient lying as quietly in the bed as the restlessness so common in the affection will permit, in order to avoid this pain, which a movement of his limbs, a light touch upon the surface of his body, even the shaking of the bed, will evoke.

It is a symptom which often, when present, interferes greatly with the examination of the sick. Its commonest seat is the anterior surface of the body, especially of the legs and thighs, though it is everywhere present. It is often associated with marked intolerance of light and sound.

Anæsthesia of portions of the surface in some instances follows the symptom just described. Stillé observed it in many cases during the epidemic which is recorded in his writings. It is sometimes marked, sometimes a mere numbness. It disappears, as a rule, during the progress of the case.

SYMPTOMS REFERABLE TO THE SKIN.

Many observers state that the skin is apt to be dry in the early days of the disease, and afterward bathed in moisture, especially the head, face, and neck. But in this respect cases furnish no constant condition. The pallor and cyanosed appearance of the surface, especially of the face, which is uniformly present, has been already more than once alluded to in the description of the disease.

Cutaneous lesions are very common and constitute a group of phenomena of great interest in the study of this affection. In some epidemics they are present in by far the greatest number of the cases, while in others no eruption whatever can be discovered in most of the instances of the disease. Were this statement not correct, it would be impossible to reconcile the conflicting accounts of many competent observers.

Ziemssen remarks that it would be difficult to understand from observations of the German epidemics why the disease has been called "spotted fever" by American physicians. Many recorded observations of epidemics in the United States, however, go to show that in not infrequent instances eruptions are altogether absent, or present in but a small number of the cases. There can be no doubt that roseola and petechiæ are more common in cerebro-spinal fever in this country than in Europe.

Herpes is the most common of the eruptions. It is usually confined to the face (*herpes facialis*), but may appear upon the trunk as shingles (*herpes zoster*), or in circumscribed patches upon the extremities. It begins generally in the region of the mouth, upon the upper or lower lip, and extends to the nose, cheeks, ears, and eyelids. In many cases one or both sides of the face are covered with a hideous mass of herpetic vesicles or crusts. This eruption is also met with on the mucous surface of the nostril and cheek, and upon the scalp. It is in many cases **an early symptom, appearing on the second or third day;** but irregular outbreaks of vesicles often **take place late in the convalescence.** No prognostic significance can be ascribed to it.

Petechiæ occur in the next order of frequency. **The mottling is more or less** distinct and widely diffused over the surface, sometimes even involving the face. Larger spots of like character resembling the eruption of purpura are likewise common. Wide effusions of blood and its coloring matter beneath the skin (*vibiics, ecchymoses*) also occur. They **have sometimes** a regular, sometimes an irregular or ragged edge or border, which may remain fixed from the time of its appearance, or **may** extend rapidly over large surfaces. **They** are sometimes light or bright red in the beginning, and grow dark or livid in the lapse of a short time; oftener, however, they are dark purple or black from the first, and have been likened to splashes of ink. They often **resemble the livid staining of** the skin in the cadaver. These extensive effusions of the coloring matter of the blood **are of ominous significance as betokening the gravest disintegration of its corpuscular elements.**

Dr. J. Lewis Smith has observed that the size and position of such spots is sometimes determined by bruises which the patient receives during his **spells of restlessness.** The purpuric spots are sometimes hard to the touch, with defined margins. Vesicles may form and superficial gangrene of the skin take place, which, if recovery follows, gives rise to permanent scarring. A cyanosed appearance of the surface and livid mottlings may also occur without distinct eruption.

Less frequent are *roseola, erythema, urticaria, erysipelas,* and *sudamina.* **Not infrequently a patient presents three or four separate forms of cutaneous eruptions.**

The symmetrical distribution of the eruptions of cerebro-spinal fever have often been made the subject of remark. It is not uncommon to find similar eruptions and patches of eruption seated in the same position, upon both sides of the body or on the two extremities. This, together with **the variety of the forms,** their frequency and the hyperæsthesia and anæsthesia, point to disturbance of innervation in the central **nervous** system directly affecting nutrition (disturbances of trophic innervation). The purpuric eruptions are chiefly due to the breaking up of the red blood-corpuscles, and the solution of their coloring matter in the serum; and perhaps in part, to other causes not yet known.

THE PHENOMENA OF THE FEVER.

The temperature during the course of the disease does not give rise, **when** depicted in the graphic method, to a typical curve.

It is above the normal in every case, except perhaps those of the fulminant variety, in which the patient falls speedily into a state of collapse. It is generally moderately high, not always on the first day, but from the second or third day. In some instances it rises rapidly after the chill, which marks the onset of the attack. After the characteristic symptoms of the disease are fully established, the temperature rarely falls below 37.5° C. (99.5° F.), and ranges in adults from 38° C. (100.4° F.) to 40° C. (104° F.) for average cases. In children it is frequently higher. J. Lewis Smith has recorded a temperature (rectal) of nearly 42° C. (107.4° F.), a few hours after the onset of the attack, in a young child who died on the third day, and in two other instances a temperature of over 41° C. (106° F.). Both of these cases also terminated in death, one on the ninth day, the other in the ninth week. In severe cases it is apt to be high, and in particular, it rises as death approaches. Periods marked by long-continued subfebrile temperatures often occur in the course of an attack, and very irregular variations, both below and above the average range, are common. During such periods of relatively low temperature the other symptoms remain unabated. There is no constant and notable difference between the morning and evening, as in typhoid and typhus. A gradual fall (lysis) marks the beginning of convalescence; an abrupt fall ushers in collapse or death. A critical fall in temperature does not occur to signalize the favorable termination of this disease.

Wunderlich[1] concluded, from a study of thirty cases, that three varieties of the fever course may be distinguished:

"(a) In some very severe and rapidly fatal cases the temperature, though not invariably very high at the beginning of the disease, reaches very striking heights in the briefest time. It remains high, rising even higher at the approach of death, till in the very moment of death it may attain 42° C. (107.6° F.), and more. In one of his cases it reached 43.7° C. (110.7° F.). It may rise some tenths of a degree after death. In the case just cited, it was 44.1° C. (111.5° F.), three-quarters of an hour after death. There were also some fatal cases in which the temperature for some time was very moderate, and rose rapidly with abruptness at the close of life.

"(b) On the other hand, relatively mild cases exhibit a fever of only short duration, although there are sometimes considerable elevations of temperature and often an interrupted course. Recovery does not take place by crisis, but happens rather with a remittent defervescence (lysis). Here and there cases occur which, after defervescing and apparently almost recovering, relapse all at once with a rapid rise of temperature and run a course like the cases marked (a).

"(c) In contrast with these brief courses of fever with either very severe or slight character, we find cases which are more or less protracted.

---

[1] The Temperature in Diseases. Trans. of New Sydenham Society. London, 1871.

The height of the temperature in these may be varied, and indeed may exhibit manifold changes in the very same case, though this chiefly depends

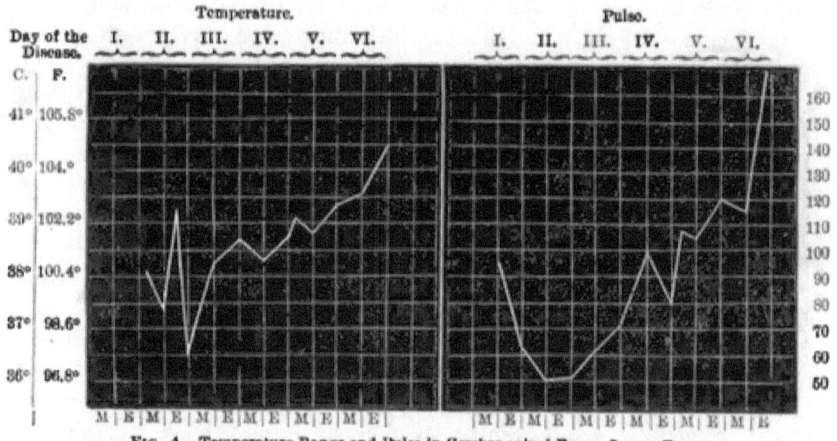

Fig. 4.—Temperature Range and Pulse in Cerebro-spinal Fever, Severe Form.

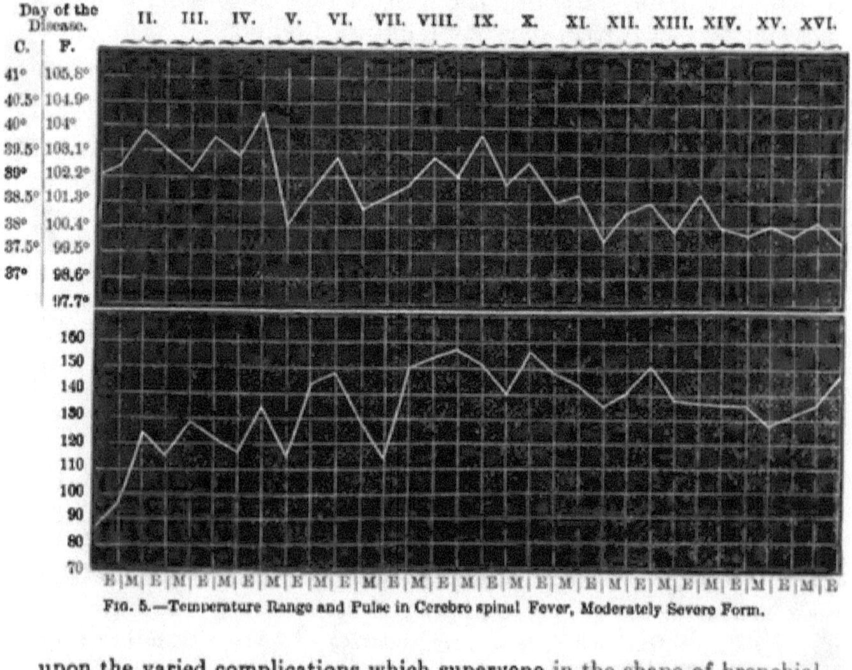

Fig. 5.—Temperature Range and Pulse in Cerebro-spinal Fever, Moderately Severe Form.

upon the varied complications which supervene in the shape of bronchial, pulmonary and intestinal affections, and affections of serous membranes."

Githens made records of the temperature in forty-four of his ninety-eight cases, with the following results. "In two cases only did the thermometer in the axilla reach 105°. In fifteen cases it was between 104° and 105°; in twelve between 103° and 104°; in seven between 102° and 103°; in six between 101° and 102°, and in two it was below 100°. The figures given are the highest points reached in each case. The difference in the temperature at the evening and morning observations was not so marked as in most other fevers—a fall of more than one degree being unusual, and frequently there was no change. A regular and gradual descent indicated the beginning of convalescence ; a rapid fall was the sure precursor of collapse."

Sanderson found that exacerbations of pain were always accompanied by a rise in temperature of from two to three degrees Fahrenheit.

One of the more notable characters of the febrile phenomena of this malady is their extreme irregularity. The temperature ranges not only do not coincide, they also do not even approach an ideal type.

Finally, I transcribe from the pages of Ziemssen diagrams illustrating the curves of the temperature and pulse in severe, mild, and the so-called intermitting cases. They are accompanied with brief abstracts of the clinical notes of the cases:

CASE I.—*Severe form.*—" L. W., aged fifteen years, a plasterer's apprentice. Access abrupt, with chills, cephalalgia, vomiting, trismus, tetanus of cervical and spinal muscles. Conjunctivitis, hyperæsthesia of skin. After the fourth day, herpes facialis, roseola, erythema, urticaria, and petechiæ on the extremities. Moderate fever, with retardation of pulse. Furious delirium followed by sopor. With rapid elevation of temperature and pulse, death ensued on the seventh day of the disease.

Autopsy.—Purulent cerebro-spinal meningitis. Remains of old pleurisy and perihepatitis. Partial atelectasis of lungs, and lobular pneumonia. Cadaverous softening of stomach and diaphragm (five hours after death). Suppurative tendo-synovitis in the left hand. Areas of degeneration in the spinal and recti abdominis muscles. Ulceration of cornea " (Fig. 4, p. 80).

CASE II.—*Moderately severe form.*—" M. V., aged twelve years, daughter of a stocking-weaver. Access abrupt, with chill and vomiting, which lasted during the first four days. Frontal headache. Stiffness and pain in the spine, jactitation, urgent thirst. Mind at first clear, afterward delirium and somnolence. Petechiæ on the second day, herpes on the face on the seventh, on the thumb on the tenth day. Effusion into the right wrist-joint. Conjunctivitis and keratitis. Aphthæ. Temperature at first high, but gradually diminishing, while pulse became very rapid. Tedious convalescence. Duration of the disease about six weeks " (Fig. 5, p. 80).

CASE III.—*Mild form.*—" C. H., aged ten years, daughter of an umbrella-maker. After a prodromal stage of two days' duration, patient was taken ill with pains in the head, extremities and epigastrium, nausea, vomiting, stiffness of neck, delirium, premature menstruation ; herpes on the fourth day, conjunctivitis, and transient fever. Improvement at the end of the first week. Duration of the disease, three weeks " (Fig. 6).

CASE IV.—*Intermittent form.*—" Th. M., aged nineteen years, student. Prodromata for eight days. Access abrupt, with cephalalgia, vomiting, slight convulsions, unconsciousness. Neck somewhat stiff. No fever during the first few days. Exacerbation on the fourth day, with fever of short duration, followed by apyrexia and disappearance of the malaise. On the fifth, seventh, and eighth days, the exacerbations recurred with marked spinal symptoms. Then followed daily exacerbations, but of less intensity and shorter duration. No eruption. Complete cessation of febrile attacks after the eighteenth day. Recovery. Duration of disease six weeks; of convalescence, four weeks" (Fig. 7).

In these cases the thermometer by no means shows the regularity that characterizes malarial fevers, or that a superficial study of the symptoms

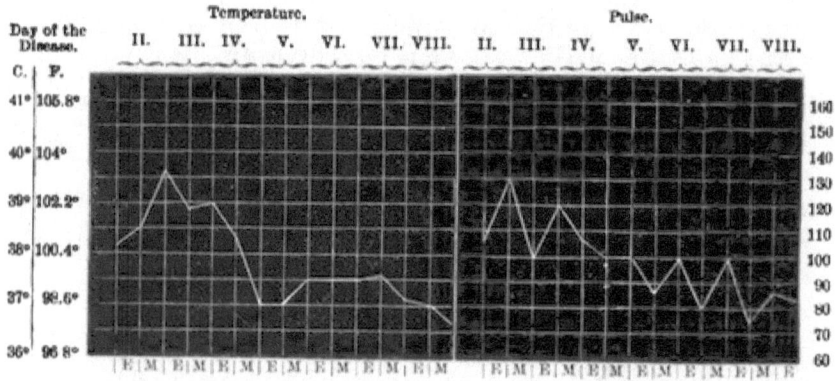

FIG. 6.—Temperature Range and Pulse in Cerebro-spinal Fever, Mild Form.

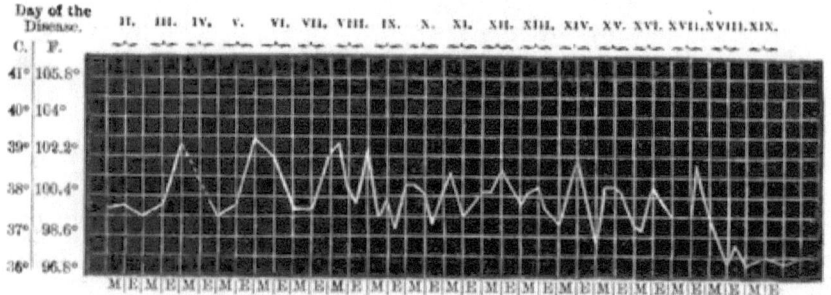

FIG. 7.—Temperature Range in Cerebro-spinal Fever, in so-called Intermittent Form.

would indicate. True intermittence, in the sense that the term carries when applied to ague, does not belong to cerebro-spinal fever. Ziemssen regards the exacerbations of fever as due to the irregular progress of the inflammation at the beginning and during the course of the attack; as due to slight returns of the inflammation when they occur during convalescence; and finally, as in the retrogressive stage, presenting the character of the absorptive fever, often met with during the retrograde metamorphosis of purulent exudations in other serous membranes (pleura, peritoneum).

The pulse is as variable as the temperature. Diminished heart-power, and a tone so impaired that slight causes give rise to extreme depression as manifested in a rapid, feeble, and compressible pulse, characterizes the circulation in this fever.

The frequency of the pulse by no means constantly corresponds to the intensity of the febrile action and the gravity of the other symptoms. It is in many cases scarcely increased in frequency beyond the normal, in others moderately quickened, and again it may be very frequent indeed. In children it is constantly accelerated. It is rarely retarded, in this respect differing from the pulse in tuberculous basilar meningitis. A slow pulse is in some instances present in the beginning of the disease, ere the temperature has risen; but, as a rule, it quickly, with the onset of marked fever-symptoms, rises in frequency. In fatal cases it is often so rapid that it cannot be counted. Perhaps the most constant character of the pulse is its variations in rapidity. Within a few hours it often varies from forty to fifty beats per minute, and a difference of twenty or thirty beats may be counted within the lapse of a few minutes. A very rapid pulse, which continues so, is to be regarded as unfavorable.

In quality, the pulse may be normal, or its fulness and tension may be augmented. When depression comes on, it becomes small, weak, and often intermittent. The feebleness and rapidity of the pulse in cases tending to a fatal issue, and particularly in those patients who rapidly approach death, is notable. To quote Githens: "The pulse varied from normal to 150 beats per minute in uncomplicated cases, and as high as 160 in two puerperal women; it was in all very weak, with dicrotic tendency; sometimes entirely imperceptible in the radial artery, and always interrupted by a very slight pressure." Da Costa[1] has observed well-marked blood-murmurs in the heart, even early in the course of the disease.

The *nutrition* of the patient generally suffers seriously. The wasting is very rapid and extreme in severe cases. When death takes place after a long illness, the corpse presents a high degree of emaciation. This is due not alone to the fever and grave inflammatory lesions of the nervous system, but the loss of appetite, obstinate vomiting, restlessness and pain, also contribute a large share in bringing it about.

An early, sudden and great loss of strength is a frequent and prominent feature of this malady. Syncope sometimes occurs at the beginning of the sickness. The patient is not only the victim of an extreme de-

---

[1] J. M. Da Costa : Medical Diagnosis. Philadelphia, 1864.

bility during his illness, but he comes out of it thoroughly exhausted, and is a long time in regaining his strength. The prostration which is so prominent a symptom of cerebro-spinal fever cannot be said to be characteristic of this disease as distinguished from some of the other continued fevers, but is notable for the frequency with which it occurs, the high degree which it attains, and the early period in the course of the attack at which it appears in the affection under consideration. It is to this character of the disease that it owes the old misleading names of "Sinking Typhus," "Typhus Syncopalis," etc.

SYMPTOMS REFERABLE TO THE ORGANS OF RESPIRATION.

In mild, uncomplicated cases, the respiration is for the most part quiet and easy, though slightly accelerated. Its rhythm is undisturbed. If cough be present, it is usually slight and accidental. In the grave cases the respiration is more or less disturbed. It is sighing, labored, or interrupted. As the case draws to a fatal termination, the breathing becomes more and more embarrassed; it grows very rapid, arhythmic, and often presents that alternation of respiration with respiratory pauses, known as the Cheyne-Stokes respiration.

It is probable that pressure upon, or œdema of, the medulla oblongata, gives rise to the interrupted respirations. There can be no doubt, however, that the tonic contraction of the spinal muscles, and other groups more directly concerned in the performance of the acts of respiration, has much to do with the embarrassment of breathing, which is still more common.

The organs of digestion are deranged. In addition to the vomiting which has already been described, there is nausea and more or less complete loss of appetite.

The tongue is moist and coated with a light or thick, white fur. In cases attended with great prostration, and in collapse, the tongue is dry and brown. If the patient rally from the prostration, the tongue quickly becomes moist again with the reappearance of the whitish fur. It is sometimes clear at the tip and edges. A moderate degree of retraction of the belly is sometimes present. Constipation is the rule. Diarrhœa sometimes occurs. The latter is more frequent in children, and in some cases precedes the attack. If constipation be present, it readily yields to the action of purgatives.

Thirst is almost constantly a tormenting symptom. It is unappeasable, and frequently persists till convalescence.

Jaundice occurs in a few cases. No other symptom of disturbance of the liver is noted.

The spleen is very rarely sufficiently enlarged to occasion an increase of its area of dulness discoverable during life.

The urine is sometimes normal in quantity, oftener increased. It may be much increased, even during active fever with high temperature. Urates, as in all fevers, are often thrown down as the urine cools. The reaction is usually acid. A moderate amount of albumen is occasionally to be detected, and more rarely cylindrical casts and blood-corpuscles. Phosphates may be present. In delirium and in coma, retention of urine may be overlooked, and if catheterization be delayed, cystitis may result.

Polyuria is very common in children, and has been observed in rare cases as a symptom persisting for years after convalescence. Transient albuminuria has also occurred.

Inflammation of the joints, resembling rheumatism, is occasionally met with. It is commonly slight, but in rare instances may run on to a suppurative arthritis. The wrist-joints are most frequently affected.

Swelling of the parotid glands is an infrequent accident of the disease. It may be slight, or it may run on to suppuration. Tourdes saw suppurative parotidites in two fatal cases. Githens met with it in two out of ninety-eight cases, both of which recovered, and Stillé saw two or three cases.

DISTURBANCES OF THE ORGANS OF THE SPECIAL SENSES.

The eye and ear are frequently involved, and are often the seat of serious lesions. It is not known to what extent the taste and smell may be affected. In mild cases the perception of odors is normal; the taste is perverted, as it is apt to be in the catarrhal state of the mouth and stomach which belongs to fever. In grave cases the condition of the patient precludes the investigation of this point. J. Lewis Smith ascertained that in one nostril the sense of smell was lost altogether, in a case under his observation.

To return to the consideration of the eye and ear.

The pupil is often normal during the whole course of the disease. In other instances it dilates toward the end; again it is frequently contracted in the beginning, and dilates after some days' sickness; not infrequently the pupils differ in size, one being contracted, the other dilated, and the two responding differently to the same light. Feeble response to light is a common symptom.

Intolerance to bright light is an almost constant symptom.

Nystagmus may occur in consequence of clonic spasm of the muscles of the eyeball, and spasm of particular muscles or groups of muscles may give rise to transient strabismus, which may appear and disappear several times before convalescence sets in. Paralysis of certain of the ocular muscles also causes squint, which may last several weeks, or even be permanent. Paralysis of the various cranial nerves depends mostly upon the lesions consequent upon the extension of the meningeal inflammation to their trunks, and arises either from the pressure exerted by the surrounding exudation, or from contraction of the hyperplastic connective tissue of the nerve-sheath.

The further lesions of the eye consist of inflammatory affections of the organ of sight itself, and are: (*a*) inflammatory hyperæmia of the conjunctiva. This is of frequent occurrence. There is a uniform diffused redness of the conjunctiva, not so dusky as in typhus. It is an early symptom. At times it amounts to an intense conjunctivitis, with œdema of the eyelids and a free muco-purulent secretion. When it is severe the cornea becomes opaque and the seat of ulceration. Ziemssen has pointed out the fact that this form of destructive keratitis is frequently due to the exposure of the cornea to the action of the air, as a result of partial palsy of the orbicularis palpebrarum muscle, and consequent imperfect closure of the eyelids: (*b*) severe, suppurative irido-choroiditis, or panophthalmitis. The media grow cloudy, the iris discolored, the pupils become irregular and are blocked with inflammatory exudation. The storm subsides, leaving distorted pupils, the lens cataractous, the retina detached, and ultimate atrophy of the globe ensues; or, in rarer cases, the eye is destroyed by perforating ulceration of the cornea and the formation of anterior staphyloma: (*c*) optic neuritis terminating in atrophy of the nerve.

Disturbances of hearing are noticed within the first few days. The patient is annoyed by loud sounds; humming and ringing in the ears are speedily followed by more or less complete deafness. These manifestations are usually bilateral. They are due to two processes: (*a*) inflammation of the middle ear. If it be catarrhal in character and mild, as is most commonly the case, it subsides without loss of hearing; if it be purulent and severe, perforation of the membrana tympani occurs, and an otorrhœa of variable duration ensues: (*b*) suppurative inflammation of the labyrinth, with destruction of the membranous labyrinth. The patient loses his hearing without otorrhœa, otalgia, or other local symptom. The loss of hearing does not always come on at the same period of the disease; the majority of cases are observed to be deaf as soon as the stupor goes off and full consciousness returns, while in rarer instances those who become deaf are able to hear more or less distinctly at this time, but lose this function in the course of the convalescence. This form of deafness is complete and permanent. It affects both ears, and has been observed in some instances to be associated with a staggering gait.[1] It is probable that the inflammation makes its way within the sheath of the auditory nerve (A. Heller).

---

[1] See Proceedings of Philadelphia Pathological Society, Philadelphia Medical Times, January 31, 1874.

Serious lesions of the eye and ear, resulting in the permanent and complete loss of sight and hearing, occur in some cases that run a mild course as regards the general phenomena of the affection. Dr. Schaffner[1] records a case in which a boy aged six, after a sickness of two weeks with symptoms of mild character, complained of blindness. On examination the loss of sight was found to be due to optic neuritis.

COMPLICATIONS AND SEQUELS.

Some of the complications—those involving the eye and ear, the joints and the parotid glands—have already been considered in the analysis of the symptoms.

Catarrhal and croupous pneumonia, bronchial catarrh, pleurisy, endocarditis, pericarditis, also occur as complications in some cases of almost every epidemic. Atelectasis and broncho-pneumonia are more common in those patients who for a long time have suffered from orthotonus (Ziemssen). The combination of croupous pneumonia with cerebro-spinal fever has been observed to be of common occurrence in some of the recent German epidemics. This serious complication has been encountered with greater frequency at the close than at the beginning of the epidemic, "as if the infectious poison had then lost its violence, and was able to resume its activity only when aided by the force of other diseases."

Intestinal catarrh also occurs as a complication. Malarial and enteric fever, and measles, scarlet fever and cholera, have been met with as intercurrent affections.

Convalescence is irregular and uncertain. After severe cases it is apt to be tardy. Relapses are not uncommon, and are often fatal.

The sequels are: (*a*) prolonged debility and emaciation, dyspepsia, boils, carbuncles due to the blood-changes that take place in this as in other infectious diseases; and (*b*) those due to the lesions resulting from the inflammation of the brain and cord, and their membranes, and its extension to the organs of the special senses, namely: pareses and paralyses, impairment of intelligence in consequence of chronic meningitis and chronic hydrocephalus, and more or less complete deafness and loss of vision.

General motor weakness and *paralyses of single extremities or particular nerves* are not very infrequent. They depend upon lesions of the brain or spinal cord, or on the results of injury to the parts in consequence of the pressure exerted by the contraction of the organized inflammatory exudation.

Feebleness of the intelligence and weakness of memory, with defects of

---

[1] Philadelphia Medical Times, May 16, 1874.

speech, are often sequels. In most cases, they gradually disappear in the course of some weeks or months, and when permanent are the result of chronic inflammatory processes affecting the brain.

One of the most important of the cerebral affections left by this fever is chronic hydrocephalus.

The **symptoms** are paroxysmal; they consist of severe headache, intolerance of light and sound, vertigo, pains in the neck and limbs, vomiting, involuntary discharges, convulsions, loss of consciousness. The attacks occur either at long and irregular intervals or in rapid succession. The mental and bodily condition of the patient during the intervals is sometimes such as to lead to delusive hopes of his recovery. If partial recovery take place the mind remains weak, and the limbs paralyzed and deformed. In the rarest of instances has an approach to complete recovery been recorded.

In a majority of instances the condition in the intermissions is such as to preclude all expectation of recovery, the mind being irritable and unsteady, the limbs slightly palsied, muscular movements inco-ordinate, and the development of the body in the young retarded. The head is large, the skull thin, and the eyes prominent. Headache is a common symptom.

Ziemssen gives the following account of the successive anatomical changes which attend the development of this process, as the result of his autopsies during the epidemic and in following years:

"During the second week the meningeal exudation, which has hitherto been little changed, or perhaps somewhat thickened, undergoes fatty degeneration of the cells and fibrin, and is thus slowly or rapidly absorbed, or ultimately shrinks into caseous matter, if absorption does not occur; the connective tissue of the arachnoid and pia mater proliferates, the hyperæmia of the substance of the brain disappears, and the purulent effusion in the ventricles increases. From the twenty-seventh to the thirtieth week the arachnoid and pia mater exhibit a pulpy hyperplasia or already a cicatricial thickening; the caseous remains of the meningeal exudation are still more shrunken; the ventricular effusion has become more moderate in amount, but quite clear, owing to the inspissation of the cellular elements into small, caseous flakes on the dependent parts of the ventricles. The earlier hyperæmia of the brain is completely gone; the brain is anæmic, even œdematous; the ependyma of the ventricles thickened and distinctly granulated, and the choroid plexus bloodless. Unless the hydrocephalic effusion be moderate, the brain-substance is atrophied sometimes to a very considerable degree. In a boy two years of age we found the medullary and cortical layers of the cerebrum together only seven and a half lines in thickness, while the central ganglia were much flattened."

The same author states that the interval of apparently progressive convalescence which usually occurs between the acute stage of the meningitis and the appearance of the symptoms of hydrocephalus, renders probable the supposition that the increase of the ventricular effusion may be due to the shrinking and thickening of the pia mater.

The various lesions of the eye and ear that give rise to defects of or loss of sight and hearing, as sequels, have already been pointed out. These lesions are either the result of the extension of the inflammatory process from the pia mater along the sheath of the optic and auditory nerves to the respective organs, or of a simultaneous localization of the inflammation in the pia mater and the eye and its tunics, and in the pia mater and the labyrinth and tympanum, as effects of a single disease-producing cause.

Complete deafness in young children who have not yet learned to talk, and even in those who have more or less perfectly acquired the power, results in deaf-mutism. In those who have learned to talk, speech is, after several months, understood with difficulty, and gradually, in the course of a year or more, becomes quite unintelligible. It is necessary for such children to be sent to institutions for the education of deaf-mutes.

Some observers have noticed that deaf-mutism is an uncommon result of this fever, even when complete deafness occurs, and Hirsch has called attention to the fact that impairment of speech, and even aphasia, may arise coincidently with the loss of hearing as a co-effect of the meningeal inflammation. When the loss of speech is a result of the deafness, it is preserved for a time after the meningitis subsides, and gradually grows more and more imperfect, till it is, as articulate language, lost altogether.

Pathology, Morbid Anatomy.

The essential pathological processes in cerebro-spinal fever are twofold: (a) the constitutional disturbances due to the direct action of the infecting poison upon the blood, giving rise to the group of symptoms constituting fever; and (b) the local inflammation. As is seen by the foregoing study of the clinical phenomena of the disease, one or the other of those processes may predominate, and the course and symptoms of the attack vary accordingly. If the phenomena of infection are most conspicuous, and the symptoms of the local inflammation are but slightly developed, the affection presents striking resemblances to some of the other infectious diseases, while on the other hand the latter symptoms may be so prominent as to overshadow the infectious nature of the affection, and present the appearance of a simple inflammation with attendant symptomatic fever. The latter form is met with during the epidemic prevalence of the disease, but is most common in the sporadic cases.

Between these two extremes every variety of combination of the two processes is to be encountered, but in all a careful study of the course and symptoms of the attack will reveal the manifestations of both.

In like manner the morbid anatomy reveals the lesions due to the infective character of the disease, and those resulting from the local inflammation which is its constant attendant.

These lesions are constant. They vary only in the degree of their development.

The emaciation, in cases of long duration, is extreme.

Cadaveric rigidity is marked and long continued.

Extensive discolorations of the dependent parts rapidly show themselves. Large patches of a livid hue may even appear elsewhere upon the body.

Stillé has published the account of a case in which the whole body became rapidly almost black, during the two hours before death, but the countenance afterward nearly regained its natural hue. As a rule the purpuric spots on the anterior surface, the redness of the eyes and the like, fade as the staining of the posterior parts of the cadaver deepens.

The skin shows the vesicles and crusts of herpes, the mottlings and staining of petechiæ. Patches of superficial gangrene, and bed-sores are sometimes seen.

The muscles are dry, soft, brownish red, sometimes pale, and atrophied. They are found to have undergone granular degeneration. These changes especially affect the muscles extending along the spinal column.

The heart is often flabby, and contains dark, thin fluid blood, with loose soft coagula, or less frequently it contains firm fibrinous clots. The cardiac muscle shows the same histological changes as the voluntary muscles. In the fulminant variety it is unchanged.

Klebs[1] found the condition of the blood very variable. In rapidly fatal cases it was very fluid and the clots were soft and scanty. Dr. Levick[2] states that the blood is in all cases fluid. Upon microscopical examination, the red corpuscles are shrivelled, crenated, not formed into rouleaux, and "numerous white corpuscles are found in the field."

Multiple abscesses have been found both in the subcutaneous connective tissue, and in that of the intermuscular planes. When the joints have been swollen and tender during life, they have been found the seat of sero-purulent effusions.

The lungs show frequent changes. Hyperæmia, hypostatic congestions, œdema, bronchitis with a tenaceous secretion, are often met with. The infiltrations of catarrhal and, less frequently, of croupous pneumonia, are also encountered.

---

[1] Zur Pathologie der epidemischen Meningitis. Virchow's Archiv, xxxiv.
[2] See Report of the Committee on "Spotted-Fever, so called," by James J. Levick, M.D. Transactions American Association, vol. xvii., 1866.

The pleuræ and pericardium are sometimes inflamed, ecchymosed, and contain purulent exudation.

Recent endocarditis is rare.

The liver is congested, but rarely enlarged. Its cells show a granular, albuminoid, or fatty cloudiness (Klebs).

The spleen is very variable in size. It is usually small, but sometimes moderately, never greatly enlarged. It is usually softened.

The intestinal mucous membrane is usually normal. It is sometimes injected and thickened. The solitary and agminate glands are enlarged and sometimes ulcerated.

The lymphatic glands nearly always present a reddened appearance.

The kidneys are generally congested. The tubules are sometimes blocked with fat-granules and fibrinous casts.

The lesions thus far described are for the most part those met with after death, from infectious diseases in general, and are not distinctive. Those which we now come to discuss are as characteristic as the intestinal lesions of enteric fever. They are the results of the inflammatory processes which have their seat in the cerebro-spinal axis and its enveloping membranes.

The calvarium in many instances shows no change; it is most frequently congested, especially in the line of the sutures.

The dura mater of the brain is often tense, smooth on the outer surface, at points firmly attached to the inner table of the skull, and showing scattered punctiform hemorrhages or small effusions of blood; the inner surface hyperæmic, and more or less closely adherent to the arachnoid.

The *sinuses* are distended with thin fluid blood, and contain soft *post-mortem* clots or firm thrombi.

The *arachnoid* is often found quite normal, especially in cases that have run a rapid course; it is sometimes hyperæmic or stained with blood, or again it may be dry, lustreless, and opaque. The space between the dura and the arachnoid has been observed to contain a considerable quantity of serous effusion, or more rarely of pus.

After a protracted illness, when the exudation has begun to become organized, the arachnoid is rough and thickened.

The *pia mater* is hyperæmic, with intense, diffuse capillary injection and points of capillary hemorrhage. It is adherent to the surface of the brain, from which it can be separated with difficulty, and often only by tearing the brain-substance. In those cases which end fatally in the course of a few hours, this hyperæmia is, as a rule, the only change in the meninges discoverable by the unaided eye. Free exudation is absent. But upon microscopical investigation the pia mater is found to be densely infiltrated with cells, especially along the line of the vessels. If the case have been a more protracted one, this membrane becomes œdematous from the transudation of serum into its meshes, and in one or two days, a cloudy serum, or a thin, yellowish exudation accumulates in the sub-arachnoid space. By the second or third day the exudation is found to be distinctly purulent, of a butter-like, gelatinous, or firmer consistence, and from one to four lines in thickness. It is of a yellowish or greenish color, or may be deeply tinged with blood. It is at times distributed in a broad layer over considerable spaces, both on the convexity and at the base, most abundant in the sulci, along the course of the vessels, over and around the optic chiasm, over the pons Varolii, the cerebellum, medulla oblongata, and in the great fissures of the brain. In rare cases the exudation uniformly covers the whole surface of the brain. It sometimes extends in strips along the vessels and in the integral spaces, at others it is scattered in detached, island-like plaques. The extent of the exudation and its amount vary greatly in different cases. No part of the pia of the brain or cord may be free, or it may be limited to patches or strips on the convexity, at the base, or on the cord. It occupies the subarachnoid space; the arachnoid space is free. The thickening of the visceral arachnoid is due to purulent infiltration. The seat of the primary inflammation is the pia mater. The exudation consists of fibrin, mucine, pus-cells, and free granules.

The membranes of the spinal cord present similar anatomical changes. The *dura* is often separated from the vertebræ by collections of extravasated blood, its inner surface smooth, or in many cases injected or slightly adherent to the arachnoid; or finally, collections of serum or pus occupy regions of the space between these two membranes. The *arachnoid* is often normal, in other cases cloudy and infiltrated with pus. The *pia* is, as in the brain, but, as a rule, less deeply and less extensively hyperæmic. It is also roughened, thickened, and intimately adherent to the substance of the cord. The exudation here also appears early as a cloudy serum, but a little later in bands or strings of fibrino-pus, which often assume an irregular, net-like appearance, and later still as thick layers of pus, resembling in all its character the exudation described above. Its seat is almost exclusively upon the posterior surface of the cord, very rarely, and never wholly, in the cervical portion, but commonly extending from the cervical to the dorsal enlargement of the cord downward to the cauda equina, and it is most abundant in the lumbar region. The roots of the spinal nerves are frequently imbedded in it. The anterior surface is much less rarely the seat of the exudation, and when this is the case, the whole cord is surrounded.

According to Hirsch, the accumulation of the exudation upon the lower portion and the posterior surface of the cord, is chiefly due to the fact that it flows there by gravitation whilst fluid, and Ziemssen observes that in the rare cases where the whole cord is imbedded, the variation from the rule depends mainly upon the viscidity of the exudation from the beginning.

The *brain-substance* is frequently congested, with numerous "puncta vasculosa" upon the incised surface, and the secondary development of local areas of softening, which are most abundant in the neighborhood of the purulent exudation and about the ventricles. The nerve-elements are more or less disassociated as a result of the imbibition of fluids. Occasionally the entire brain is somewhat softened. In rare instances it is œdematous, even after an illness of only a few days.[1] This condition is more common in cases that have been very acute or very long-continued. The latter class of cases present a brain with a smooth, level surface, and a watery appearance on section. More rarely the consistence of the brain is firmer than normal. In most cases, and in particular in those in which the illness has been protracted, the ventricles contain more or less turbid serum, and in some cases they are distended with pus. The choroid plexus and the ependyma are deeply congested, or even ecchymosed, and covered with pus and lymph. The same anatomical changes are found in the third and fourth ventricles. In cases terminating after a long illness, the effusion may reach an enormous amount, and give rise to atrophy of the brain-substance with flattening of the convolutions, and œdema of the brain and spinal cord.

The retrogressive changes consist of resorption of the serous effusion, shrinking and organization of the exudation between the arachnoid and pia mater, with opaque thickening of these membranes or caseous degeneration of the exudation. In rare instances, diffuse purulent encephalitis takes place and purulent infiltration of the brain-substance or deposits of pus are found at the necropsy.

Like changes are met with in the *substance of the spinal cord*, namely, hyperæmia, serous infiltration, and softening. They are less marked, as a rule, and less uniformly distributed.

In a girl, aged fourteen, who died on the fourth day, the autopsy disclosed a large, serous effusion, purulent exudation into the ventricles, including the fourth, and dilatation of the central canal of the cord, which was filled with pus (Ziemssen).

In cases of the fulminant variety, where death quickly follows the onset of the sickness, it is probable that the subject is overwhelmed by the poison ere the characteristic anatomical changes have time to develop, as occurs in rapidly fatal cases of other epidemic and infectious diseases, as variola, scarlatina, etc.

The amount of the exudation and the extent of the secondary changes in the substance of the brain and cord, are not always proportionate to the intensity of the symptoms or the duration of the case.

---

[1] Hutchison: American Journal Medical Sciences, July, 1866.

Diagnosis.

The direct diagnosis of epidemic cerebro-spinal fever usually presents but little difficulty if the attack be primary and occur during the epidemic prevalence of the disease. Under certain circumstances, as when it develops as an intercurrent affection in pneumonia, typhoid fever, or other acute diseases, or when very young infants are the subject of the attack, and when **sporadic cases** occur either beyond the limits of the territory in which the disease is rife, or at the beginning of the outbreak, the diagnosis is attended with the greatest difficulty, and often cannot be made until some days have elapsed.

The character of an epidemic outbreak of the disease may be inferred from the suddenness of its oncoming and the rapidity of its spread.

The diagnosis of individual cases is based upon the presence in varying combinations of the characteristic symptoms of the affection. Most prominent among these are, furious headache, with acute pains in the neck, spine, and extremities, faintness, with a sinking sensation in the epigastrium, and vomiting which is uncontrollable; and contraction, first of the cervical muscles, later of those of the spine, with general cutaneous hyperæsthesia. Add to these morbid phenomena the abruptness of the attack, with or without prodromes; extreme restlessness; delirium alternating with periods of quasi-consciousness and merging into stupor or coma; the occasional convulsive spasms; the eruptions, especially herpes; the irregular temperature, and the extraordinary variations of the pulse in frequency and volume, and the case presents a picture not difficult of recognition.

The uncertainties which beset the diagnosis of *sporadic cases* of the affection arise from its less abrupt onset, its less acute course, the frequently indistinct spinal symptoms and the great rarity of its occurrence. In these cases, also, the pains in the back and limbs, the orthotonus and the hyperæsthesia of the skin and soft parts, are often altogether wanting, and the stiffness of the neck is less perfectly developed than in the epidemic disease.

Tuberculous basilar meningitis is to be distinguished from cerebro-spinal fever by the long duration of the period of prodromes, which is rarely absent, by the less abrupt and less violent onset, by its slower course marked with remissions, its slow pulse, the great irregularity of the respiration, and the absence of eruptions. Furthermore, there will usually be elicited some history of scrofulous and phthisical affections, or of a hereditary tendency to tuberculous disease. But in children, or during the prevalence of an epidemic of cerebro-spinal fever, or in those cases in which the tuberculous process extends to the membranes of the spinal cord (Hirsch), the diagnosis is far from easy.

Pernicious intermittent fever, with its fulminant manifestations, its speedy collapse and fatal coma, may be confounded with the fulminant variety. The diagnosis rests upon a consideration of the etiological factors of the two diseases. The season of the year, the nature of the country, which is usually in the highest sense insalubrious, and the endemic or epidemic prevalence of ordinary intermittent or remittent fever, tend to clear up the obscurity arising from any accidental resemblance of the symptoms. Moreover, an attack of intermittent fever rarely declares itself as pernicious or malignant in the first paroxysm; it is only after one, two, or more seizures, differing not at all, or but slightly, from the common manifestation of the disease, that it discloses its true character.

Scarlet fever in some instances may resemble, in its sudden onset, high febrile movement, vomiting, convulsions and stupor, cerebro-spinal fever as it occurs in children. The presence of the peculiar redness of the palatine half-arches, which is characteristic of the former disease in its earliest stage, may aid in the diagnosis. In a few hours the efflorescence will clear up any uncertainty.

Enteric fever in its typical form presents marked points of difference from the fever under consideration. Yet it has in more than one local outbreak presented symptoms that have for a time rendered it doubtful which of the two diseases was present.

The following brief tabular arrangement of the prominent symptoms will serve to contrast the two diseases:

CEREBRO-SPINAL FEVER.	ENTERIC FEVER.
Abrupt, overwhelming onset, with or without prodromes.	Gradual approach with marked prodromes, and often obscure beginning.
Headache, acute, agonizing.	Headache, dull, heavy.
Vomiting, constant.	Rare.
Muscular contraction within the first two or three days.	Absent altogether.
Constipation the rule.	Diarrhœa the rule.
Active delirium, alternating with stupor, or stupor deepening into coma.	Mental hebetude, muttering delirium, stupor.
Curve of temperature extremely irregular and atypical.	A typical thermal line.
Attack reaches its maximum within four or five days.	Develops slowly to its maximum.
Various eruptions, chiefly herpetic and petechial; they appear early.	A characteristic lenticular, rose-colored eruption, which does not appear until the end of the first week.

Much confusion has arisen from the fact that not a few among the older writers, and some of a later date,[1] have confounded cerebro-spinal

---

[1] See Murchison: London Lancet. April, 1865.

fever with *typhus*, or regarded it as a *variety of typhus*. These diseases are not only, as is at this day universally admitted, unlike in every respect save their infectious nature, but—as has been pointed out by Stillé, whose learned treatise has done much to finally settle every question of doubt concerning the identity of these two diseases in this country—they are also in strong contrast in respect of their causes, symptoms, course, lesions and sequels, and all physicians who have witnessed epidemics of both affections agree in pronouncing them to be radically different.

I venture to transcribe, from the pages of the last named author, a table of the important phenomena of these two affections, believing that, by so doing, their essential independence and the striking points of differential diagnosis between them will be most clearly demonstrated:

Epidemic Meningitis (Cerebro-spinal Fever).	Typhus Fever.
A *pandemic* disease. Occurs in places remote from one another, and without intercommunication.	Essentially an *endemic* disease. Always due to local causes. Spreads by intercommunication only.
Attacks all classes of society. Is never primarily developed by squalor and deficient ventilation.	Attacks primarily the poor, filthy, and crowded alone.
Is not contagious.	Contagious to a high degree.
More males than females attacked.	The two sexes equally affected.
More young persons than adults.	More adults than young persons.
Generally occurs in winter.	Epidemics are irrespective of season.
Eruptions are wanting in at least half of the cases; they occur within the first day or two.	The eruption is rarely absent, and disappears between the fourth and the seventh days.
The eruptions are very various, including erythema, roseola, urticaria, herpes, etc. Ecchymoses are common.	The eruption is uniformly roseolous and then petechial. Ecchymoses are rare.
Headache, acute, agonizing, tensive.	Headache, dull and heavy.
Delirium often absent; often hysterical, sometimes vivacious, sometimes maniacal. Generally begins on the first or second day.	Rarely absent; usually muttering. Rarely begins before the end of the first week.
Pulse very often not above the natural standard; often preternaturally frequent or infrequent. Is subject to sudden and great variations.	A slow pulse exceedingly rare. Its rate is pretty constantly between 90 and 120.
"The temperature is lower than that recorded in any other typhoid or inflammatory disease." It is also very fluctuating.	The temperature is always more or less elevated, and it does not fall until the close of the disease. "The skin is hot, burning and pungent to the feel."
The body has no peculiar smell.	The mouse-like odor of typhus is characteristic.
The tongue is generally moist and soft; sordes of the teeth, etc., rare.	The tongue is generally dry, hard, and brown; and the teeth and gums fuliginous.

CEREBRO-SPINAL FEVER.

EPIDEMIC MENINGITIS (CEREBRO-SPINAL FEVER).	TYPHUS FEVER.
Vomiting, generally of bilious matter, is an almost constant and urgent symptom, especially in the first stage.	Vomiting is rare and not urgent.
Pains in the spine and limbs, of a sharp and **lancinating** character, are usual and evidently **neuralgic**.	Pains are dull, heavy, and apparently muscular.
Tetanic spasms in a very large proportion of cases, and within the first two or three days. They are due to an inflammatory exudation within the spinal canal.	Tetanic spasms are unknown in typhus. Convulsions sometimes occur, due to "pyæmia."
Cutaneous hyperæsthesia is a common symptom.	The sensibility of the skin is generally blunted.
Strabismus common.	Rare.
The eye, if injected, has a light red or pinkish color.	The blood in the conjunctival vessels has a dark hue.
The pupils are often unequal.	Always equal.
Deafness often complete and permanent.	Hardly ever permanent or attended with signs of the disorganization of the ear.
Duration very indefinite, but generally from four to seven days.	Duration from twelve to fourteen days.
Relapses are common.	Relapses are rare.
The blood is often highly fibrinous.	The blood is never fibrinous.
The lesions, unless in the most rapid cases, consist of a fibrinous or purulent exudation in the meshes of the cerebro-spinal pia mater.	There are no inflammatory lesions whatever.
Mortality from twenty to seventy-five per cent.	Mortality from eight to forty per cent. STILLÉ.

Certain cases of cerebro-spinal fever occurring in nervous females at the close of epidemics, or sporadically, have presented a delirium so peculiar and an array of symptoms so little characteristic, that they have been looked upon as the manifestations of *hysteria*. It is to be hoped that a period has been reached in the progress of medicine, in which such an error can no longer arise — a period in which this term shall be used with a degree of circumspection proportionate to the vagueness of its meaning. The use of the thermometer will clear up any uncertainty of diagnosis between this fever and most cases of functional disturbance of the nervous system.

PROGNOSIS AND MORTALITY.

The course of the disease is very variable. In individual cases the prognosis can never be made with certainty. The abortive cases and those of the fulminant variety run the most rapid course. Hirsch has emphasized the fact that certain cases, which at the onset present the symptoms of cerebro-spinal fever, promptly recover after an illness of a few hours, which terminates in free sweating. The most intense cases, on the other hand, prove fatal in a few hours—as few as five, and constantly as early as the second or third day. The course of the moderately severe cases continues one or two weeks, to the beginning of convalescence, but it in other cases extends over months. Sad examples of the ravages of the disease are to be encountered after every epidemic. In the fulminant cases death is by far the most common termination; in cases of average severity it is still frequent, and it not seldom occurs in cases that have run an apparently mild course, in consequence of complications or sequels. The first week is the time of greatest danger; if the patient survive that, hope of his recovery may be entertained. The symptoms that render the prognosis unfavorable are: a very high degree of excitement, the early appearance of depression, return of the vomiting, intense headache, continuous coma, recurring convulsions, and irregular respiration.

In cases of average severity, and in mild cases, a guardedly favorable prognosis may be based upon the uniform gradual amelioration of all the symptoms within the first or second week, and the establishment of convalescence without the occurrence of grave complications or sequels. It is to be borne in mind that relapses are not infrequent, and that they are often fatal.

The high death-rate that attends cerebro-spinal fever places it among the most dreaded of epidemic diseases. The mortality varies greatly in different epidemics; in the mildest, it is about thirty per cent., in the most severe, seventy-five per cent. The average may be stated at about forty per cent. A comparison of the statistics of the epidemics of the early and middle periods of the century with that of those that have prevailed within the last two decades, suggests the probability of a gradual diminution of the violence of the disease. This difference in the death-rate is, however, without doubt due in part to the fact that the energetic depletory measures of treatment formerly extensively in vogue are now wholly abandoned.

Mode of death.—Death occurs in a majority of cases by failure of the respiratory nerve-centres, in some instances from asthenia, and in the fulminant variety probably from necræmia.

TREATMENT.

Our ignorance of the precise etiological conditions of the disease, limits *prophylaxis* to general sanitary measures for the purification of houses and localities, and attention to personal hygiene. This, as all epidemic diseases, assumes, as a rule, its worst form, and numbers the most victims, where anti-hygienic conditions most abound. Attention to the cleanliness of streets and dwelling-places, to the condition of drainage and sewerage, the prompt removal of accumulations of refuse matter, and the avoidance of overcrowding, cannot fail to diminish the severity and mortality of the disease.

The evidence that the fever-poison, in some instances, spreads among the different members of a household, either from the individual first attacked, from his personal effects, or in consequence of some unknown favoring condition of the surroundings, renders it advisable that, where practicable, the dwellings in which the disease has made its appearance, should be abandoned until after the close of the epidemic (Ziemssen). It is recommended that all the linen and other articles used by the patients should be carefully disinfected, or perhaps burned.

The use of plain and wholesome food, the avoidance of unusual fatigue, both bodily and mental, and of excesses of every kind, are important. Moderate exercise, quiet, and regular living, may afford some, but by no means complete security during the epidemic.

Nervous persons and those in feeble health, should, when possible, leave an infected district upon the outbreak of the disease.

The treatment of the disease has been almost as various as its various physiognomy. In different epidemics and at different periods, divergent and even opposite methods of treatment have been adopted. On the one hand, a vigorous tonic and stimulant plan has been pursued by those to whom the disease has presented, in an extreme degree from the onset, the symptoms of depression; again, the urgent symptoms of an intense inflammatory process localized in the membranes of the brain and spinal cord, have seemed to some to indicate the energetic use of depletory and other antiphlogistic remedies, including the administration of mercurials, while others have been content with a modified expectant plan of treatment, in which a careful regimen and efforts to combat the symptoms as they arise, play the chief part.

In the present state of knowledge, it is impossible to decide whether or not any plan of treatment yet resorted to is capable of so affecting the mortality as to lower the death-rate in particular epidemics, while there is reason to believe that the extreme fatality characterizing some of the older epidemics has been in part due to the repeated and copious bloodlettings, and other depressing measures entering into the treatment. The difficulties connected with the consideration of the treatment of this disease are partly inherent to the subject of the treatment of the infectious diseases in general, in which, the cause being beyond our reach and its nature unknown, we are compelled to direct our therapeutic efforts alone against the consequences of its action. They are, however, in a much greater degree dependent upon the variable and diverse forms in which this disease presents itself. Efforts to deduce, from statistics, conclusions in regard to the success of different modes of treatment in an epidemic disease in which the mortality ranges between thirty and seventy-five per cent., must yield unsatisfactory, if not fallacious results. It is not only impossible to compare the results of treatment in different epidemics, but, from the capricious nature of this affection and its various manifestations, it is even impossible to compare cases in the same epidemic, or indeed, to compare the cases which occur during the rise, the maximum, or the decline of the same epidemic. We have to do with cases of this fever to which the term average cases may be aptly applied, as qualifying the intensity of the morbid phenomena and the rate of mortality which attends them, which yet differ among themselves by as many shades as there can be various combinations of the infectious or blood element and the local inflammatory element which jointly underlie its manifestations. Cases are far from rare in which the attack is of the mildest form, only to be recognized by the lurid light of the outbreak in which they occur, cases requiring no treatment, sometimes not even compelling the subject to take to his bed. In strong and terrible contrast to such cases are those in which, in the midst of health, while at his ordinary occupation or on awaking from sleep, the patient is overwhelmed by the poison as by an avalanche, and passing rapidly from agonizing suffering to coma, perishes in the course of a few hours. Here the brevity of the course and the nature of the lesions alike show the powerlessness of our efforts to control the attack. Medicine, with all its resources, is neither adequate to combat it, nor responsible for its result. As Stillé has said, "the first symptoms of the disease are the first phenomena of death."

We are driven then, in estimating the results of treatment, to restrict our observations to the effect of remedies upon the individual patients, the immediate influence upon their symptoms, both objective and subjective, and the permanence of that influence.

A judicious treatment must be based upon the broad general principles of therapeutics.

Antiphlogistic treatment would seem to be indicated by the prominence of the symptoms of inflammatory congestion of the meninges at the onset of the attack, by the nature of the lesions constantly found after death, and by the relief it affords in a large proportion of the cases. But, in view of the infectious character of the cause of the affection, its rapidly disintegrating effect upon the blood, the early and often alarming debility in some cases, the marked depression that in others follows the active symptoms, the great emaciation and the tedious convalescence, measures of depletion must be employed with the greatest caution, and are in all but the sthenic cases contraindicated. In the young, and particularly in children, the abstraction of even small quantities of blood is liable to be followed by alarming symptoms of depression. Dr. J. Lewis Smith reports a case in which the application of a leech to each temple in a child aged four years was followed by extreme and almost fatal exhaustion.

General bloodletting is in no case admissible. It is to be borne in mind that the pulse is almost always, even from the onset, such as would contraindicate the abstraction of blood, and if the urgency of the symptoms of the local inflammation and the critical state of the patient seem to call for the employment of energetic measures, the clinical history of the disease reminds the physician that a no less marked depression is speedily to follow and calls for a thoughtful regard for the future. Even in the sthenic cases the local application of cut cups to the nape of the neck and along the spine is to be employed with caution. Leeches may be applied to the temples and in the neighborhood of the mastoid processes. These measures are of great value in mitigating the headache and spinal pains which form so prominent a symptom in many cases.

If such local abstractions of blood be contraindicated by the state of the patient, dry cupping may be employed with advantage.

The direct application of *cold to the head and spine* by means of ice, snow, or a freezing mixture in rubber bags made for the purpose, and to be had at the apothecaries' shops, is not open to the same objections as bloodletting, and at the same time is attended with satisfactory results as regards the symptoms of inflammation. If the bags cannot be procured, a bladder filled with cracked ice mixed with bran may be substituted. In children gentle cold affusions may be practised. The application of cold by these means is in most cases followed by very marked mitigation of the pains, and often by quietude or sleep. It should be continued as long as the patient is comfortable, and repeated upon the return of the symptoms. Patients frequently require the continuous application for hours at a time. A hot mustard foot-bath, or a general hot bath, 38°—39° C. (100.4°—102.2° F.), should be employed as early as possible, care being taken that the strength of the patient be in no wise taxed. This may be followed by gentle frictions with some stimulating liniment, or with oil of turpentine, if the surface be cold and the circulation depressed. A stimulating enema may at the same time be administered. The patient should also be covered with warmed blankets, and *artificial heat* applied to his sides, thighs, and extremities. In all cases it is well, while using the cold to the head and spine, to counteract its depressing effect by the application of moderate heat elsewhere. This may be accomplished by means of hot flannels, bags of hot sand or salt, bottles filled with hot water, or heated billets of wood well wrapped up. At the same time, if necessary, sinapisms are to be applied to the extremities and the præcordium.

Bartholow holds that the application of ice to the head and spine may do mischief by the depression of the circulation which it causes. He advises, instead, the use of hot water applied by a sponge passed over the spine every two or three hours. The best modern American authorities agree in advising the continuous use of external heat, to anticipate and counteract the early depression which is so grave an element of the disease, a practice very general in the early epidemics in this country, but for a long time strangely overlooked here, and altogether neglected abroad.

Blisters upon the occiput and upon the nape of the neck are not only to be advised upon theoretical grounds, but they are of great practical value in relieving pain and in diminishing delirium, spasm, and coma. They should be applied early in the course of the disease.

The use of *mercury*, except at the onset of the attack, in the form of a dose of calomel as a purgative, is to be discountenanced. No single drug has been employed to a greater extent than mercury in the treatment of cerebro-spinal fever, but almost all authorities at this time regard with disfavor the employment of the preparations of this metal for its supposed antiphlogistic or antiplastic effect, or its absorbent effect upon the exudation. Among the most recent German writers, Ziemssen, however, recommends its use in the form of mercurial ointment or calomel, "for the purpose of preventing the extension of the meningeal inflammation and exudation." He employs free inunctions and the internal use of calomel "in almost every case," but admits that when used in connection with other remedies, it is difficult to ascertain its share in the common effect, and that even when used alone its efficacy is by no means clearly established.

The antipyretic treatment by cold baths and enormous doses of quinia, as practised by the Germans in diseases attended by hypyrexia, can be rarely necessary, for the reasons that in most cases the fever is moderate, and in those cases characterized by an excessively high temperature, the fatal event is due to other causes than the fever. Quinia has no control over the intermittent variety of the disease. In the report of the Committee of the American Medical Association, written by Dr. Levick, above quoted, the use of quinia in large doses, *at the very beginning of the disease*, is favorably spoken of; but its administration in the later period, when the phenomena all point to intra-cranial exudation, is said to be of no use and liable to prove even hurtful, except in small doses as a tonic to an enfeebled system.

The statement that this drug has appeared to abort the disease in some instances is not borne out by sufficient evidence.

There is no abortive treatment.

Opium, by the concurrent testimony of observers in all countries, now holds the highest place in the treatment of this disease. It was used in this country in the early part of this century, adopted as a treatment in France at a later period, and has recently found favor in the eyes of the physicians of Germany. Ziemssen says of morphia, that it "may be regarded as one of the most indispensable remedies in the treatment of epidemic meningitis."

All the distressing symptoms, the headache and spinal pains, restlessness, the spasm, the hyperæsthesia, and the inability to sleep, call for the administration of this drug. At the same time our knowledge of the nature of the lesions suggests its use. Opium slows the heart and increases arterial tension. It is to be employed at the earliest moment possible, and in full doses. By this means we may anticipate the occurrence of exudation, or limit it.

Experience has shown that a remarkable tolerance for this drug exists in most cases of cerebro-spinal fever. Some of the older physicians gave large doses. Strong[1] in one case "gave sixty drops of laudanum every hour till half a fluid ounce was taken. The whole of it was retained, and it subdued the excitement and relieved the pain, but produced no sleepiness or other apparent effect of opium. Others among the early American writers gave enormous doses, 16 c.c. or half an ounce of the tincture, or from 2—4 grammes (thirty to sixty grains) in substance, in the course of twelve hours, being necessary to control the urgent symptoms. Such cases recovered.

Chauffard,[2] to whom Hirsch erroneously ascribes the first advocacy of the opium treatment, gave it in doses of from 0.2—1.0 gramme (three to fifteen grains). Boudin[3] frequently gave up to 0.45—1.0 gramme (seven to fifteen grains) at a single dose at the commencement of the attack, and afterward 0.065—0.13 gramme (one to two grains) every half-hour. As soon as the symptoms abated or the patient became drowsy, the dose was diminished. Stillé gave 0.065 gramme (one grain) every hour in very severe cases, and every two hours in moderately severe cases without narcotism, or even an approach to that condition. He adds that "under the influence of the medicine, the pain and spasm subsided, the skin grew warmer and the pulse fuller, and the entire condition of the patient more hopeful."

The remedy must be given for its effect, and the quantity necessary is to be prescribed. Its action is to be carefully watched. Its greatest usefulness is to be reached only by its administration early in the course of the disease. After the symptoms indicative of effusion appear, it must be given in lessened doses, and its utility is greatly diminished. It is among the most notable facts respecting the use of opium in this disease, that the early American physicians did not hesitate to employ it when coma, a condition usually thought to preclude the use of narcotics, threatened, nay, Strong and others have recorded their opinion that it is a powerful agent in removing such comas as are not "absolutely irrecoverable."

When the condition of the patient is such as to render its administra-

---

[1] Quoted by Stillé.
[2] Revue médicale, LXXXVI. 1842.
[3] Histoire typhus cérébro-spinal, par C. M. Boudin. Paris, 1854.

tion by the mouth impracticable, or when the repeated vomiting prevents its absorption, it may be given in the form of enemata or suppositories, by the rectum, or one of the morphia salts may be substituted in hypodermic injections. The latter is in most cases the best plan of treatment.

In view of the fact that children are peculiarly susceptible to the action of this drug, the dose must be regulated with caution. A boy aged six, under the care of Dr. J. Lewis Smith, was quieted by the subcutaneous injection of $\frac{1}{32}$ of a grain (0.002 gramme) of morphia sulphate.

Ergot and belladonna have been used upon theoretical grounds, on account of their influence in diminishing vascularity of the nervous centres, but the evidence of their value is not satisfactory. Rosenthal[1] urges great caution in the administration of belladonna and in the hypodermic use of atropine.

Cannabis indica, the fluid extract of gelsemium (Bartholow), *zinc oxide*, large doses of *chloral hydrate* and *inhalations of chloroform*, have been employed in the management of the excitement. Chloral is to be emphatically condemned in the treatment of a disease attended with vomiting so continued as often to interfere with the assimilation of food, and characterized by a tendency to extreme exhaustion; and chloroform inhalations, when from the outset we often have to do with a feeble and irregular action of the heart, showing itself in extreme weakness and irregularity of the pulse, and a tendency to syncope upon assuming the upright posture; of the others it may be said that they are useful auxiliaries to treatment, but that they do not in severe cases constitute an efficient medication.

The last remark holds true also of the *potassium bromide*, a remedy, which has, however, great value in the treatment of mild cases, and in the treatment of children. It may be advantageously combined with opium or morphia.

In cases of extreme urgency, the inhalation of Squibb's ether may be resorted to for the purpose of securing temporary relief from the torturing pain, the jactitation, and the spasm.

Upon the approach of depression, *excitants* and *stimulants* are to be resorted to. Among the more useful are *ammonium carbonate, spirits of chloroform, turpentine,* and the *preparations of alcohol.* Cold affusion, practised several times a day, is recommended by German writers. It is a remedy scarcely likely to be widely used in this country. Quinine may be given in moderate doses.

Alcoholic stimulants are required at some time in the course of the ma-

---

[1] Rosenthal: Diseases of the Nervous System. American edition. New York, 1879.

jority of cases. Their use as a remedy in the treatment of this fever, independently of the indications which govern their use in the general management of diseases, has not been followed by satisfactory results. They are to be promptly resorted to when symptoms of depression of the nervous system show themselves, whether it be at the onset of the attack or later in the progress of the case. Their amount must be regulated by the effect which they produce. The pulse and the first sound of the heart are the best guides. If the pulse, after the free administration of alcohol becomes less frequent, stronger and fuller, and the first sound more distinct, it is beneficial; but if the pulse increases in frequency, the heart's action being excited, the tongue grows dry, and the excitement augments, the alcohol must be given in decreased doses or abandoned altogether. If the need be urgent, and the patient unable to swallow, brandy should be given hypodermically.

During the convalescence the vegetable tonics and iron are to be employed. *Arsenic*, and especially *potassium* arsenite, are also useful at this period. The latter has been praised as a remedy of value in the management of the acute disease. These praises are unfounded. *Cod-liver oil* is of use, and in proper cases *potassium iodide* is of proved service in promoting the resorption of the exudation. Its use should be long-continued, and at the same time flying blisters, daily hot affusions, and, after all acute symptoms wholly cease, mild continuous currents should be employed.

The potassium iodide is not of use in the treatment of cerebro-spinal fever during its acute course. Ziemssen states that he has not found it of the slightest benefit in the chronic hydrocephalus occurring as a sequel —a result which the nature of the lesions in that affection would lead us to expect.

Diet.—A generous alimentation is to be given from the beginning of the sickness. During the continuance of the febrile phenomena, milk, eggs, meat-juice and broths should be given at regular intervals, and continued in severe cases during the night. If food cannot be taken by the mouth, an attempt should be made to administer nutritious enemata.

As soon as he is able, the patient should be allowed an abundance of solid food. The appetite is often excellent, even in the early days of convalescence.

When there is *thirst*, the desire for water must be freely gratified. This symptom is often very distressing.

Constipation may be relieved by a dose of calomel with or without jalap, by other simple drugs, or by enemata. Neither constipation nor diarrhœa are, as a rule, difficult of relief.

When there is much prostration, and, indeed, in most cases, the patient should be guarded against assuming the **erect** posture, or, in truth, against even sitting upright in bed, on account of the danger of **syncope.**

The room should be darkened, and all noises **and other disturbing** influences avoided.

Delirium, spasm, and irritability of the stomach, too often, in the **severe cases,** render the administration of medicine **and** food impracticable.

IV.

ENTERIC OR TYPHOID FEVER.

DEFINITION.—An acute, endemic, febrile disease, of long duration, due to a poisonous principle associated with certain forms of decomposing animal matter. It is characterized by a gradual and often insidious commencement ; dull headache, followed by stupor and delirium ; a red tongue, occasionally becoming dry and brown ; in most cases tympany, abdominal tenderness, and diarrhœa ; an eruption of isolated, slightly elevated rose-colored spots, disappearing on pressure and developed in successive crops ; increased splenic dulness ; epistaxis ; late prostration and tardy convalescence. After death, constant lesions of the solitary and agminate glands of the ileum, with enlargement of the mesenteric glands and of the spleen, are found.

SYNONYMS.—Typhus nervosus; Typhus mitior; Abdominal typhus; Darmtyphus; Synochus and typhus with abdominal affection; Typhus gangliaris vel entericus; Ileo-typhus; Typhia; Typhus; Fièvre typhoïde ; Typhoid fever ; Mild typhoid fever.

Febris non-pestilens ; Endemic fever; Autumnal or fall fever.

Remittent fever ; Infantile remittent fever.

Febris lenta; Slow or lent fever; Febris chronica; Chronic continued fever ; Fièvre continue.

Nervous fever; Slow nervous fever; Irregular low nervous fever; Low fever ; Nervenfieber; Fièvre nerveuse.

Febris putrida ; Febris putrida nervosa; Sepimia ; Entérite septicemique.

Febris hectica; Infantile hectic fever.

Febris gastrica; Febris acuta stomachica aut intestinalis; Febris glutinosa gastrica; Febris gastrica acuta; Gastrisches Fieber; Fièvre gastrique; Epidemic gastric fever; Gastric fever; Febris biliosa; Bilious fever; Bilious continued fever ; Febris biliosa putrida ; Synochus biliosa ; Bilio-gastric fever ; Gastro-bilious fever.

Febris colliquativa ; Febris stercoralis ; Febris mucosa ; Febris pituitosa; Morbus bilioso-mucosus; Febris pituitosa nervosa; Schleim-Fieber; Fièvre muqueuse ; Mucous or pituitous fever.

Febris mesenterica maligna ; Febris intestinalis vel mesenterica;

Febris mesenterica acuta; Enteritic fever; Gastro-entérite; Entero-mesenteric fever; Dothiénentérie; Muco-enteritis; Fever with affection of the abdomen; Fever with ulceration of the intestines; Gastro-enteric and gastro-splenic fever; Entérite folliculeuse; Enteric fever; Febris tympanica; Intestinal fever.

Night-soil fever; Cesspool fever; Pythogenic fever.

Rock fever; Mountain fever.

The above long list of the terms under which this fever has been described at various periods and by many different authors, is taken, with few exceptions, from Dr. Murchison's great work upon the "Continued Fevers of Great Britain." They are variously derived from its supposed relationship to typhus, its mode of prevalence, its remittent character, its long duration, its supposed nervous origin, the occurrence of septic or putrid symptoms, its hectic phenomena, the presence of symptoms denoting disturbance of the stomach and liver, the intestinal symptoms, the morbid anatomy, its mode of origin, and localities in which it has prevailed. The term "abdominal typhus" and its equivalents, in general use in Germany and elsewhere upon the Continent, are open to the objection that they suggest a relationship with typhus that is now acknowledged on all sides not to exist. They are, in fact, due to the opinion formerly generally entertained, that there existed between the two affections an essential pathological identity—that they were, in fact, two varieties of a single species of fever. This opinion is no longer tenable. "Typhoid," suggested by Louis in 1829, is open to the same objection, since the labors of pathology during the past half-century have shown with increasing clearness, not that the fever in question is *like* typhus, but that it is *unlike* it. This term has, however the sanction of very general acceptance in France and among English-speaking physicians. The strongest objection to its use for any purpose whatever, lies in its common employment as an adjective to designate a condition or group of symptoms that may appear in the course of any acute disease—a use that has given rise to endless confusion of thought and vagueness of description. The term "*enteric fever*," proposed by the late Professor George B. Wood, possesses the advantage of designating at the same time the anatomical seat of the constant primary lesion, and, by a now accepted usage of the word fever in combinations of this kind, the infectious nature of the disease. It was adopted in the "Nomenclature of Diseases," in 1869.

Historical Sketch.

Enteric fever has been separated from the general group of the fevers as a substantive disease only in the present century. It is probable, however, that it has come down to us from a remote antiquity. The description of a continued **fever mentioned by** Hippocrates as prevalent in the autumn, and characterized by diarrhœa, bilious **vomiting, tympany,** abdominal pain, red rashes, bleeding at the nose, sleeplessness or a tendency to coma, delirium **and** subsultus, irregular remissions, a long duration and great emaciation, doubtless refers **to this disease.** It **has been** thought likewise **that Galen** described **it under the name of** "*hemitritæus*," a name applied to **a disease resulting from the grafting** of a tertian upon a quotidian intermittent. **Dr. Murchison thinks there** is little **doubt that the** "*febris semitertiana*" of the writers **of the seventeenth** century **was** true **enteric fever.** Spigelius (1624), **Panarolus** (1654) and Baglivi (1696) in Italy, **Thomas** Bartholin (1641) in Copenhagen, **and** Willis (1659) and Sydenham (1685) in England, recorded their observations of cases of fever, which both in the symptoms and the post-mortem appearances corresponded with enteric fever as we know it.

During the eighteenth century many accounts of **enteric** fever were published, and the difference between its **symptoms and those of** typhus, as well as the prominence of the **intestinal lesion, attracted** the growing attention of British and Continental physicians.

Strother (1729) distinguished **the epidemic fever of** 1727–29 in London, which we **know to have been typhus, from the** *slow fevers,* one variety of which, "the lent fever, is a **symptomatical fever,** arising from an inflammation, or an ulcer, fixed on some of the bowels."

Gilchrist (1734), Languish (1735), and Huxham **(1739), called** attention to the differences between the *Nervous Fever*, **or the** *Slow Nervous Fever,* **and the** *Malignant Continued Fever,* or *Putrid Malignant Petechial Fever,* generally prevalent **in England and Scotland.**

Sir Richard Manningham **(1746) published an account of the** "Symptoms, Nature, etc., of the Febricula or Little Fever." This fever **was** described **as of insidious origin, and apt in the** beginning to be disregarded; but at length conspicuous and very terrible symptoms arose, upon which the physician was sent for **in** the greatest haste, and "the little, neglected fever **proves of** very difficult **and** uncertain cure, and too often becomes fatal **in the end."** Its prominent symptoms were a red tongue, often dry, abdominal pains, diarrhœa, hemorrhage, a quick pulse, loss of memory, and sometimes slight delirium. It was known popularly as the *Nervous or Hysteric Fever, Low Continued Fever, Fever on the Spirits, Vapors, Hypo or Spleen.*

Others, at this period and **later, regarded the** *febris nervosa* as a very different disease from the *febris carcerum*, and in particular, Willan (1799) "observed **that Cullen had** improperly comprised under the term **typhus the slow or nervous fever** described by Gilchrist and **Huxham,** which may rather be **considered as a** species of hectic, **and is not received** by infection." It is at this point worthy of remark that the term hectic, thus employed by Willan, seems singularly appropriate in view of the modern doctrine of a primary and secondary or septic fever in the course of the disease, and that Manningham's observation that the "little, neglected fever" might prove at length "very difficult and uncertain of cure," and "often fatal in the end," is in full accord with the now well-known fact that cases untreated in the beginning are apt to be very serious and often fatal.

The Irish physicians of the last quarter of the eighteenth century also make frequent mention of the *febris nervosa*, a continued fever of three or four weeks' duration, more frequently occurring in the autumn, and attended by diarrhœa and by hemorrhages.

Upon the continent, during the same period, many accounts of enteric fever were published. De Haen, of Vienna (1760), describes it as *Miliary Fever;* Stoll (1785) as the *Pituitous* or the *Slow Nervous Fever;* others as *Febris Intestinalis*.

About the beginning of the present century the pathologists of France began to study the pathological anatomy of fever with great earnestness. Prost (1804) announced that mucous, gastric, ataxic and adynamic fevers have their seat in the mucous membrane of the intestine. Broussais (1816) advocated similar views. He regarded it as useless to distinguish between the ulcerations found in fever and frequently having their seat in the intestinal glands, and inflammations of other portions of the intestines. He believed that the symptoms were due to the inflammation, *gastro-entérite*, and upon this opinion he based his advocacy of copious depletion.

Petit and Serres (1813) described enteric fever as the *Fièvre entéro-mésentérique*. They called attention to the fact that the disease differed from ordinary enteritis, and were the first to look upon it as specific. They regarded the lesions as the result of the introduction of a poison, and as of an eruptive nature, like the pustules of variola, failing, however, to localize the processes of the disease in the solitary and agminate glands. To these observers is due the credit of having first pointed out the fact that the intestinal lesions are limited to the ileum, and principally to its lower parts.

In 1818 Bretonneau began at Tours the series of anatomical researches that enabled him to prove that the solitary and agminate glands of the ileum are always implicated in the processes of this fever, and that it differs essentially from all other inflammations of the bowels. He maintained that the disease was due to a poison communicable from the sick to the healthy, and regarded the intestinal lesion as analogous to the cutaneous eruptions of the exanthemata. He pointed out the fact that the severity of the general symptoms bears no relation to the intensity of the eruption. He named the disease *dothiénentérie*, or *dothiénentérite* (δοθιήν, a small abscess, boil, and ἔντερον, intestine), the "Dothinenteria" of the translators of Trousseau. These views were made known in Paris in 1820, but were first published by Landini[1] and Trousseau,[2] pupils of Bretonneau in 1826, and by Bretonneau[3] himself in 1827.

Louis's elaborate work, "Recherches sur la maladie connue sous les noms de gastro-entérite, fièvre putride, adynamique," etc., appeared in 1829. It not only contained an admirable and exhaustive description of the fever, but also an analysis of the symptoms and pathological phenomena so accurate and full, to use the words of Gerhard, "as to surpass any other description of individual diseases." Enteric fever was so well studied by Louis, and its symptoms so well set forth in this work, that it served from that time as a standard of comparison for other affections less thoroughly understood. It may be said then, with truth, that the appearance of this work marked an important epoch in the history of the continued fevers. Louis gave to the disease the name *fièvre typhoïde*, which was a few years later adopted by Chomel (1834), and soon passed into general use in France and among English-speaking physicians.

At the period of these investigations into its pathology, enteric fever was very prevalent in Paris and elsewhere in France. There was, accordingly, abundant opportunity for its clinical and anatomical study. Typhus fever was, on the other hand, unknown. An epidemic of typhus, brought back by the retreating armies of Napoleon after the disastrous campaigns of 1813–14, in Germany and Eastern France, had prevailed extensively in Paris and elsewhere in the large cities, and had been everywhere extremely fatal. But, from the date of the subsidence of this outbreak, typhus had not occurred within the borders of France, and was, to the French physicians of the time of Louis and Chomel, practically unknown. These observers therefore fell into the error of regarding the contagious fever of camps and armies, and of the British writers, as identical with the prevalent fever known to them as *dothiénentérie* or *typhoïde*.

At the same time English physicians were not idle in the study of the pathology of fever. As a result of their investigations, however, it was discovered that in by far the greatest number of fatal cases the intestines showed no evidences of disease.

To this general statement there were, nevertheless, numerous exceptions. Sutton (1806), William (1801), Muir (1811), Bateman (1819) and others, published accounts of outbreaks of fever, prevailing principally in the autumn, and attended by diarrhœa, in which, after death, the intestines were found to be "inflamed and gangrenous." Edmonstone (1818) recorded an extremely interesting history of an outbreak of enteric fever at Newcastle in 1817. This fever presented striking contrasts to

---

[1] Thèse inaugurale sur la dothiénentérie. Paris, 1826.

[2] De la maladie à laquelle M. Bretonneau a donné le nom de dothiénentérie ou de dothiénentérite. Archiv. gén. de médecine, Sér. I., Tome X., 1826.

[3] Notice sur la contagion de la dothiénenterie. Archiv. gén. de médecine, Sér. I., Tome XXI., 1829.

the epidemic fever then prevalent in various parts of the kingdom, and which at a later period fell upon Newcastle itself. Many of the first cases occurred among the better classes, and among servants residing in the best-aired parts of the town. Children and young adults in the vigor of life were almost exclusively affected. The duration of the attack was from fourteen days to a month. The disease was thought not to be contagious, and several members of a family were seized at the same time. It was almost unknown in the portions of the town inhabited by the poor, and in which typhus, upon its appearance, chiefly prevailed. The symptoms included vomiting, purging, bleeding at the nose, and hemorrhages from the intestines. Abercrombie (1820), Hewett (1826), and Bright (1827) recorded cases of fever in which the lesions of enteric fever were found after death. Alison (1827) stated that he had encountered in Edinburgh the intestinal affections described by French authors; but he maintained that they were not found after death from the ordinary typhus. Tweedie and Southwood Smith (1830) recorded a number of cases that had fallen under their observations in the London Fever Hospital, in which, after death, the intestines showed ulceration and the mesenteric glands were enlarged; in other cases, however, these parts were unaffected. These lesions thus came to be regarded by the English and Scotch pathologists as accidental complications of fever, and one of the earliest results of the awakened interest in the study of the morbid anatomy of fever was that the clinical distinction between the slow nervous fever and the malignant fever of camps and jails was lost sight of, both in France where the former only was prevalent, and in the British Isles, where the two fevers were constantly met with side by side.

In Germany, however, this distinction was recognized. Hildenbrand (1810) pointed out the difference between the contagious typhus and the non-contagious nervous fever. From this period the *typhus exanthematicus* and the *typhus abdominalis* were regarded as well-marked varieties—a view which is not yet finally abandoned at all hands in Germany, and which, while it was in advance of the doctrines held in France and England at the period of which we are writing, nevertheless has since had great influence in retarding the spread of the doctrine that they are essentially distinct, separate, and independent infectious diseases.

The distinction between the two fevers, based upon their clinical differences, that have arisen in the eighteenth century, had been lost sight of; that resting upon the differences in their morbid anatomy failed of recognition because of the confusion of the symptoms. It remained to study at the same time the symptoms during life and the appearances after death, and to compare them; in other words, to apply to this epidemic fever of Great Britain, typhus, the analytical method of study that Louis had applied to enteric, the endemic fever of France. This done, the two fevers were no longer to be regarded as the same disease: when it was thoroughly done, they were not even to be looked upon as varieties of the same disease; they were to unprejudiced eyes clearly seen to be separated by their causes, their symptoms, their course, their duration, and their anatomical characters, and no more closely related to each other than that they are both acute, specific, infectious diseases.

The process of accumulating the necessary facts upon which to base a convincing demonstration of the non-identity of typhus and typhoid fever was a slow one. From the appearance of the first edition of Louis' great work in 1829, in which the *fièvre typhoïde* and the *typhus* of English writers were spoken of as identical, till the issue of the second edition in 1841, the question of the identity or non-identity of these two fevers attracted the widespread interest of medical men both in England and France.

Prominent among the names of those engaged in the discussion which this question called forth are those of Drs. Peebles, A. P. Stewart, Perry, Barlow, Lumbard, Messieurs de Clautry, Montault, Rocheux, and Dr. Staberoh, of Berlin.[1]

To Drs. Gerhard and Pennock, of Philadelphia, belongs the honor of having first in America clearly set forth the distinction, between the two fevers, that was gradually taking form in the minds of the British and continental physicians. These gentlemen had studied enteric fever both in France, with Louis, and afterward in America, and had arrived at the conclusion that the *dothiénentérie* or *fièvre typhoïde*, of the French, and the prevalent continued fever of this country are identical. Upon the appearance of typhus in Philadelphia, in the spring of 1836, they recognized it as a different disease and after a careful study of the epidemic, they were enabled to point out the most important points of difference between the two affections, and to classify the epidemic fever among the continued fevers, "distinguished by the terms typhus, typhus gravior, petechial fever, etc."

"By diagnosis," Dr. Gerhard wrote, "we mean the comparison of all the symptoms appreciable by us in disease. This comparison requires a careful examination of the symptoms presented during life, and of the phenomena observed after death, in such cases as terminate unfavorably. We do not base our classification of diseases solely upon their anatomical lesions, although those lesions are oftentimes more constant than any other single symptom whatever; but we group together lesions and symptoms whenever they occur together with sufficient frequency to admit this process of generalization."

Proceeding to compare the two fevers upon this plan, they showed that the lesions of Peyer's patches and of the mesenteric glands invariably present in enteric, were never found in typhus; and that English obser-

---

[1] For a detailed account of the conclusions reached by these observers, see Murchison. It is to his work on the continued fevers that I am indebted for the outline and for most of the facts of this historical sketch.

vers were in error in regarding these lesions as merely complications of typhus; that there was a "marked difference between the petechial eruption of typhus and the rose-colored spots of **typhoid fever**;" that the train of symptoms associated with the intestinal lesions was very different from those of typhus, and that "the distinctive characters of the two diseases were such as in practice would not allow them to be confounded." They pointed out the fact that typhus is very contagious, whilst they were convinced that "dothinenteritis is certainly not contagious under ordinary circumstances," although in some epidemics, they said, "we have strong reason to believe that it becomes so."

Their observations and conclusions were published by Dr. Gerhard in February and August, 1837.[1]

Dr. Shattuck, of Boston (1839), strongly insisted, after watching some cases in the London Fever Hospital, upon the existence of two fevers in England, and pointed out, in a paper communicated to the Medical Society of Observation in Paris, the distinctions between them with considerable minuteness.

During 1840 several able papers, setting forth the differences between the two fevers, made their appearance; and in 1841, in the second edition of his work, Louis declared that "the typhus fever of the English is a very different disease from that with which we are occupied."

Other French and English physicians adopted similar views; but the doctrine of non-identity met with general opposition, and the opposite view continued to be taught in most of the medical schools.

In America, Bartlett, in his work on the "History, Diagnosis, and Treatment of the Fevers of the United States,"[2] treated of typhus and typhoid fevers as distinct diseases.

Sir William Jenner, in a series of papers upon "Typhus Fever, Typhoid Fever, Relapsing Fever, and Febricula, the diseases commonly confounded under the term Continued Fever"[3] (1849–52), contributed greatly to the final overthrow of the doctrine of the identity of the two fevers first named. He not only confirmed and extended the distinctions between the symptoms and post-mortem appearances, pointed out by previous observers, and in particular by Gerhard and Pennock, supporting his statements by the histories of carefully recorded cases and elaborate analyses of the symptoms and anatomical lesions of many cases of both fevers, but he also demonstrated the non-identity of the causes of the two fevers, and showed by an analysis of all the cases admitted to the London Fever Hospital in two years, that they did not prevail together and that the one did not give rise to the other; and he called at-

---

[1] W. W. Gerhard, M.D.: On the Typhus Fever which occurred at Philadelphia in the Spring and Summer of 1837. American Journal of Medical Sciences February and August, 1837.

[2] Philadelphia, 1842.   [3] Medical Times. November, 1849, to March, 1851.

tention to the fact that an attack of one of them mostly confers immunity from subsequent attacks of the same, but not of the other fever.

From the period of the appearance of these papers, the doctrine of the specific distinctness of enteric and typhus fevers was gradually accepted; it is now generally entertained in all parts of the world. If there be those who are exceptions to the rule that competent observers regard these two diseases as essentially distinct, they are very few, and their protests no longer retard the progress of knowledge.

The geographical distribution of enteric fever is wide. It has been observed in all countries and in every climate. It is endemic in the British Isles, all parts of Europe, and in North America. Hirsch[1] has reached the conclusion that its general prevalence in Europe and America dates no farther back than the second and third decades of the present century—that is, from the period at which typhus (der Petechialtyphus) became less common, and in part disappeared altogether.

Enteric fever is, according to Murchison, common in Scotland, more common in Ireland, and most common in England, but everywhere prevalent within the United Kingdom. Dr. Cayley, in his Croonian Lectures,[2] declares that upwards of eighty per cent. of the cases, if properly nursed and fed, that is, if treated upon the expectant plan, will recover. Some idea of the extent of the prevalence of enteric fever in England may be formed from his statement that upwards of 73,000 persons have died of it during the past nine years in that country alone.

There is, in medical literature, abundant evidence that this fever is also endemic in France, Spain, Italy, Turkey, Switzerland, Germany, Russia, Norway and Sweden, and in Iceland.

In North America it is endemic from Hudson's Bay to the Gulf of Mexico. In new and sparsely settled districts, where the land is being gradually, strip by strip, so to speak, brought under cultivation, the malarial fevers prevail; after a time, as populations increase, the malarial diseases and typhoid fever occur side by side, the one often modifying the symptoms of the other and complicating its course; and finally, when the land has been generally taken up and drained and tilled for some generations, and when the population has grown dense and villages and cities abound, the malarial diseases, true agues and remittents, come to impress communities but faintly, or they disappear altogether; but enteric fever grows very common, and asserts itself as the predominant endemic disease in proportion to the neglect of the sanitary measures by which alone it can be kept in check in populous localities.

---

[1] Handbuch der historisch-geographischen Pathologie. By Dr. A. Hirsch, Erster Band. Erlangen, 1860.
[2] On Some Points in the Pathology and Treatment of Typhus Fever. By Wm. Cayley, M.D., F.R.C.P. London, 1880.

It is far from uncommon in tropical and subtropical countries. Many observers have met with it in India. It has been reported as occurring in Egypt, Algeria, the west coast of Africa; in the West Indies, Mexico, and upon the Pacific slopes; Central America has not escaped it, and it is said to be extremely common in Brazil and Peru. Enteric fever has also been encountered in the British settlements of Australia, New Zealand, and Van Diemen's Land.

In tropical countries it has doubtless been frequently confounded with remittent fever.

ETIOLOGY.

I. PREDISPOSING CAUSES.

Climate, not of itself, but indirectly as determining the mode of life in communities, has a manifest influence upon the extent of the prevalence of enteric fever. This, like many other widely prevalent infectious diseases, is met with, as has been just indicated, in all parts of the world, but manifests, at the same time, a decided preference for certain broad areas or belts of the earth's surface. It is especially frequent and constantly present everywhere in Europe, Great Britain, and in the United States, and Southern Canada. These countries lie within the limits of the northern temperate zone, in which enteric fever possesses a fixity of tenure.

The season of the year is a predisposing cause of great importance. Epidemics of enteric fever commonly occur during the last half of the year, and the number of cases in localities where it is endemic is usually greatest from August to November, decreasing in December; and is lowest from February to May, again increasing in June.

Hirsch found that 519 epidemics of typhoid fever were distributed among the seasons as follows: in the spring, 29; in the summer, 132; in the autumn, 168; and in the winter, 140; and of 116 circumscribed epidemics occurring in France between 1841 and 1846, recorded by de Claubrey, 20 began in the first quarter of the year, 21 in the second, 39 in the third, and 36 in the fourth.

The following table shows the relative frequency of typhoid fever in the different seasons:

Number.	Locality.	Observer.	Date.	Spring.	Summer	Autumn	Winter
488 Cases	Lausanne.	Delaharpe	1851	44	122	211	111
74 Cases	Geneva.	Lumbard.	1834–37	7	24	28	15
355 Deaths	Geneva (Canton).	D'Espine.	1838–45	70	75	115	95
14,547 Cases	Nassau (Duchy).	Franque.	......	2,597	3,095	4,827	4,028
645 Cases	Lowell (Mass.).	Bartlett.	1840–47	102	163	250	130
2,826 Deaths	Massachusetts.	Curtiss.	1846–48	429	671	1,182	544
183 Cases	Strasbourg.	Forget.	1841	38	49	60	36
131 Cases	King's College Hosp.	Todd.	1860	21	25	51	34
5,988 Cases	London Fever Hosp.	Murchison	1848–70	759	1,490	2,461	1,278

This fever is so much more common in the latter part of the year that it has received in some districts of the United States the popular names of "Autumnal" or "Fall Fever."

The development and spread of enteric fever is favored by the high temperature of summer, and checked by the low temperature of winter. The maximum of temperature and the period of greatest prevalence of the fever are separated by an **interval of two or** three months, the former occurring in July, the latter in September and October; and the minimum of temperature, occurring in January, precedes the period of the least prevalence of the fever, in February or April, by a like interval, so that if the curves of temperature and of the frequency of enteric fever be projected diagrammatically, as has been done by Liebermeister,[1] they will be seen to nearly correspond.

The interval of about two months is not accounted for by the supposition of Murchison that the cause of the disease is called into action by the *protracted* heat of summer and autumn, and that the *protracted* cold of winter and spring is required to impair its activity or destroy it; but this time is probably consumed, as Liebermeister suggests, in the penetration of the warmth to the places where the poison is elaborated, its development outside the body, the stage of incubation, and the period from the beginning of the attack to the admission of the patient to hospital or his death. On the other hand, the time between the lowest temperature and the least prevalence of the disease is to be accounted for by the stage of incubation and the length of the patient's illness before admission to the hospital, or death as the case may be, the poison having already been introduced into his body, and by the infection of new cases from sources of contagion within dwellings, where it remains unaffected by the outside temperature.

Closely connected with the subject of the temperature as influencing the prevalence of enteric fever is *the state of the weather as regards dryness and moisture.* Hot and dry summers favor the development of the disease; cold and wet summers check it. This statement is supported by the concurrent testimony of observers in all countries. In England the summers and autumns of 1865, 1866, 1868, and 1870 were remarkable for their great heat and prolonged drought, and for an unusual and early increase of enteric fever. On the other hand, there have been few years in which the summer and autumn have been more cold and wet than in 1860, while the remarkable diminution of the prevalence of enteric fever over the whole country in that year, and in London during the wet autumn of 1872, was a subject of general observation. The admissions into the London Fever Hospital for 1860 fell to one-half of the average of the previous twelve years, and this diminution was due to the absence of the ordinary autumnal increase (Murchison).

---

[1] Ziemssen's Cyclopædia of Medicine, vol. i.

An analysis of the outbreaks of enteric fever which occurred in Stuttgart from 1783 to 1837, made by Cless, shows that all arose at the end of the summer or in the autumn, and that all had been preceded by unusually hot seasons. Virchow also found that, in Berlin, the years in which the rainfall was small were attended with severe epidemic and typhoid affections, while in wet years the mortality from enteric fever was decreased.

Dryness of the atmosphere alone does not, however, lead to an increase of enteric fever. In cities and other localities supplied with a system of underground drainage, warm damp weather often leads to an outbreak of the disease, while heavy rainfalls, by flushing the drains, remove the causes to which its origin and spread are chiefly due. On the other hand, outbreaks of enteric fever may be traced to the influence of abundant rains in washing the germs of the fever into water used for drinking purposes, particularly where the water-supply is derived in part from manured fields.

Pettenkofer and Buhl have shown that the prevalence of enteric fever in Munich is dependent upon changes in the height of the deeper springs of water. When the water steadily rises, typhoid decreases; when the water sinks, it increases. This observation corresponds with the statement just made, that enteric fever is much more frequent after hot and dry summers than after cold and wet ones. These observers explained the varying prevalence of enteric fever in connection with changes in the ground-water by the assumption that the causes of typhoid fever lie deep in the earth. When the water-level sinks, the layers of earth, containing moist organic substances and exposed to the air, undergo changes which lead to the development of the fever-poison. When, on the contrary, the water rises, these layers of earth are again covered and the development of the germs arrested. The explanation advanced by Buchanan and Liebermeister, namely that the lower the water is, the greater must be the proportion of solid matters suspended in it, and that therefore in localities where typhoid fever is endemic and the specific cause is in the earth, or soaks from privies into the earth, this poison must be relatively more abundant in the water the lower it is, is probably correct.

Age is of great importance among the predisposing causes of enteric fever. It is pre-eminently a disease of adolescence and early adult life. Of 5,911 cases admitted to the London Fever Hospital during twenty-three years (1848–70), nearly one-half, or 46.55 per cent., were between fifteen and twenty-five years of age, and more than one-fourth, or 28.58 per cent., were under fifteen. Less than one-seventh, 13.3 per cent., were above thirty, and only 1 in 71 exceeded fifty, (Murchison). The mean age of 1,772 cases was 21.25, that of the males being inconsiderably higher than that of the females. It may be stated that the greatest predisposition is between the ages of fifteen and thirty, and that it diminishes progressively both above and below these limits. Cases in the first year of life are exceedingly rare. The same is true of old age, although well authenticated cases of enteric fever in persons seventy, eighty, and even ninety years of age, are reported. The infrequency of the attack in the latter periods of life is doubtless to be accounted for, in part, by the fact that many persons, having already passed through the disease, are insusceptible to its poison.

Sex exerts little influence as a predisposing cause. The statistics of enteric fever, almost exclusively collected from the reports of hospitals, show a marked preponderance in the number of males. This preponderance is to be explained, not by an increased liability on the part of men, nor, in truth, to increased exposure to the causes of the disease, but by the fact that in most places more men than women seek treatment in hospitals.

Of 138 cases observed by Louis in Paris, 106 were males. This excess, however, is accounted for by the circumstance that a large number of males were strangers in Paris, and could not be treated at their lodgings.

Occupation exerts no influence whatever as a predisposing cause of enteric fever.

The mode of life of the individual is also without influence. Enteric fever is as common in the houses of the affluent as in the most crowded and destitute localities. In fact, the presence of stationary wash-stands in bedrooms, and the arrangement of bathrooms and water-closets near sleeping-rooms, expose the well-to-do to dangers of infection that the less fortunate escape. Enteric fever attacks by preference strong and healthy persons, passing by those, for the most part, who are the subjects of previous severe or wasting disease.

There is no relationship whatever between enteric fever and variola,' and enteric fever is not, as has been suggested, at all more prevalent in communities protected by general vaccination than in those less fortunate in this respect. The suggestion of Dr. Harley that scarlatina and enteric fever are different manifestations of the same poison, or that enteric fever is an abdominal scarlatina, is untenable. The two diseases are essentially different in their causes, course, symptoms, duration, and lesions after death.

Habitual exposure to the poison of enteric fever confers an immunity from the disease. Instances are recorded where successive visitors at the same house, at intervals of months, or even years, have been seized shortly after their arrival with enteric fever, or intestinal catarrh, from which the ordinary inhabitants were exempt. Persons changing their residence, from one part of a city to another, have not unfrequently been attacked with enteric fever, and persons coming from the country into cities very frequently become the subject of the disease. The French observers strongly insist upon recent residence as a predisposing cause. Of 129 cases Louis found that 73 had not resided in Paris more than ten months, and 102 not more than twenty months.

It has been suggested that one of the causes of the frequency of typhoid fever in the early autumn in our American cities, among well-to-do people, is to be found in the circumstance that, during an absence of two months or more in the mountains or by the sea, they have to some extent lost the immunity acquired by habitual exposure to sewer-emanations, and return to the atmosphere of the city unprotected.

Severe mental disturbance, fear, sorrow, care, and *great fatigue,* doubtless render individuals less able to resist morbid influences, and therefore act as accidental predisposing causes; but that they can give rise to enteric fever, as was held by the older authors, is a view wholly at variance with modern theories of the cause of the disease.

Pregnant and *lying-in women,* and those who are nursing infants, enjoy a relative immunity from enteric fever.

II. THE EXCITING CAUSE.

Up to the present time the exciting cause of enteric fever has eluded all attempts to demonstrate its nature, either by chemical analysis or microscopical examination.[1] It is known to us, as are the causes of most of the infectious diseases, only by its effects.

---

[1] Prof. C. J. Eberth, of Zurich, has recently examined the lymphatic glands, spleen, the affected parts of the intestine, and the liver, kidneys, as well as other organs, with a view to discovering the character of the lower organisms said to be the exciting cause of typhoid fever. Of twenty-three cases examined, micro-organisms were found in twelve—twelve times in the lymphatic glands and six times in the spleen. They were much more numerous in the lymphatics. Eberth does not regard these organisms as micrococci. They usually assume the rod shape, and are about the size of the bacilli found in decomposing blood, only with the difference that they usually take a narrow, oval, or stumpy spindle shape rather than a cylindrical outline. They are slightly rounded at the end, not cut off sharp. Together with these rods, small egg-shaped forms are met with, resembling micrococci. Undoubted spherococci were not observed. The peculiar delicate outline of these bodies serves to distinguish them from the bacilli of putrefaction. They contain one to three spore-like bodies, and are not so easily stained in methyl violet as the ordinary micrococci and bacilli. Eberth gives several interesting facts regarding the number of organisms found at different stages of typhoid, and concludes by asserting the probability that they stand in some relation to the essence of the disease.—*Virchow's Archiv,* Band LXXXI., 1880.

Professor Klebs, of Prague, also believes that he has discovered the micro-organism which constitutes the specific agent of typhoid fever, and develops his views in a paper entitled "Der Ileotyphus eine Schistomycose," published in the *Archiv für Experimentale Pathologie,* T. XII., 1880. Professor Klebs has for a long time, assisted by his pupils, been making researches in this direction. He writes that he has been able to find, at the necropsy of twenty-four persons carried off by dothinenteritis, microbes in various organs: in the intestinal mucous membrane, in the thickness of the cartilages of the larynx, in the pia mater, in the foci of lobular pneumonia, in the mesenteric ganglia, in the parenchymata of the liver, and generally diffused in the organs which showed the most decided lesions. These micro-organisms showed them-

The view that it is an organic poison is tenable only when the term poison is understood in the broadest sense. A poison produces sickness and destroys life, but it cannot infect. Much less is it capable of indefinitely reproducing itself either within or outside of the body, or of a prolonged continuous existence, during which it successively affects an endless series of individuals exposed to its influence in precisely the same manner and without exhausting its noxious power. The ingenious suggestion that it may be some derivative of albumen capable of setting up, in other albumen and albuminous compounds, chemical changes by "catalysis," and of thus inducing the series of changes in the body, which, taken together, constitute enteric fever, lies wholly within the domain of hypothesis. Without entering upon a detailed discussion of its improbability and its inadequacy to explain the well-proved facts of the pathogenesis of the disease, it is only needful to state that no derivative of albumen possessing such "catalytic" properties is known, and that organic compounds of the kind indicated are unstable; so that it is improbable that they would remain undecomposed, in such localities as are the favorite lurking-places of the germs of enteric fever, for any length of time, much less during the lengthened periods that such places retain their power of distributing the infection.

Without doubt the fever-producing principle is an organized germ, a micro-organism, a protomycete, a *contagium vivum*.

It is by this theory alone that we can understand the known facts bearing upon the origin and transmission of the disease.

Although the nature of the germ which produces enteric fever is unknown, many of its properties are established.

1. It is invariably derived from a previous case of enteric fever.
2. When introduced into the human body, it is, under favorable circumstances, capable of indefinitely reproducing itself.
3. It is eliminated with the fecal discharges.
4. It is not capable of producing enteric fever in other persons at once, but must undergo certain changes outside the body before it acquires this power.
5. It retains its activity, when it finds its way into favorable situations, for a lengthened period after it has passed out of the body, the

---

selves in the form of rods, about eighty micrometres in length and 0.5 to 0.6 micrometre in thickness. They have been constantly observed in the bodies of dothinenteric patients since the attention of Professor Klebs was drawn to the subject, and they are always absent from the organs, and specially the intestines, of subjects who have died from any other disease than typhoid.—*British Med. Journal*, Oct. 16, 1880.

Further researches are necessary to establish the causal relation between particular forms of protomycetes and enteric fever, but it may be confidently predicted that ere long the specific cause of this and many of the other infectious diseases will be demonstrable.

requirements to this end being decomposing animal matter, especially fecal discharges and moisture. Hence, cesspools, sewers, drains, dung-heaps, wet manured soils, are its usual habitat.

6. There is reason to believe that in such situations it is capable of reproducing itself.

7. It remains suspended in, and may be conveyed by, water used for drinking purposes, and usually finds access to the body by this means.

8. Suspended in the atmosphere, it also reaches the blood by means of the inspired air.

These statements are supported by the following facts and observations:

1. From the assertion that *the specific cause of enteric fever is invariably derived from a previous case of the disease,* many observers, even among the most recent authorities, strongly dissent. Among them Dr. Murchison is most prominent.

It had long been held that air and drinking-water, polluted with decomposing sewage and other kinds of putrefying organic matter, were capable of causing fever; but it was Murchison who first, in 1858, pointed out that the fever thus produced was different from that arising from other causes. He showed that the fever thus caused is always enteric, and never typhus or relapsing fever; that its origin in substances of this kind accounted for its endemic prevalence and the occurrence of circumscribed epidemics; that it also accounted for its attacking the rich as well as the poor, its occurrence in isolated country-houses as well as in towns and cities, and for its increased frequency in autumn and warm seasons. He adduced many conclusive facts in support of these statements. It is now universally admitted that the cause of enteric fever is traceable to air or drinking-water defiled with decomposing organic matter, and, in particular, with the emanations from sewage.

The name "pythogenic," signifying, as it does, "produced by putrefaction," is based upon this generally received opinion. It was first suggested by Dr. Murchison.

But Dr. Murchison and his followers go much farther than this. While admitting the now unassailable doctrine that the poison of enteric fever finds its way into drains, sewers, and the like, by means of the dejections of persons ill of the disease, and that a single case may in this manner give rise not only to other cases, but even to extensive epidemics, they also insist that the specific cause of this disease may be generated *de novo* in sewage, without the presence of the enteric excreta.

In support of this opinion two principal arguments are adduced. The first rests upon the well-established fact that persons may be exposed to recent typhoid stools in their most concentrated form, and fail to contract the fever unless decomposing sewage be present. Thus, in hospitals the disease rarely extends to the medical officers, the attendants upon the sick, or to the other patients. In nine years there were treated in the London Fever Hospital 3,555 cases of enteric fever, in the same wards with 5,144 patients not suffering from any specific fever. Not a single case of enteric fever arose among the patients suffering from other maladies in the whole course of this time, although it was a common practice for them to use the same water-closets and night-stools, and the use of disinfectants was exceptional. In the same hospital, during a period of twenty-three years up to 1870, 5,988 cases of enteric fever were treated, and seventeen of the resident medical officers contracted the disease, but of this number only five were in communication with the enteric fever cases, and twelve occurred at a time when serious defects existed in the drainage of the house. Since 1871, 1,447 cases of enteric fever have been received and treated in the same wards with 692 patients suffering with other diseases, and in this period only three nurses, and not a single patient, have contracted the disease. On several occasions, however, cases lying in other wards have been infected.

Liebermeister states that up to 1865, in the hospitals he had visited, namely, at Greifswald, Tübingen, and Berlin, he had never seen a single patient, nurse or physician attacked by enteric fever, although such cases were placed in the general wards.

Similar observations have been made in the hospitals at Paris and elsewhere on the Continent.

In this country it is customary to treat typhoid cases in the wards of general hospitals side by side with other patients. I have never known of the transmission of the disease to other occupants of the wards, nor to the attendants. When the disease has appeared as an epidemic in hospitals, it has seized upon persons occupying separated wards or rooms, and has almost invariably been traceable to defective water-closets or leaking drain-pipes.

There are, nevertheless, observations of an opposite character. Epidemics have on many occasions appeared in hospitals, and particular hospitals have suffered repeatedly from local outbreaks; but these extensions of the disease may be traced to local causes. Thus, Liebermeister states that, in the hospital at Basle, during his service of six years from 1865 to 1871, such hospital infections occurred repeatedly. During this period one thousand nine hundred cases were treated, of which, in forty-five, the disease originated in the hospital. In addition to these a number of cases of slight febrile affection, probably due to slight infection, also occurred, and cases of afebrile intestinal catarrh, which were to be imputed to the typhoid infection, were very numerous. Of the forty-five cases of the developed disease, many had never been brought in contact with the fever patients. For example, a patient, who had gone through an attack of variola in the isolated wards set apart for that disease, was attacked immediately after his discharge, with fatal typhoid fever. In the wards for syphilis, also isolated, and in the surgical wards, some cases arose. Many of the officers of the house, and the washer-women in particular, who never entered the wards, also contracted enteric fever. But there were facts that clearly indicated the existence of foci of infection within the hospital. Cases occurred, for example, with notable frequency among the patients and nurses in two rooms, one directly above the other. A wooden pipe, leading from the main sewer to the roof, passed by both these rooms. The sewer was faulty in construction at this point, so that matters accumulated there. It was, in fact, liable to become choked. Attention to this defect and its correction in part was followed by satisfactory improvement. It would appear that the hospital at Basle, where enteric fever is very prevalent, is saturated with the poison of this disease, and that its drainage is far from efficient. Not only have numerous cases of infection occurred since the observations of Liebermeister, but it is stated that almost all the new attendants have suffered from abdominal catarrh without fever.

Observations of this kind, which appear to show that enteric fever is not transmitted from the patient to those about him, and that it occurs promptly in those who are subjected to the emanations from choked drains and otherwise defective sewerage at a distance, lose their value in view of the fact, now generally admitted, that the specific germs cannot produce the disease in their fresh state, in the recently voided dejections of the typhoid patients. In truth, these very observations are evidence in support of this view. If the water-closets in connection with the wards occupied by fever patients are in order and the drains free, no infection takes place; the excreta are swept away before there is time for them to develop their poisonous properties. But, if the closets are in bad order and the dejections remain and undergo decomposition, other cases arise among those who use them; or, if the drains are choked, infection arises—not in the neighborhood of the patients, but at distant parts of the hospital, at the point of obstruction, that is, where the emanations from the arrested, or leaking and decomposing excreta, escape into the atmosphere.

The well-known fact that patients taken ill at a distance rarely transmit the disease to those about them on being removed to their own homes, is without value in support of the doctrine of the independent origin of the fever from decomposing sewage without the previous introduction of the poison; but the fact that such cases sometimes do give rise to epidemics, and that in such instances there is always either defective drainage, or direct or indirect contamination of drinking-water by soakage or otherwise, from the dejections of the patient, is of the most convincing force in support of the opposite view. Persons may be exposed to the direct emanations from decomposing human excreta, and drink water rich in the leakage from neglected privy-wells with impunity, as regards the danger of enteric fever, for an indefinite period; but the day a case of the disease appears upon the scene, the danger becomes direct and enormous, and, unless it is at once appreciated and provided against, other cases arise. That which was foul and indecent, injurious yet incapable of occasioning a specific disease, becomes a nidus for the growth of a poison and a focus of infection.

The following case, which very fully illustrates this statement, came under the observation of Dr. Flint:[1]

"In 1843, in a little settlement called North Boston, situated eighteen miles from the city of Buffalo, consisting of nine families, all being within an area of a hundred rods in diameter; but the few houses in which the disease occurred were closely grouped together around a tavern, the house farthest removed from the tavern being only ten rods distant. A stranger from New England, travelling in a stage-coach which passed through this settlement, had been ill for several days, and, on arriving at this stopping-place, was unable to proceed farther. He remained at the tavern, and, after a few days, died. He was seen by several physicians of the vicinity, and there can be no doubt that his disease was the same as that with which others were subsequently affected. Up to this time typhoid fever had never been known in that neighborhood. The sick stranger was seen by the members of all the families in immediate proximity to the tavern, with a single exception. One family named Stearns, having quarrelled with the tavern-keeper, had no intercourse with the family of the latter, and very little with the other families, all of whom were tenants of the tavern-keeper. No member of the family of Stearns saw either the sick stranger or any of those who were taken ill after the stranger's death. Members of the family of the tavern-keeper were the first to become affected, the first case occurring twenty-three days after the arrival of the stranger. Other cases speedily occurred in the surrounding families. In a month more than one-half the population, numbering forty-three, had been affected, and ten had died. Of the families immediately surrounding the tavern, that of Stearns alone escaped; no case occurred in this family.

"The occurrence of the disease produced great excitement in the neighborhood; poisoning was suspected, and Stearns was charged with having poisoned a well used in common by all the families except his own. A fact which encouraged this suspicion was, the common well, being owned by the tavern-keeper, he had refused permission to use it to Stearns, who had, in consequence, been obliged to dig a well for his own use. An examination of the water from the common well showed it to be perfectly pure.

"The disease was undoubtedly typhoid fever. Visiting this settlement during the prevalence of the disease, and recording the symptoms of several cases then in progress, the clinical history furnished abundant evidence of the nature of the disease. Moreover, I made an examination of the body of one of those who had died with the disease, and found the Peyerian patches ulcerated and the mesenteric glands greatly enlarged."

Dr. Flint was of the opinion that the spread of the fever was due to personal intercourse with the sick stranger, but it is beyond doubt that the water from the well served as the means of transmission.

The second argument is based upon the not uncommon observation

---

[1] A Treatise on the Principles and Practice of Medicine. By Austin Flint, M.D. Second edition. Philadelphia, 1867.

that enteric fever has broken out in isolated localities in which it had not hitherto been known, and the inhabitants of which, as far as could be discovered, had had no communication with any place in which the disease existed.

The following instances are taken from a large number of observations of like character collected by Dr. Murchison.

"In August, 1829, twenty out of twenty-two boys, at a school at Clapham, within three hours were seized with fever, vomiting, purging, and excessive prostration. One other boy, aged three, had been attacked with similar symptoms two days before, and had died comatose in twenty-three hours; another boy, aged five, died in twenty-five hours; all the rest recovered. Suspicions were entertained that they had been poisoned, and a rigorous investigation ensued. The only cause which could be discovered was that a drain at the back of the house, which had been choked up for many years, had been opened two days before the first case of illness, cleaned out, and its contents spread over a garden adjoining the boys' play-ground. A most offensive effluvium escaped from the drain, and the boys had watched the workmen cleaning it out. This was considered to be the cause of the disease by Drs. Latham and Chambers, and by others who investigated the matter, and also by Sir Thomas Watson. The morbid appearances in the two fatal cases were described as 'like those of the common fevers of this country.' Peyer's patches and the solitary glands of the small and large intestines were enlarged like 'condylomatous elevations,' and, in one case, the mucous membrane over them was slightly ulcerated. The mesenteric glands were enlarged and congested."

In June, 1861, a case similar to those at Clapham came under Dr. Murchison's observation.

"A girl, aged nine, was seized with febrile symptoms, vomiting, purging, and intense headache, followed by acute delirium, and died forty-seven hours from the commencement of her illness. After death the characteristic lesions of enteric fever, in an early stage, were found in the bowels. Accompanied by Dr. Stewart, I visited the rooms, over a stable, occupied by this girl's family. The privy was in the stable, and drained into a cesspool near the door, which had become choked up. Over the cesspool was an open grating, by which the stable drained into it, and from which the most offensive smells had issued since the beginning of the warm weather—so offensive that the horses had sometimes to be removed. The girl had been playing close to this grating at the time of her seizure. The cesspool did not communicate with the public drain, and no other cases of fever had occurred in the mews."

"About Easter, 1848, a formidable outbreak of fever occurred in the Westminster School and the Abbey Cloisters, and for some days there was a panic in the neighborhood respecting the 'Westminster Fever.' No case of fever had occurred in the Abbey Cloisters for three years, and there was no evidence of its having been imported. Within little more than eleven days it affected thirty-six persons, all of the better class, and in three instances it proved fatal. Shortly before its first appearance 'there occurred two or three days of peculiarly hot weather,' and a disagreeable stench, so powerful as to induce nausea, was complained of in the houses in question. It was found that the disease followed very exactly in its course the line of a foul and neglected private sewer or immense cesspool, in which fecal matter had been accumulating for years without any exit, into which the contents of several smaller cesspools had been pumped immediately before the outbreak of fever. This elongated cesspool communicated by direct openings with the drains of all the houses in which it occurred; the only exception was that of several boys who lived in a house at a little distance, but who were in the habit of playing every day in a yard in which there were gully-holes opening into the foul drain."

These observations are open to the serious objection that in none of them has the *possibility* of the presence of germs derived from previous cases of enteric fever been excluded.

The account of the outbreak at the school at Clapham is not sufficiently explicit as to the condition of the drain at the back of the house, which had been choked up for many years. It is not stated whether or not it connected with drains with the neighboring houses, nor whether there had been, some time before, cases of fever in the house or neighborhood, nor how many years the drain had been choked up. If this case has value at all as illustrating the subject of the etiology of enteric fever, it seems to me that it is in this, that it shows that the germs of the disease may retain their vitality, under favorable circumstances, for a long period—*many years*, and that, when so long imprisoned, it becomes highly infectious and capable of producing the most profound disturbances of the functions of the body with great rapidity.

In the second example, the possibility of the child's infection from an entirely different source, distant from her home, while visiting or at school, must be excluded in order that the observation may have weight in the argument.

In the third example it is distinctly stated that there had been no case of fever for three years, and that there was no evidence of its having been imported. There was, however, "a foul and neglected private sewer or immense cesspool, in which fecal matter had been accumulating for years without any exit, into which the contents of several smaller cesspools had been pumped immediately before the outbreak of the fever." It is to be remarked that fever probably had occurred in the neighborhood three years before, as the expression used in the account indicates. Now, there is reason to believe that the poison retains its activity outside the body for a long time under favorable circumstances, and farther on examples will be given to prove that it does actually retain it for many months. Is there a limit to the time? Where then is it? If the contagion remains active many months, why not three years? In this "immense cesspool" the conditions for its survival, perhaps also for its multiplication, were complete. Moreover, there was no exit. Immediately before the outbreak of fever the contents of this pool had been agitated by the pumping into it of several smaller depots of ordure! Who can be sure that a person suffering from typhoid fever, in a mild or even grave form, had not used some one of these numerous wells during the period preceding the occurrence of this epidemic?

These observations are certainly inconclusive, and they are neither better nor worse than the others of a long list adduced in defence of a view that, in spite of the ablest advocacy, is gradually giving way before the overwhelming force of accumulating facts that need no logic to render them convincing.

In addition to the example observed by Dr. Flint at North Boston, and given above, the following facts are cited by Dr. Cayley to show that the contamination of drinking-water by fecal matter may exist for an indefinite period without giving rise to enteric fever, but that upon the arrival of an infected person, the disease speedily makes its appearance as a local epidemic.

One is the well-known outbreak at Over Darwen.

"The water-supply pipes of the town were leaky, and the soil through which they passed was soaked at one spot by the sewage from a particular house. No harm resulted till a young lady suffering from typhoid fever was brought to this house from a distant place; within three weeks of her arrival the disease broke out, and one thousand five hundred persons were attacked."

A second took place at Calne.

"A laundress occupied the middle one of a row of three houses supplied by one well, into which the slop of the laundress's house leaked. She, on one occasion received the linen soiled by the discharges of a case of typhoid fever, and after fourteen days cases occurred in all three houses."

"At Nunney a number of houses received their water-supply from a foul brook contaminated by the leakage of the cesspool of one of the houses, but no fever showed itself till a man ill with typhoid came from a distance to this house. In about fourteen days an outbreak of fever took place in all the houses."

The record of the outbreak at Lausen, in the Canton Baselland, in 1872, is of great value as illustrating this and other facts in the pathogenesis of enteric fever. From the time of the passage of the allied armies in 1814, Lausen had suffered from no epidemic of typhoid fever. Isolated cases had never spread the infection. During the seven years preceding 1872, not a single case of typhoid had occurred.

"This village is situated in the Jura, in the valley of the Ergolz, and consists of one hundred and three houses, with eight hundred and nineteen inhabitants; it was remarkably healthy, and resorted to on that account as a place of summer residence. With the exception of six houses, it is supplied with water by a spring with two heads, which rises above the village at the southern foot of a mountain called the Stockhalder, composed of oolite. The water is received into a well-built covered reservoir, and is distributed by wooden pipes to four public fountains, whence it is drawn by the inhabitants. Six houses had an independent supply—five from wells, one from the mill-dam of a paper factory.

"On August 7, 1872, ten inhabitants of Lausen, living in different houses, were seized by typhoid fever, and during the next nine days fifty-seven other cases occurred, the only houses escaping being those six which were not supplied by the public fountains. The disease continued to spread, and in all one hundred and thirty persons were attacked, and several children who had been sent to Lausen for the benefit of the fresh air fell ill after their return home.

"A careful investigation was made into the cause of this epidemic, and a complete explanation was given.

"Separated from the valley of the Ergolz, in which Lausen lies, by the Stockhalder, the mountain at the foot of which the spring supplying Lausen rises, is a side valley called the Furlenthal, traversed by a stream, the Furlenbach, which joins the Ergolz just below Lausen, the Stockhalder occupying the fork of the valleys. The Furlenthal contained six farm-houses, which were supplied with drinking-water, not from the Furlenbach, but by a spring rising on the opposite side of the valley to the Stockhalder.

"Now, there was reason to believe that, under certain circumstances, water from the Furlenbach found its way under the Stockhalder into one of the heads of the fountain supplying Lausen. It was noticed that when the meadows on one side of the Furlenthal were irrigated, which was done periodically, the flow of water in the Lausen spring was increased, rendering it probable that the irrigation water percolated through the superficial strata, and found its way under the Stockhalder by subterranean channels in the limestone rock. Moreover, some years before, a hole on one occasion formed close to the Furlenbach by the sinking-in of the superficial strata, and the stream became diverted into it and disappeared, while shortly after the spring at Lausen began to flow much more abundantly. The hole was filled up, and the Furlenbach resumed its usual course.

"The Furlenbach was unquestionably contaminated by the privies of the adjacent farm-houses, the soil-pits of which communicated with it. Thus, from time immemorial, whenever the meadows of the Furlenthal were irrigated, the contaminated water of the Furlenbach, after percolation through the superficial strata and a long underground course, helped to feed one of the two heads of the fountain supplying Lausen. The natural filtration, however, which it underwent rendered it perfectly bright and clear, and chemical examination showed it to be remarkably free from organic impurities; and Lausen was extremely healthy and exempt from fever.

"On June 10th one of the peasants of the Furlenthal fell ill with typhoid fever, the source of which was not clearly made out, and passed through a severe attack, with relapses, so that he remained ill all the summer; and on July 10th a girl in the same house, and in August a boy, were attacked. Their dejections were certainly, in part, thrown into the Furlenbach, and moreover, the soil-pit of the privy communicated with the brook. In the middle of July the meadows of the Furlenthal were irrigated as usual for the second hay crop, and within three weeks this was followed by the outbreak of the epidemic at Lausen.

"In order to demonstrate the connection between the water-supply of Lausen and the Furlenbach, the following experiments were performed: the hole mentioned above, as having on one occasion diverted the Furlenbach into the presumed subterranean channels under the Stockhalder, was cleared out and eighteen hundred-weight of salt were dissolved in water and poured in, and the stream again diverted into it. The next day salt was found in the spring at Lausen. Fifty-six pounds of wheat flour were then poured into the hole, and the Furlenbach again diverted into it; but the spring Lausen continued quite clear, and no reaction of starch could be obtained, showing that the water must have found its way under the Stockhalder in part by percolation through the porous strata, and not by distinct channels."

It is a matter of the commonest observation that the decomposition of organic substances, and the drinking of water containing the products of such decomposition, are not of themselves sufficient to produce enteric fever. These are the conditions **favorable to the** development of the poison; but, in order that the **disease may be produced, something** more is **necessary,** and that is **the specific poison itself.**

The view that enteric **fever never originates spontaneously, but that every case is due to the continuous transmission of the poison, the sewers** or drains serving as the ordinary means of conduction, or as "a direct continuation of the diseased intestines," was first taught by von Gietl, in Munich. It was afterward ably advocated in England by Dr. Budd,[1] and **is to-day,** though not generally accepted, steadily gaining ground. If we assume **that a fever** so specific in its clinical and anatomical characters **must be due** to a specific **cause,** and that the specific cause is an organism of some kind, the view that **the** poison does not arise independently, but in every instance from a parent stock, **becomes a** logical postulate of these assumptions; otherwise, we are forced to accept the theory of spontaneous generation. **If we admit that** the decomposition **of organic and excrementitious substances in some instances can produce enteric fever—a specific disease, but in by far the greatest number of instances, even when every predisposing influence to the disease** exists, fails **to do so—we are** yet left **to grope** in the dark for the cause of the different behavior of such substances. It is conceded on all sides that when outbreaks of the disease **follow** the introduction of **a case into a locality** previously free **from it, the affection** spreads **not by direct contagion, but by the well-recognized methods of** sewage contamination **from** the dejections of the patient. **Examples of** this abound **in recent** medical literature. The accidental presence of the specific poison, and its prolonged latent existence, are capable of explaining every case of the apparently spontaneous origin **of the disease, with less violence** to our sense of the relation of cause and effect than the doctrine of independent origin. Moreover, **there are** two general truths relative to the etiology of the infectious diseases that aid us in reaching a reasonable conclusion. First, a **mild case of such** diseases may produce by infection the gravest forms of the **disease in** other persons. Thus, a walking case of **typhoid** fever, not **recognized as** such, **or a** case of mere intestinal catarrh, due to the cause of **typhoid, may** import the specific germs into a locality previously exempt, and in this manner **give rise** to an outbreak apparently spontaneous. Secondly, the **contagium is** capable of being transported in **the** bedding or clothing of **patients, and in other substances which may** serve as *fomites*. There is abundant reason to believe that the changes in the stools of typhoid fever,

---

[1] On Intestinal **Fever: its Mode of Propagation.** By W. Budd, M.D. Lancet, vol. ii. 1856.
Intestinal Fever Essentially Contagious, etc. Ibid. Vol. ii. **1859.**
On Intestinal Fever. Ibid. Vol. i. 1860.

which give rise to the infection, may take place not only in drains, sewers, and other similar situations, but also in the excrement discharged into the clothes or the beds of the patients. In this manner the germs may gain access to localities in which no case of the disease has occurred within the memory of man; and if, as is most probable, they retain their activity for a long time, all connection with any previous case disappears from the memory of those who may have known of it, and when new cases arise they present the appearance of being autocthonous. It is possible to conceive of other methods by which the germs may be imported without the importation of cases. Moreover, the people are always, the physicians often, untrained to the kind of scrutiny which alone will reveal the channel by which an infectious disease reaches a new quarter, unless it be so plain that the wayfaring man need not err concerning it.

The following observation is recorded by Dr. Cayley. It illustrates the statement that the poison, not at first active, becomes so within a short period in the bed, or the clothing, or about the person of the patient, just as in the drain of a defective water-closet:

"A boy was admitted into the Middlesex Hospital, under my care, on March 27, 1879, suffering from a very severe attack of typhoid. For several days he lay in an unconscious condition, and during this time he had very profuse diarrhœa—twelve to twenty liquid motions daily—which were, for the most part, passed in the bed. In the next bed was a boy aged six, who had been admitted on April 16th, with acute renal dropsy and bloody urine. He was kept strictly confined to bed, and never got up to go to the water-closet, down which the motions of the typhoid case were thrown. On May 11th, when he was convalescing, the dropsy having disappeared and the albumen much diminished, he was seized by typhoid fever, and passed through a moderately severe attack, with a well-marked rash and characteristic symptoms. This at first sight appeared to be a case of direct contagion, but there is no doubt that the true explanation is this: the bedding of the first patient was constantly kept saturated by his liquid motions, and, though every care was taken to change the linen frequently, it was obvious, from a distinctly fecal smell which was always present, that the bedding or mattress remained contaminated, and thus time was given for the poison to develop its infectious properties. Another patient in the same ward, admitted for acute rheumatism, was also attacked by the fever. He occupied a bed on the opposite side, and never came near the first case; but, being convalescent, he used the water-closet down which the motions of the typhoid case were thrown; and it so happened that at this time the closet was out of order, the contents were retained, and an offensive smell was constantly present. Hence, there can be no doubt but that he was infected by the emanations from the evacuations of the first case."

Murchison relates the following fact, which was communicated to him "on excellent authority." It proves beyond question the possibility of the transmission of the infecting principle of enteric fever to a considerable distance, without the direct importation of a case, and without the person who serves as the vehicle of importation necessarily becoming the subject of the disease:

"In 1859, the wife of a butcher residing in the small village of Warbstowe, situate between Launceston and Camelford, on the Cornish moors, travelled to Cardiff, in Wales, to see her sister, who was ill and soon after died of 'typhoid fever.' She brought back her sister's bedding. A fortnight after her return to Warbstowe, another sister was employed in hanging out these clothes, and soon after was taken ill with 'typhoid fever,' which spread from her as from a centre. The woman who had been to Cardiff never took the fever herself; there had been no cases in Warbstowe previous to her return; neither were there any cases in the neighboring villages, either before or after."

The frequency with which washer-women are attacked is to be explained by the fact that the dejections undergo the changes necessary to render them capable of producing the disease in the bed-linen and clothes of the patient.

The weight of evidence is decidedly against the doctrine of the independent origin of the disease from decomposing animal matter or fecal discharges.

There remains, however, another method by which enteric fever has been supposed to originate, namely, from the eating of diseased meat. The following are some of the most important of the observations upon which this supposition rests:

"On July 10, 1839, the local choral society held a festival meeting at Andelfingen, in the Canton of Zurich, after which 513 persons of all ages sat down to a cold collation, consisting chiefly of veal and ham. It was noticed at the time that neither the veal nor the ham were perfectly good. Some portions of the former had a greenish color and a disagreeable smell; the ham also is said not to have tasted well. But most of the guests observed nothing amiss, and ate heartily. Of the 513 persons who partook of this collation, 421 were subsequently attacked by an acute febrile disease, which was regarded at the time as typhoid. Thirty-four inhabitants of Andelfingen were also attacked, who had taken no part in the choral festival, but all of whom, it was ascertained, had been supplied by the same butcher who had furnished the veal and ham for the festival.

"The day after the festival there was a wedding in the neighborhood of Andelfingen, at which 15 persons were present, only one of whom had attended the choral meeting. The meat—veal and beef—for the wedding-breakfast was supplied by the same butcher. Of these 15 persons 11 were attacked.

"The period of incubation of this epidemic was very variable. A few were seized with nausea and vomiting on their way home, but this was ascribed to their having drunk too much wine. Out of 230 cases in which the incubation period was ascertained, 43 were taken ill during the first five days, 123 during the second five days, 48 during the third five days, and 16 during the fourth five days, 6 being attacked on the nineteenth day.

"The symptoms were those of severe gastro-intestinal irritation, with high fever, delirium, stupor, congestion of the lungs, and great prostration. No rose rash was observed, but in some cases there were petechiæ. The duration of the milder cases was about eight days; of the severer ones, three to four weeks. Convalescence was slow, and often the hair fell out. The mortality was slight, and on post-mortem examination, in some cases there were infiltration and ulcerations in the lower part of the ileum, with enlargement of the spleen; in others these changes were not observed.

"There can be no doubt as to the meat having been the cause of the epidemic, as only those persons who had partaken of it were attacked; while a very large number of persons from all parts of the canton were present as singers or spectators, who did not share in the collation, and they all escaped. But great doubts have been expressed as to whether it really was typhoid fever, or a form of poisoning resembling sausage-poisoning."

Liebermeister, recognizing the importance of this outbreak in reference to the etiology of typhoid fever, made a careful study of the printed accounts of it, and came to the conclusion that it was certainly not typhoid fever. Of more than five hundred persons who fell ill, only nine or ten died. He was at first led to the conclusion that it was an unusual form of trichinosis, but this opinion was not confirmed by the microscopical examinations that had been made. Liebermeister considers it probable that there is a special form of disease produced by meat-poisoning.

An epidemic, apparently due to the same cause as that which occurred at Andelfingen, but which was in part undoubtedly typhoid fever, occurred at Kloten, a place about seven miles north of Zurich, in 1878.

"On Ascension Day, June 30th, a festival was held of the united choral societies of the district, together with choirs from Zurich and Winterthur. The festival collation was furnished by the landlord of one of the inns, who himself was attacked by the epidemic. The food supplied consisted of ragout of veal, roast veal, and veal sausages.

"The meat, which came from various sources, was hung up in the meat-room of the inn, and the day before the festival was partly roasted, partly minced up for sausages, and the fragments used for the ragout were cooked on the following day.

"Nothing amiss was observed with the ragout, but the cold roast veal was in part decomposed, and the sausages were manifestly bad. In consequence of this they were largely distributed among the spectators, the children, and persons who could not afford to pay. Out of 690 persons who sat down to the collation, 290 were attacked. In all, 668 persons were infected who had partaken of the meat provided for the festival, either at the collation or at the inn, or who had been supplied with it at home; besides which, 49 secondary cases occurred—*i.e.*, persons who subsequently became affected by contagion, without having eaten of the meat. All other sources of infection could be almost certainly excluded, and Kloten was quite free from typhoid at the time. A very large number of the visitors to the festival ate no meat, but only drank wine; none of these were attacked. And it was clearly shown that the water was not the cause of the outbreak. Many persons who had drunk no water were attacked, and others who had drunk freely escaped. Several persons who drank wine to excess, and consequently vomited in the evening, afterward escaped.

"The incubation period, as in the Andelfingen epidemic, was for the most part very short. Some persons were taken ill on the second day, with loss of appetite, nausea, headache, pain and swelling of the belly, and slight fever. These early cases were the mildest, and many patients recovered in a few days. The greater number fell ill between the fifth and the ninth days. The symptoms were chills, fever, diarrhœa, great prostration, in many cases early and violent delirium. Epistaxis frequently occurred, and also profuse intestinal hemorrhage. The roseolous rash of typhoid was present in almost all the cases, and in many was remarkable for its extensive development, sometimes leading to little infiltrations forming distinct elevations, and leaving behind slight pigmentations.

"Post-mortem examination showed the characteristic appearances of typhoid fever, infiltration, and sloughing of Peyer's patches and the solitary glands, with characteristic ulcers where the sloughs were detached; not infrequently also infiltration and sloughing of the solitary glands of the large intestine, great enlargement of the mesenteric glands and spleen.

"With regard to the meat supplied, the following facts were ascertained: forty-two pounds of veal were furnished by a butcher at Seebach, taken from a calf which appears to have been at the point of death from some disease, when it received the *coup de grâce* from the hands of the butcher. All the flesh of this animal was sent to supply the festival at Kloten; but the liver was eaten by an inhabitant of Seebach, and he was attacked by typhoid fever; and the brain was sent to the parsonage at Seebach, and all the household became affected by the same disease.

"It was also ascertained that another of the calves which supplied the veal was suffering from umbilical phlebitis and peritonitis, and was at the point of death when it was slaughtered. The veal from this calf had been kept fourteen days, and was in a decomposed condition. All the meat was placed together in the meat-receptacle of the inn, which was in a horribly filthy state, and no doubt the putrefying flesh of this last calf, together with the state of the receptacle, would rapidly excite decomposition in the whole supply."

This meat was possessed of two injurious qualities: it was putrid, and it was in part infected with the specific typhoid poison (Huguenin, Cayley). In accordance with these two causes of disease, the outbreak which followed the eating of the meat was composed of two distinct groups of cases. These were, respectively, an acute gastro-intestinal catarrh arising shortly after the ingestion of the food and due to its putridity, and typhoid fever, showing itself after a more or less extended period of incubation, and due to the specific cause of that disease. In some instances the patients suffered from both of these diseases, the specific fever arising after the putrid catarrh. The mortality was slight.

In reply to the question as to how the cause of typhoid fever found access to the meat, Dr. Cayley states it as his opinion that there can be but little doubt that it was derived from the meat supplied by the butcher at Seebach. The liver and brain of this animal were eaten at Seebach by persons who had not visited Kloten, and who were attacked by typhoid fever within a short time. This calf was apparently dying of disease when slaughtered by the butcher. Huguenin regarded its sickness as typhoid fever. He believes that this fever is common among the cattle in Switzerland. It is stated that in the house of one of the persons attacked in the Kloten epidemic, while he was still laid up, two calves fell ill and were killed. Their intestines showed the characteristic lesions of typhoid fever, and it was thought that they were infected by the dejections of their owner in the early days of his sickness, while he was yet going about.

Huguenin states that the meat of cattle suffering from typhoid fever is often eaten without injurious effects, some change of decomposition being necessary to call the poison into activity. In this respect the development of the poison is the same in infected meat and in the alvine discharges of patients. It requires time and decay. It may be permitted me to quote another instance from Dr. Cayley, who has been at great pains to collect the observations bearing upon this question of the origin of enteric fever from diseased meat. This example took place recently at Kronau.

"A butcher refused to buy a calf because it was ill; the family to whom it belonged, therefore, ate it themselves, and six members were attacked with typhoid fever."

A few other instances of outbreaks due to like causes are to be found in recent medical literature.

Important as these observations undoubtedly are, they cannot be looked upon as evidence that enteric fever can be produced by the eating of putrid or decomposing meat. On the other hand, a careful scrutiny of the facts lead to the conclusion that whatever may be the disease produced under such circumstances, and however closely it may resemble enteric fever, it is not in fact that disease unless, along with the decomposed meat, the specific cause of enteric fever has been introduced into the body. These observations, therefore, so far from being looked upon as favoring the view that enteric fever can arise independently, are, when carefully looked into, confirmatory proof that it arises only by continuous transmission, and that the poison in every instance is derived from a previous case of the disease.

In confirmation of the statement of Huguenin, referred to above, that typhoid fever occurs among the cattle in Switzerland, it may be mentioned that several observers have encountered a similar disease in others of the lower animals.

Thus, Jaccoud[1] states that a disease altogether analogous has been observed among several species of animals, especially in the horse, the ass, the rabbit, the hare, much less commonly the dog and the cat. He refers to the observations of Bruckmüller, Röll, Serres, and Birch-Hirschfeld, in connection with this subject.

The last of these gentlemen studied the effects of the introduction of the blood and diarrhœal products into the bodies of rabbits at Dresden, in 1873. The animals succumbed to the injection of a certain quantity of blood subcutaneously, but the characteristic lesions of the intestines were not found. The introduction of the stools by the mouth was also without effect in producing the disease.

---

[1] Traité de pathologie interne. By S. Jaccoud. Tome II. Fifth edition. Paris, 1877.

A different result, however, followed the injection of the intestinal products into the œsophagus, and the effects were proportionate to the quantity of the matter injected and to the gravity of the disease in the patient from whom it was derived. The animals were seized with fever, the temperature reaching 41° C. (105.8°), great emaciation, diarrhœa; after death, swelling, pigmentation, and in ten cases, commencing ulceration of Peyer's patches was found; there was also enlargement of the mesenteric gland and of the spleen. In several cases there was recent pneumonia, and in one instance an intense "follicular catarrh" of the large intestine. Birch-Hirschfeld does not regard the disease thus artificially produced as identical with enteric fever, but its resemblance is very close.

Many experimental efforts to produce enteric fever in the lower animals, by exposing them to the emanations from decomposing animal substances, the effluvium from cesspools and the like, have been made without result. If the view of the nature of the cause of the disease advocated in these pages be correct, the failure of such experiments is to be looked for, in all cases where the decomposing material is not derived from, or commingled with, the dejections of previous cases.

Murchison fed a pig upon barley-meal mixed with the fresh stools of enteric fever patients for six weeks. The animal appeared to suffer 'no inconvenience,' but grew very fat, and, when killed, its intestines were found to be perfectly healthy. The same observer believes that there is no clear proof that any of the lower animals are liable to enteric fever. To this opinion are opposed the facts of the epidemic at Kloten, as well as the statement of Huguenin, and the strong probability which the results of Birch-Hirschfeld's experiments carry with them. In view of the probability that cattle, and particularly milch-kine, are occasionally subject to enteric fever, the question as to whether or not the disease is communicable by means of the milk of animals suffering from it, assumes great importance. It is a question, however, that can be settled only by future observations.

2. *When the germ of enteric fever is introduced into the body, it is capable, under favorable circumstances, of indefinitely reproducing itself.*

In this respect the specific cause of enteric fever in no way differs from that of the other acute infectious diseases. It is a result of this power of reproduction in the cause, that a single case may become the focus of a local or general epidemic.

The time between the introduction of the poison into the body and the development of the symptoms of the disease—the period of incubation—is occupied by this process, but it certainly does not come to an end with the outbreak of the attack. The causes of true relapses are involved in great obscurity, but it is probable that they arise in consequence of the multiplication of some colony of germs within the body, that have not passed through the usual phases of development at the same time with the others, and undergo such changes at a later period; or else, that they are due to reinfection from the patient's own decomposing discharges retained in his clothes, bedding, upon his person, or about the drains connected with his room.

The length of the period of incubation is variable. The difficulties in fixing it arise from the difficulty in determining, in a disease due to indirect contagion, the exact date of infection on the one hand, and, when the onset of the attack is insidious, the exact date of the beginning of the disease on the other. The indefinite prodromes are usually included in the period of incubation, and the attack is commonly dated from the commencement of the fever. It is not, however, often in practice, that the date of the first rise in temperature can be ascertained.

The beginning of the disease must, therefore, be reckoned from a much less definite circumstance, namely, the day when the patient is obliged to desist from his ordinary occupation, or to betake himself to bed. In cases where the period of invasion is distinctly marked by rigors, one of the chief difficulties in regard to the determination of the length of the period of incubation does not arise.

It is probable that in children the prodromic period is often included in the fever, for the reason that the nervous system reacts to relatively slight disturbing influences much more strongly than in adults. This may explain the fact that many of the instances of apparently short periods of incubation have occurred in children. This supposition, however, certainly does not account for the remarkable instance of the school at Clapham, in 1829. Here, twenty out of twenty-two boys were attacked, within four days of their exposure to the effluvia from materials removed from an old stopped drain, by vomiting, purging, fever and extreme prostration, and two died within twenty-four hours. The symptoms of the *stadium prodromorum* doubtless assumes in children much greater severity than at later periods of life, but the occurrence of fever and the rapidly fatal result, indicate that, in these cases, the actual disease began with the occurrence of the symptoms named, and it appears probable that the shortness of the period of incubation was due to the enormous amount of the fever-producing principle and its concentration. It is proper to state that doubts have been entertained as to the precise nature of this fever; but Dr. Murchison and others, who have carefully studied its history, are of the opinion that it was typhoid, and cases of undoubted typhoid have been repeatedly observed which were due to a similar cause and attended by a like symptom-grouping.

There is reason to believe that the period of incubation is longer when the poison finds access to the patient's system by means of the ingesta than when by means of the air; but the facts thus far advanced in support of this opinion are too few to warrant a positive conclusion.

Cases are not uncommon in which diarrhœa, vomiting, headache, and the like occur for a day or two in persons who, in the course of a fortnight or more, develop the symptoms of enteric fever. These symptoms subside, but the patient does not regain his feeling of health during the intervening period. It is probable that the poison in such instances exercises a primary irritating influence upon the intestinal mucous membrane at the time of its absorption, and thus occasions a transient gastro-intestinal catarrh, which is followed at the termination of the incubative period by the attack.

The duration of this period is commonly about fourteen days, but great variations are met with.

Murchison places it at two weeks, having, however, met with only two cases in his own experience that shed any light upon the matter. In one of these it was not longer than fourteen days; in the other not longer than twenty-one. Budd, as the result of the study of a large number of cases, was led to the conclusion that it varies from ten to fourteen days. There is evidence that the period of incubation may be much shorter. Greisinger[1] relates three instances in which the attack began within twenty-four hours after exposure to the infection. These cases are so often referred to, yet appear to me so inconclusive, that I cite them in order to show how little value they possess, rather than as illustrations of very short periods of incubation.

"One day, while sitting by the bedside of a patient very ill of enteric fever, whom he had examined for a considerable time, he suddenly felt unwell and thought he had caught the fever. The next day he was taken ill.

"One of the patients of his clinic had gone, perfectly well, to nurse a case of enteric fever. She slept the first night in the sick-room, and the next day was taken ill.

"A man passed a quarter of an hour in a building where there were cases of the fever (in einem Typhus-hause), and he fell ill the following day."

From what is now known of the nature of the cause of enteric fever, it is in the highest degree improbable that the fever was contracted in any of these cases, at the time Griesinger supposed. If it could be shown that these persons had not been exposed to the contagion before, the cases would be startling examples of short incubative periods. As it is, Greisinger himself had undoubtedly been constantly exposed to it, and the presumption that the others had also been previously exposed amounts almost to a certainty, in view of the overwhelming evidence that the period of incubation is always longer than in these instances.

The following examples of unusually short periods are related by Professor Quincke, of Bern:

---

[1] Virchow's Handbuch der speciellen Pathologie und Therapie. Band II., Abtheil. II. Erlangen, 1864.

"Three boys played on successive days, from March 13th to 16th, with some straw from a mattress soiled with the discharge of a fatal case of typhoid. All three were infected; the first was taken ill on March 22d. Here the maximum limit was nine, the minimum three days."

"Another case was that of a woman, who came from a distance to an infected house, where she stayed two or three days. She felt ill on her way home, and after a few days took to her bed, and died on the fourteenth day. Here the maximum limit was six days."

Here the probability that the beginning of the period of incubation was determined within the limits indicated is very great, but it is impossible to exclude a previous infection.

Instances of a longer duration are much more common. An incubation of twenty-one days is far from rare. Liebermeister places it at three weeks as an average, and states that it sometimes reaches four. A longer duration than this is mentioned, but such protracted periods may be ascribed to errors of observation.

The variations are doubtless in part to be attributed to constitutional peculiarities on the part of the patient, rather than to differences in the poison or the mode of its introduction. Thus, several persons may be infected at the same time and in the same manner, yet fall sick at variable intervals of time afterward. The following instance of the infection of a number of persons by drinking-water at the same time illustrates the foregoing statement. It is related by Professor Quincke:

"On June 22, 1873, the Federal Gymnastic Festival was held at Münsingen, a village situated about seven miles from Bern, and there was a large gathering of visitors from all parts of Switzerland, most of whom left Münsingen the same day. It so happened that the wife and son of the landlord of one of the inns, which was close to the ground where the gymnastic meeting was held, were ill with typhoid fever, and only ten feet from the leaky soil-pit of the privy of the inn was a well from which water was supplied to the persons taking part in the festival, as well as to those who took refreshments at the inn itself. An epidemic of typhoid fever broke out among persons who had attended this festival, and the particulars of fourteen cases, occurring among visitors who came from places free from typhoid, have been ascertained. In one case the attack commenced eight days after the festival, in three cases on the twelfth day, in one case on the thirteenth day, in two cases on the fourteenth day, in two cases on the fifteenth day, in two cases on the sixteenth day, in two cases between the sixteenth and eighteenth days, and in one case some time between the fourteenth and twenty-second days, when the patient first came under medical observation."

The germ may find access to the body without thereafter undergoing the changes and indefinite reproduction necessary to give rise to the fever. In order to this certain favoring conditions are requisite. What those conditions are we do not know, but we know that they are prevented in many instances by acclimatization, and in most by a previous attack.

Thus it is well known that persons are apt to be attacked in removing to a locality where the disease is endemic, as from one quarter of a city to another, or from the country into the city. Jaccoud states that, even to-day, Paris presents in this respect exceptional danger—a danger that is imminent for several months, or even a year and more. After a time, if the patient escape the attack, the danger diminishes, and those who have passed their lives in such localities often escape.

The immunity from a second attack rests upon the statements of patients who, in almost all cases, state that they have not previously suffered from the disease, and upon several remarkable observations, when the disease has become epidemic a second time in the same house or locality, after the lapse of several years, and has attacked those who escaped before, and spared the others who had suffered in the prior outbreak.

No suggestion adequate to explain the immunity conferred by a previous attack of enteric fever, or the immunity from a second attack in other infectious diseases, has yet been brought forward. The fact that second attacks do occasionally occur, adds not a little to the obscurity of the question. Murchison met with several instances of well-marked second attacks occurring after puberty, and many more of attacks of enteric fever in persons who had passed through "infantile remittent fever." Instances are also recorded by Trousseau, Bartlett, Budd, and others.

I attended a gentleman through an attack of enteric fever, with well-developed eruption, in 1873, who died of the same disease, in the spring of 1880, on the twenty-first day of his sickness. At the necropsy there were found infiltration of Peyer's patches without ulceration, enlargement of the mesenteric glands corresponding to the affected intestinal tract, and enlargement and softening of the spleen.

That a certain individual susceptibility to the poison is requisite in order that it may produce the disease, is evident from the fact that in house-epidemics due to contaminated drinking-water, or to the pollution of the atmosphere from defective sewers and the like, the whole household, although exposed to the same influences, rarely sicken at once, but commonly two or three only suffer, or, if others contract the disease, it is at varying intervals; while a considerable proportion usually escape altogether.

3. *It is eliminated with the fecal discharges.*

The alvine dejections appear to constitute the sole means of the communication of the disease. In this respect enteric fever resembles cholera and dysentery. Typhus fever and the exanthemata appear to be communicable by the emanations from the surface of the body and by the exhaled air, the specific virus being given off from the cutaneous and respiratory surfaces; but all the evidence bearing upon this point goes to show that enteric fever is not communicated in this manner. The attendants upon the sick do not contract the disease unless they are also exposed to the decomposing excrements of the patients, or to the continuing influences to which the patient's sickness is due ; the garments worn by the patient, and his bedding, do not communicate the disease to others unless they are defiled by his dejections; finally, persons may be in close relation to the patient without contracting the disease, while others in the same house, who have had no communication whatever with him, are occasionally attacked.

Piedvache mentions a remarkable instance of enteric fever, in a boys' school at Dinan, that strikingly illustrates the last statement :

"The boy first attacked was nursed by his fellow-pupils, more than twenty of whom passed the night with him during his illness, and used no precaution against the contagion. Not one of the boys thus exposed took the fever; but the second case occurred nineteen days after the death of the first, in a boy who had had no communication with the first patient, who had never entered his room, and who slept in a remote part of the building."

On the other hand, there are very plain observations to show that the disease is communicable by substances contained in the excrements of the patient. It is unnecessary to adduce particular examples of this fact. Almost every instance of the transmission of the disease cited in these pages in illustration of the various points of the etiology of enteric fever, is also an illustration of this point. Persons who do not breathe an atmosphere charged with the emanations from sewers, privies, dung-hills, which contain the dejections of typhoid-fever patients, nor drink water polluted with soakage from similar sources, nor handle linen, bedding, or other substances defiled with such dejections in a decomposing state, are not liable to contract the disease, no matter to what extent they are brought into personal relations with patients suffering from it. Enteric fever is therefore indirectly, but not directly, contagious.

4. *The germ of enteric fever is not capable of producing the disease in other persons in its fresh state, but must undergo certain changes outside the body before it acquires this power.*

Hospital experience, as well as that of those physicians in private practice who have given their attention to the matter, conclusively proves that the fresh stools are incapable of communicating the disease. Neither the attendants who empty the vessels, nor medical men who examine the stools, nor the patients in adjoining beds, who are of necessity exposed to the effluvium from them in their recent state, are attacked, save exceptionally and under circumstances which warrant the supposition that they have also been exposed to excrement undergoing decomposition. Yet, as has already been shown, all these classes of persons are liable to infection if the removal of the dejections be delayed by reason of defective sewerage or other causes. The changes take place in **soiled linen**, the bedding, or upon the surface of the patient's body, if the liquid dejections are not removed; but they take place with greater activity in **drains, privies,** or upon ground **saturated with** organic substances, where the dejections are collected **together**. It is in this **manner that** a single person ill of enteric fever, brought to a house or locality previously free from the disease, may become the means of indirectly infecting many others.

At this point there arises a question of the greatest practical importance. It is this: Within what time do the germs, innocuous at first, acquire their infecting properties? Unfortunately, the facts upon which to establish a positive reply to this question are wanting. It is probable that the time is very short. The fact that washerwomen frequently contract the disease from handling the clothing of patients—a fact confirmed by numerous observers—would indicate that the development by which the germ assumes its power of infecting is very rapid, for such articles are rarely retained any considerable length of time.

Dr. Cayley states that at the Middlesex Hospital it was formerly the custom to keep the stools of cases of enteric fever, which the physician wished to inspect, in pans which were kept in the water-closets of the wards. The time during which these stools were kept rarely exceeded twelve hours; nevertheless, several instances occurred in which patients using these closets contracted the disease. If these reserved stools were the cause of the infection, the period within which the contagious properties became developed in them could not have been longer than twelve hours.

The time is certainly short, and it is impossible to overrate the importance of prompt measures to render the stools inert by disinfection, or by the action of "powerful decomposing chemical agents," as suggested by Budd, and their efficient removal to localities in which they are little likely to do harm.

Enteric fever, in view of the fact that it is communicable from the sick to the well only after its specific germ has undergone certain changes outside the body, belongs to the class of miasmatic-contagious diseases, as defined by Liebermeister.[1]

5. *The germ of enteric fever retains its activity, in favorable situations, for a long time.*

Murchison informs us that he has met with several instances where single cases of enteric fever have originated in the same house, year after year, without traceable importation of the poison.

---

[1] See Ziemssen's **Cyclopædia**, vol. i., p. 27 et seq. I have employed this term in reference to catarrhal fever and cerebro-spinal fever, in a somewhat different and more familiar sense.

"For instance, six cases were admitted from a single house into the London Fever Hospital: one in June, 1849; one in October, 1851; one in February, 1854; one in November, 1855; one in November, 1856; and a sixth in July, 1857."

Dr. Cayley mentions an instance in which an interval of two years passed without any fresh importation of the poison, yet a new case arose.

Such cases as these speak strongly in favor of the prolonged continuous existence of the poison; but they are to some extent open to the objection that infection may have occurred in some other than the suspected manner, and have been overlooked. To the following example, cited from Dr. von Gietl, this objection cannot be urged.

"A villager, who had contracted typhoid fever at Ulm, returned to her native village, a place where typhoid had not existed for many years. The excrements of this person were thrown on the dunghill. Several weeks later five persons were employed to remove this dunghill. Of these five, four were attacked with typhoid fever, and one with gastric symptoms and swelling of the spleen. The excrements of these five persons were buried deep in the dunghill. Nine months later two persons were employed in completely removing this dunghill; one of them was attacked with typhoid and died of it."

Here, certainly, is a continuous existence of many months, during which the poison fully retained its activity. I am disposed to regard the well-known outbreak at the school at Clapham, already related, as a further illustration.

6. *There is reason to believe that the poison of enteric fever is, in favorable situations, capable of reproducing itself outside of the human body.*

Such situations are those where decomposing animal, and particularly fecal matters, are massed together, as cesspools, sewers, drains, dungheaps, or wet manured soils. In such places the excrement of a single patient may establish the focus of a local epidemic. It seems impossible to explain this fact except upon the hypothesis of the reproduction of the infecting germs in such situations. Indeed, there is good reason to believe not only that they grow and multiply enormously in localities of this kind, but that they also undergo a like increase, under some circumstances, in water itself.

In the epidemic at Lausanne, the dejections of two cases (for the boy is said to have sickened in August, probably at about the time the epidemic arose) mingled with a running stream, the waters of which were used to irrigate extensive meadows, and which thence percolated by a long underground way to the spring which supplied the village, were capable of infecting a large number of persons. The germs must, in this instance, have undergone an enormous increase in numbers, or else the poison of typhoid fever is capable of retaining its activity in an extreme state of dilution.

Compare also a recent epidemic at Caterham and Redhill, of which the particulars are related by Dr. Cayley:

"This outbreak was clearly shown by Dr. Thorne to have been caused by the contamination of the water of the Caterham Water Company by the alvine discharges of a single workman suffering from ambulent typhoid, who was engaged in the construction of a new adit. If any necessity arose for the workmen to relieve themselves during their spell of work below, which lasted from eight to twelve hours, and there should be any difficulty or delay in their being drawn up to the surface, it was arranged that they should use the buckets which were employed in raising the chalk.

"This man, it appears, had very copious diarrhœa, and had to relieve his bowels two or three times during each spell, but he positively denied ever having passed his motions in the adit without waiting for a bucket. But, nevertheless, there were undoubted means by which his evacuations could have found their way into the water, for, as the buckets were drawn up, their oscillations caused them to strike against the sides of the shaft, and some of the contents would so be shaken out and fall over a stage in the water below. And he also stated that his motions were so liquid that the buckets, which were also used to lower materials used in the construction of the adit, must have been stained with them. Here, then, we have, in all probability, only some splashings of typhoid stools mixed with a very large body of water—a proportion of the most extreme minuteness—and yet the water so contaminated gave typhoid fever to three hundred and five persons."

7. *That it remains suspended in water, and may be conveyed in it, and that it usually finds access to the body in that way*, has already been abundantly demonstrated. The following examples are peculiarly instructive:

"Epidemic in Stuttgart, 1872. The meadows from which a portion of the Stuttgart aqueduct receives its supply were, in the beginning of the winter of 1871-72, thickly manured with the matters taken from the city sewers. In January there was a thaw with rain, and the water of this aqueduct became of a yellow color, with an offensive smell. This was not produced by inorganic substances, and examination showed the presence of large quantities of organic matter. The water reduced a permanganate solution as much as would a mixture of pure water with one-half per cent. of urine. In February an epidemic broke out, in the portion of the city supplied by this aqueduct, so severe that there was an average of one typhoid patient for every two houses. In a neighboring district, partly supplied with water from the same aqueduct, there was an average of one patient to every ten houses. In the rest of the city the disease was not more frequent than at ordinary times, averaging one case to every one hundred and forty-four houses."

"Epidemic in the 'Soherenfabrik,' in Basle, 1867. In a collection of houses situated at some distance from the city, of which the inhabitants numbered about one hundred and fifty, mostly girls of thirteen to seventeen years old, there were no cases of typhoid during the severe epidemic in Basle, in 1865-66. In the year 1867, when the epidemic had subsided in the city, a single case appeared in January, a second case in February, and in May a large number, so that, within twenty-two days, thirty-six persons were attacked with typhoid fever, and many others with febrile and afebrile abdominal catarrh. It was shown that the well from which the drinking-water was drawn was fed from a canal into which emptied the privy. Eighteen days after the use of this water was forbidden there were no more new cases. A little later, three more cases occurred in persons who had probably disobeyed and drunken of the water. After the well was completely closed, there were no more cases."

A careful study of these and other cases already given shows that it is chiefly, perhaps only, in water that is confined in close situations, that the poison retains its activity for a considerable length of time, or is capable of reproducing itself. In large bodies of open water, and in running streams, it is probably speedily rendered inert.

Within recent years a number of epidemics of enteric fever have been traced to milk, supplied from dairy-farms where cases of the disease have led to the contamination of the wells from which the water used for washing the milk-cans, and perhaps also for the dishonest dilution of the milk, has been obtained.

The following is among the more recent of such outbreaks:

"At Southport, one case of typhoid after another was announced to the municipal authorities, until in about two weeks a total of twenty eight was reached. Such a rapidly invading epidemic demanded, of course, energetic measures for its repression, and a careful inspection of the various dwellings in which the victims had been attacked was undertaken. The health officers found, however, to their surprise, that with two trifling exceptions these premises were all in good sanitary condition; but further investigation disclosed the fact that in every instance milk had been served, to the families in which the typhoid fever had occurred, from a particular dairy some miles distant, and upon the grounds of this dairyman was discovered a well horribly polluted with soakage from a filthy cesspit near it. In the words of the chairman, "Chemical analysis showed that it was nothing but liquid sewage, and calculated to spread disease wherever its influence extended," and the proof that this foul infecting material had been accomplishing the work for which it was so well "calculated" is met with in the circumstance that, on stopping the milk-supply from this dairy, the epidemic ceased to spread, although not before two of the cases previously attacked had resulted fatally."[1]

It is not here expressly stated that enteric fever cases existed upon the farm in question, but the argument thus far followed has been in vain, if a specific germ was not in fact the cause of the specific disease which the tainted milk produced.

8. *The poison of enteric fever is also propagated by the atmosphere, and reaches the blood by means of the inspired air.*

The common experience of all physicians sustains this proposition. Liebermeister states that in the hospital at Basle, where infection by drinking-water could be excluded, he often saw cases of typhoid arise under conditions which seemed to exclude every other mode of infection except by the air.

The case of the villager narrated above (p. 143), whose dejections, buried in a dung-hill, gave rise to the disease in those who disturbed it some time afterward, has been cited with reason to prove that the germs

---

[1] Medical News and Abstract, January, 1881.

may be inhaled. It is in this manner that the disease is contracted by washerwomen from soiled linen. Dr. Murchison narrates the particulars of an outbreak, in which twenty-eight boys out of thirty-six were attacked in a school in the succession, and with an intensity, corresponding to the degree of their exposure to the emanations from an untrapped drain in a passage-way leading to the school-room.

In cities where the character of the water-supply precludes infection by means of drinking-water, as in Philadelphia and New York at the present time, sporadic cases of enteric fever are doubtless, in the great majority of instances, due to the inhalation of the poison. The origin of such cases is usually difficult, often impossible, to trace.

It is probable that the poison, being in the form of solid particles of extreme minuteness, is arrested upon the tongue and pharyngeal mucous membrane, and swallowed with the saliva. Some writers upon the subject entertain the opinion that it may also reach the blood by way of the respiratory mucous surfaces—a view that is rendered very probable by the readiness with which relatively coarse particles of matter of a different kind find their way into the tissues of the lung, in the case of miners and other workmen, although the constant intestinal lesion has been adduced to uphold the view that the poison enters the system by the way of the alimentary canal. Whatever the channels by which it finds access, it manifests a constant predilection for the lymph-follicles of the ileum.

Enteric fever is pre-eminently an endemic disease. The larger cities of the moderate temperate zone are never free from it. The poison is propagated continuously. From the patient it finds its way into localities suited to its growth and reproduction, and from such localities it gains access, by means of the air, the drinking-water, or other ingesta again into the human body. The dangers of infection by drinking-water are reduced to a minimum where the supply of water is drawn directly by means of distributing pipes led into the houses, from large common reservoirs adequately guarded against defilement and placed at such a height as to escape contamination by soakage from the neighboring cesspools. These dangers arise, however, wherever cisterns or tanks are introduced into houses, and especially into such houses as are provided with water-closets which are supplied, even indirectly, from the same cistern as the common wash- or drinking-water. It very often happens that the overflow pipe of the tank acts as a ventilating shaft to the sewer. Notwithstanding all the precautions of the most advanced knowledge of the draining and ventilating of houses, it is manifest that the complex system of continuous drainage necessary in cities is occasionally conducive to house and local epidemics of enteric fever, as well as, to a less degree, of others of the acute infectious diseases. The efficiency of such arrangements, for the most part, explains the much more common occurrence of single cases rather than groups of cases. A case of the disease is incapable of infecting those about it, if the dejections be promptly disinfected and swept away into well-constructed sewers; but it becomes a focus of infection if the excrement be neglected or retained. Hence, local epidemics are much more common in villages and small towns, while sporadic cases are constantly present in crowded neighborhoods and large cities. A sporadic case may, however, become the means of infecting persons at a distance, who drink the water or breathe the air polluted with the outpourings of the sewer into which the typhoid stools are thrown. It is not uncommon for the disease to arise on board vessels lying in rivers at points where the sewers of large cities empty. It is in this sense that the sewers may be spoken of as a continuation of the diseased intestine.

It would follow, from the foregoing considerations of the nature of its cause, that enteric fever is a disease essentially localized in its distribution. Such is in fact the case. The specific cause is to be found everywhere, and is readily capable of transportation from place to place, but it lurks in dark, neglected corners and about the foul ways of men's dwelling-places, and creeps along with oozing filth, crawling into wells and springs, and hiding itself in the ground, choosing now a victim, and again a group of them, but never giving rise to pandemics, or, in the wide sense, even epidemics, as do the poisons of typhus, cholera, or relapsing fever. The most extended epidemics of enteric fever extend over certain quarters of a city, or a large town, or a limited section of country, and are always made up of other distinct, local, circumscribed outbreaks or endemics.

CLINICAL HISTORY.

The course of enteric fever has been variously divided by those describing it into artificial stages in accordance with the relative importance they have attached to the symptoms or the lesions. The older authors distinguished a *stadium prodromorum*, a *stadium incrementi*, extending over the first week, a *stadium acmes vel fastigii*, extending over the second week and part of the third, and a *stadium decrementi* beginning at the end of the third week.

At a later period it became customary to divide the disease into two stages: the first covering the time of the development of the attack; the second, the period of improvement and convalescence. Under this arrangement the first three weeks of the disease are embraced in the first stage, the last week and the whole period until the restoration of health in the second.

Others, again, have divided the course of the disease, after the stage of incubation, into a stage of invasion, a stage of glandular enlargement, referring to the intestinal lesion, a stage of ulceration or sloughing, a stage of lysis, and the convalescence.

Wunderlich determined, by extended observations of the temperature in enteric fever, that the febrile movement has, in uncomplicated cases, a typical course, and that the deviations from this course, in obscure and complicated cases, are not such as to prevent a diagnosis from the temperature-curve alone, if the physician be familiar with the ordinary deviations. In well-marked uncomplicated cases the entire duration of the fever is from three to four weeks. This time may be divided, with reference to the temperature, into four periods, to each of which belongs a special fever-curve, and each of which lasts, in general terms, about a week. In accordance with these facts, the German observers divide the course of the disease into six periods, namely, the stage of prodromes, the first, the second, the third, and the fourth weeks, and the period of convalescence. It is to be borne in mind, however, that these periods may be modified by complications or by treatment, and, that by sequels, or relapses, the duration of the disease may be indefinitely prolonged.

A *stage of prodromes* usually precedes the onset of the fever, which is so insidious that the patient, in most instances, is unable to designate the day of its commencement, although his mind and memory may be still alike unimpaired. The patients are weary and complain of a general feeling of malaise with vertigo, headache, especially in the forehead, often increasing toward night. The sleep is broken and unrefreshing. Muscular pains may also be the subject of complaint. Sometimes there are uneasy abdominal pains with diarrhœa, and if diarrhœa be not at first present, it is often induced by the purgative medicines to which the patients are apt to have recourse. At the same time the patient is silent and indisposed to exertion; his expression is dull; his appetite is poor; his tongue swollen, and often heavily coated. This period has a duration of from five to ten days, sometimes it is even longer. It gradually merges into the declared disease. Sometimes slight, irregular chills, or repeated attacks of chilliness, mark the beginning of the fever. In other cases the fever is preceded by an attack resembling intermittent fever. The course of the fever, however, soon becomes remittent, and the characteristic symptoms of enteric fever are developed. Such cases are oftenest encountered in malarial districts. In rare instances the disease begins abruptly without prodromes, being ushered in with a chill followed by high fever. If there be no definite phenomenon of this kind from which to date the beginning of the attack, it is customary to reckon it from the day on which the patient was obliged to discontinue work or take to his bed. Hence, as Liebermeister has pointed out, the beginning of the disease may, in consequence of the determined character of the patient, be dated several days too late.

The first week.—The attack is to be regarded as beginning with the first chilliness or the first temperature rise. The fever steadily increases, but is distinctly remittent in type, the exacerbations occurring in the afternoon or evening, and the remissions in the morning. The rise in temperature, although steady, is gradual, the morning remission of the first four or five days being decidedly less in extent than the evening exacerbation of the previous day, so that, by the evening of the fifth day, the temperature reaches the neighborhood of 40° C. (104° F.). **The skin is** usually dry and hot; sometimes, however, especially in the early part of the day, it is moist or even bathed with sweat. The patient, with the **increased fever of** the latter part of the day, not rarely experiences sensations of chilliness, which are followed by flushing of the face **and an increased** sense of feverishness. The symptoms of the prodromic stage are intensified. The headache is violent. It is sometimes confined to the frontal regions, sometimes felt in all parts of the head. More rarely the focus of head-pain is in the occipital region. Epistaxis not infrequently occurs; it is usually slight, often not exceeding a few drops; at other times the amount is considerable. The patient feels tired, and at this period of the disease is usually obliged to keep his bed. If he rises for any purpose, his limbs tremble from sheer weakness, and he is often so dizzy that he is obliged to be supported. His expression is dull, but by no means approaches the heavy, stupid look that is characteristic of typhus. He is silent, disinclined to **mental exertion; but, when roused, his mind is** usually **clear and his memory good.** It is not, however, always **easy** to fix his attention. He sleeps restlessly and is disturbed by disagreeable dreams. **Between sleeping and waking** there is slight delirium. The lips are parched and dry; the tongue is at first moist, its mucous membrane swollen, covered with a **whitish yellow fur,** sometimes thin, sometimes thick and creamy; its margin and tip **are red.** After a time it becomes drier and is no longer swollen; the coating flakes off altogether, or remains only in irregular streaks or patches. When protruded it often trembles. The tonsils and the pharyngeal mucous membrane are **in many cases** also swollen and red. Appetite is lost, thirst is augmented. In many instances diarrhœa continues from the prodromic period; in the majority of cases, however, the bowels are at first confined, and diarrhœa sets in some time in the course of the first week. There are, in the course of the twenty-four hours, several thin, brownish stools, feculent in character, unattended by pain, and usually without tenesmus. Not infrequently diarrhœa is absent in **the first week,** or even slight throughout the attack. In such **cases,** however, laxative drugs act with **unusual energy.** Toward **the close of this period** there is some fulness **of the abdomen, or it may be decidedly distended, and even tense.** It is **tender to** pressure, particularly in **that** part of the surface corresponding to the ileo-cæcal region, and upon palpation gurgling is produced. This sign is not, however, of diagnostic importance, as it is encountered in other diseases attended by intestinal catarrh. At this time the spleen **can** be shown in most cases, upon physical examination, to be enlarged.

Sometimes its border may be also discovered upon palpation, but the abdominal distention usually renders this impossible. The urine is commonly diminished in quantity, and occasionally shows a faint reaction upon testing for albumen; the urea is increased, the chlorides diminished. In some cases the characteristic eruption appears upon the last day of the first week. The conjunctivæ are not injected, nor is the face dusky, as in typhus, but commonly there is a circumscribed pink flush over one or both cheek-bones, like the flush of hectic, and, like it, deeper toward the latter part of the day. At this period, upon auscultation, a few scattered, coarse, mucous râles may be detected posteriorly.

The second week.—The fever remains continuous at about the same height reached at the close of the first week, although in severe cases it may, at this time, rise slightly higher. The skin is hot and dry, the expression duller, the flush deeper and more continuous, the countenance sometimes slightly dusky. In the course of this period, and usually about the tenth day, the headache greatly diminishes or ceases altogether. The mental condition is now peculiar and characteristic of the disease. The patient is somnolent, but has no sound sleep, either in the night or day. He is indifferent, apathetic. It is difficult to rouse him. Partly on account of his mental condition, and partly on account of his deafness, which is now more or less marked, he must be spoken to loudly in order to attract his attention. When asked how he is, he commonly replies that he feels well. As a rule, if his attention is fixed, he answers questions correctly, but in as few words as possible. Muscular movements are feeble, tremulous, and uncertain. The tongue is protruded with difficulty, partly on account of its dryness, partly on account of the patient's indifference to what is said to him. It is red, fissured, and crusted with sordes. The patient lies upon his back, with his eyes half closed, motionless, except that he picks at the bedclothes, or makes feeble, wandering movements with his hands. There may be subsultus tendinum, or convulsive twitchings of special groups of muscles. At times, and particularly toward night, he mutters incoherently, and in the night, or when roused, there is wandering delirium. The urine and fæces are often passed involuntarily, or the urine may be retained. In other cases, and particularly in those attended by greatly elevated temperature, the mental condition is irritable. The delirium is active; it may be furious. The patients are disturbed by vivid hallucinations, they shout, attempt to get out of bed, and are restrained with difficulty. Sometimes the delirium alternates between these two forms with great rapidity, and the instances are not few, in which patients, passing rapidly from a passive, indifferent condition, with muttering delirium, into active delirium, have unexpectedly sprung from their beds and thrown themselves from the window. At this period of the disease patients should never be left, even for a moment, unattended.

The belly is now more swollen; tenderness and gurgling, especially in the ileo-cæcal region, are more marked. The diarrhœa increases—it may become profuse; the stools are watery, of a yellowish brown or ochre color, or they are of a greenish color and flocculent. The latter appearance has been said to resemble "pea-soup." The urine often contains a small amount of albumen. The spleen increases in size, but its borders can rarely be made out by reason of the abdominal distention. The eruption, which may appear upon the seventh day, is more commonly met with in the first half of the second week. It is characteristic of the disease. It consists of isolated, slightly elevated, rose-colored spots, which disappear upon pressure, and come out in successive crops and gradually fade away. These spots vary in size from one to three lines in diameter; they are of an irregularly oval shape, and, upon their first appearance, indistinctly marginate. They vary in number from a few, scattered over the abdomen and lower portion of the thorax, to many hundred, distributed generally over the body. As a rule from five or ten, to twenty or thirty may be counted in well-developed cases. They are absent altogether in some of the milder examples of the disease.

Upon physical exploration of the chest, in addition to the mucous râles heard at an earlier period of the disease, scattered sibilant and subcrepitant râles, evidences of the extension of catarrhal processes to the smaller bronchial tubes, may be detected; and, upon percussion, we find impaired resonance at the base of the chest posteriorly, commonly more marked, both in extent and degree, upon one side than upon the other. Cough is often present, but it is not proportionate to the pulmonary lesions. It is attended with a scanty mucous or muco-purulent expectoration.

The third week.—The fever changes from the continuous to the remittent form. The morning remissions become more marked, although the evening exacerbations continue to be nearly as high as during the second week. The symptoms of the second week, however, remain unabated, or they even increase until toward the end of this period, for it is not until that time that the morning remissions begin to affect the general condition of the patient. It often happens that the symptoms which, taken together, make up what is known as the "*typhoid state*," and which belong to the latter part of the second and to the third week, do not attain their full development until some time in the course of the third week. The stupor deepens, so that it is often with the utmost difficulty that the patient can be aroused. If he can be prevailed upon to protrude his tongue, he often fails to withdraw it until his attention is again aroused, and he is directed to do so. He does not ask for drink; but, when fluid is placed in his mouth, he mechanically swallows it. The fæces and urine escape involuntarily, or the urine is retained and the bladder may become enormously distended. From day to day he loses flesh and strength. He is unable to raise himself or even to turn in bed. His tissues are wasted;

his muscles soft; his cheeks hollow; the skin drawn tightly over his nose and brow; his eyes sunken; his face dusky and faintly flushed over the cheek-bones. The pulse, frequent and wanting in force throughout the case, now becomes yet more frequent and more feeble. New crops of the eruption continue to appear. At this time, over the neck, chest, and abdomen, copious outbreaks of sudamina make their appearance.

It is during this period that bed-sores are apt to form over the sacrum and other parts of the body subjected to pressure, and that other complications, especially those of the respiratory organs, become developed.

The fourth week.—The fever is now decidedly remittent. The morning and evening temperatures are, from day to day, progressively lower, while the range between the morning and evening of each day, though still greater than during the first week, tends steadily toward the normal. As the defervescence draws to a close, the type of the fever becomes distinctly intermittent, complete apyrexia being present in the morning, and a rise of a degree or more (centigrade) taking place late in the day. With this decrease in temperature the condition of the patient gradually ameliorates. The stupor disappears and the other nervous symptoms improve. The nights are more tranquil. There is natural sleep, out of which the patient awakes refreshed, while the somnolence of the previous weeks gives place to wakefulness. He begins to be conscious of his condition and to take an interest in what transpires about him. The apathy and silence of the previous weeks are replaced by questionings and complaints. The tongue and gums become clean, the mouth moist, the difficulty in swallowing ceases. The distention of the belly diminishes, the stools are less frequent, darker in color, and formed. In most cases constipation takes the place of diarrhœa. The appetite returns and thirst diminishes. Upon percussion, there is found to be progressive diminution in the area of splenic dulness. The urine is limpid and more abundant, and not infrequently the skin is bathed in perspiration; especially during sleep. The pulse gradually becomes less frequent and fuller, particularly in the early hours of the day. The emaciation, however, continues until the diurnal temperature-range becomes coincident with the normal, the patient often losing as much as the sixth or seventh part of his body-weight during the course of the disease.

Convalescence is established with the disappearance of the fever. The appetite is now good; it may be even keen. Strength gradually returns. The patient rapidly gains weight, often several pounds in the course of a week, and experiences at the same time a sense of returning mental and physical power; but, in spite of this, he is easily fatigued by exertion, and the convalescence is tedious. It is liable to be disturbed by complications and sequels. Even in uncomplicated cases it is often months before the patient fully regains his old powers of endurance for mental and bodily effort. During the early part of the convalescence the patient is prone to transient febrile or subfebrile states, which may be induced by trifling causes, such as overexertion in the sick-room, the visits of friends, mental effort, or solid food. Liebermeister states that he has frequently seen the first meal of meat followed by an increase of temperature in the evening. These recurrences of fever often arise without discoverable cause.

Occasionally the convalescence is interrupted by a true relapse.

The foregoing sketch is descriptive of a severe case, unmodified by treatment or complications, and terminating in recovery. The natural history of enteric fever calls, however, for many modifications of this description. Many cases run a very mild course, so that the patient is scarcely confined to bed, and the physician remains in doubt as to whether the disease is due to a specific cause. Others are at first severe, but speedily become mild in character, and terminate in restoration to health; while others still, in their middle course, undergo a sudden or even fatal aggravation. Striking modifications of the course of the disease are occasioned by the prominence of certain symptoms or groups of symptoms, or by the presence of complications which may become so serious as to throw the original disease into the background.

ANALYSIS OF THE PRINCIPAL SYMPTOMS.

THE PHENOMENA OF THE FEVER.

The temperature.—The consideration of the febrile movement in enteric fever is of primary importance on account of its determining the diagnosis, prognosis, and treatment.[1] As has been indicated, in well-marked, uncomplicated cases the course of the fever is typical. It may be separated into four periods, each of which is characterized by a special fever-curve. These periods are usually of about a week's duration, but they do not exceed five days in many cases, and more rarely they are extended to eight or even nine days (Fig. 8).

The typical course of the fever is frequently disturbed by complications, and prolonged by sequels and relapses.

Hence, while the average duration of the fever is from three to four weeks, it is often longer, and the febrile movement then differs in a remarkable manner from the more regular course of the other acute infectious diseases.

During the first period there is a rapid, progressive increase of the fever. This increase is not, however, steady. The rise in temperature is broken by daily morning remissions corresponding to the morning fall in the diurnal temperature-cycle in health. It takes place in a gradual zig-

---

[1] Griesinger: Das Fieber beherrscht zu grossem Theil die Situation.

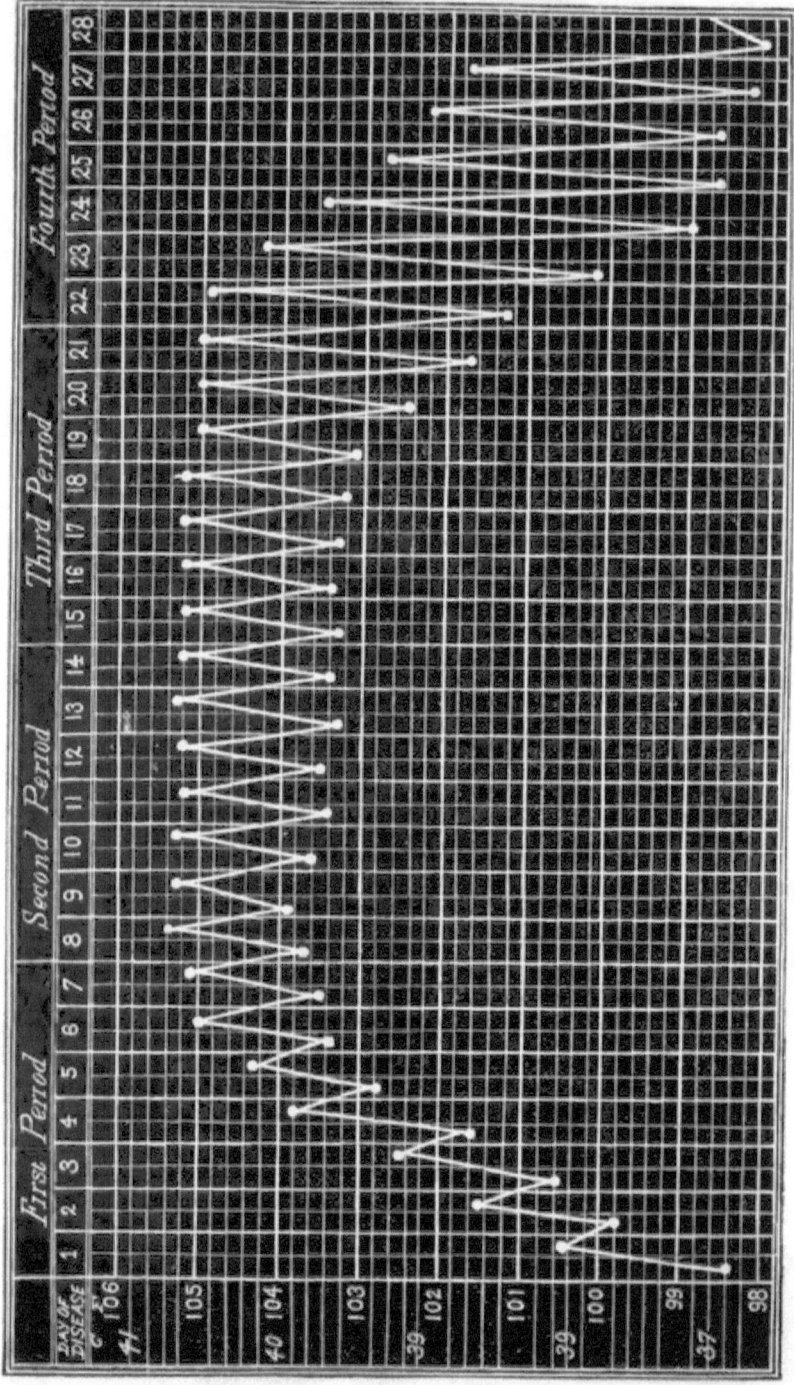

Fig. 8.—Schematic Representation of the Temperature in severe uncomplicated Enteric Fever.

zag, in such a manner that during the time occupied in attaining the maximum it rises from 1°–1.5° C. (1.8°–2.7° F.), from each morning till evening, and falls again from the evening to the following morning .5°–.75° C. (.9°–1.3° F.). On the third or fourth evening a temperature of 40° C. (104° F.) may be reached or even exceeded. The daily rise begins about noon, and is completed some time between seven and eleven o'clock in the evening. The fall begins about midnight, and the temperature is lowest, as a rule, between six and eight o'clock in the morning. In order to make satisfactory records of the temperature in any case, it is therefore necessary to take two observations daily, one as nearly as possible at 8 A.M., the other about 9 P.M. Critical studies of the temperature of enteric fever have shown that a single daily maximum is the rule, but that in many cases the diurnal curve presents two maxima—one early in the evening, and the other about midday, usually of less intensity. The highest evening temperature in the typical course of the fever is usually attained at the close of the first, or in the beginning of the second period, and is as a rule somewhere between 40°–41.5° C. (104°–106.7° F.). This maximum is commonly observed upon one day only, sometimes on two days, rarely on three.

It is very seldom that an attack of enteric fever occurring in a healthy man, or even an invalid, provided he be free from fever, does not approximate to the above type in its initial stage. It is still more rare, for any other form of disease except enteric fever, to show a similar initial stage. This course in the first week thus of itself alone possesses very great value for diagnostic purposes (Wunderlich[1]).

The second period is characterized by a fever that has been described as continuous. This term is not to be understood as meaning that the temperature remains the same throughout. If we employ it in so literal a sense, there is no such thing in the whole domain of pathology as a continuous fever. As is seen in the schematic representation of the course of the fever, the temperature is higher in the evening and lower in the morning, but the diurnal variations are not greater than in healthy persons in a state of quiet. No distinct remissions occur. It is only in cases in which the fever is very high that the temperature-curve shows a morning fall decidedly less than that which takes place at the same hour of the day in health. Toward the close of this period the evening rise, except in very severe and protracted cases, often falls a little, and at the same time the morning fall is a trifle greater.

The third period is marked by morning remissions, which from day to day become more distinct, while the evening exacerbations attain the height reached toward the close of the second period. The change from

---

[1] On the Temperature in Diseases. By Dr. C. A. Wunderlich: Sydenham Society's Transactions. London, 1871.

the continuous to the remittent type during this period of the disease is usually a gradual one, but it is not unfrequently sudden, and is then ushered in by perturbations of temperature—often an unusually high evening exacerbation, followed by a decided morning remission. It may take place as early as the fourteenth day, and when sudden it suggests the critical perturbations which occur toward the close of relapsing and typhus fevers.

During the fourth period the fever gradually changes from the remittent to the intermittent type. The morning fall is each day lower and the evening rise a little less decided, but the range between them is considerable; so that, for several days after the morning temperature has become normal, the evening shows marked fever, and upon the whole we find a defervescence of the most gradual character (Fig. 9).

Convalescence is not established until the evening temperature ceases to rise above the normal standard. In the early days of convalescence the temperature is often subnormal, especially in the morning, and it is liable to decided fluctuations in consequence of slight causes, such as overexertion, even within the limits of the bedroom, excitement, the visits of friends, or animal food. Griesinger narrates the case of a girl, whose fever had fallen to 37.3° C. (99.1° F.) in the morning, and to 38° C. (100.4° F.) in the evening, who ate sausage. Her temperature rose that evening, with general aggravation of the symptoms, to 40.5° C. (104.9° F.), and did not fall to its former level again until after three days. Jaccoud also records a case in which a lad, eighteen years of age, suffering from abortive enteric fever, was allowed, on the thirteenth day of his sickness, an egg. The temperature of the previous evening was 38.3° C. (101.5° F.), and on the morning in question 37.3° C. (99.1° F.). The same evening it reached 40° C. (104° F.). Two days later it had fallen to 37.6° C. (99.6° F.) in the morning, and the patient was allowed to eat a chop; that evening the temperature rose to 40.8° C. (105.4° F.), and, while it fell to almost the previous level on the morning of the sixteenth day, it rose that evening to 40° C. (104° F.), and only resumed the regular curve of the gradual defervescence again on the seventh day. Here the *febris carnis*, as this distinguished clinician terms this transient fever, lasted two days.

Those cases must be looked upon as severe in which the evening temperature steadily rises, and the morning fall diminishes, in the latter half of the second period; so also those in which the morning temperature does not, from day to day, fall below 39.5° C. (103.1° F.), or in which it reaches 40° C. (104° F.). Recovery rarely takes place after a morning temperature exceeding 40.5° C. (104.9° F.), or an evening temperature exceeding 41.75° C. (107.15° F.), although occasional exceptions to both these statements have been recorded. A persistently high temperature, in which the difference between the morning and evening range is slight,

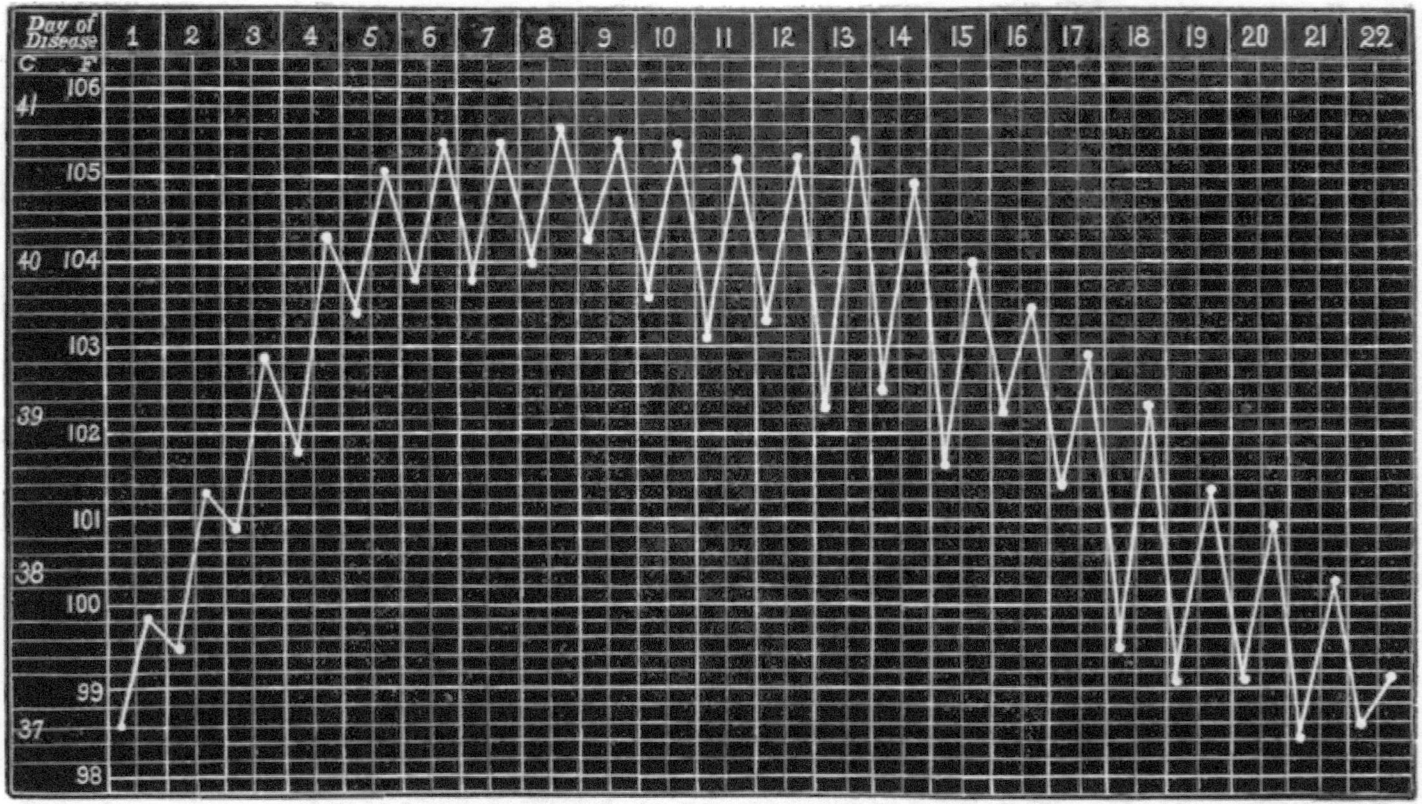

FIG. 2.—Temperature in mild Enteric Fever from First Day of the Attack. (Wunderlich.)

is much more unfavorable than a temperature characterized by high evening exacerbations and considerable morning remissions. In other words, the greater the regular daily fluctuations of the fever, the less severe is it likely to prove.

A persistent elevation of temperature, after convalescence is established, can only arise from some complication or sequel, or from the occurrence of a relapse.

A close study of the temperature, of enteric fever in its relation to the symptoms and the lesions found after death, impresses us with two facts of great practical importance. Of these the first is this: that the fever, like that of scarlet fever and of small-pox, is made up of two distinct febrile movements—first a primary fever, resulting from the infection of the tissues of the body by the specific virus, and later a secondary, irritative, or hectic fever caused by the localized ulceration of the intestines, the formation of slough, and the resorption of septic materials.

The second practical fact with which we are impressed by a near examination of the temperature-range in a considerable number of cases, is that the balance between the heat-production and the heat-elimination in enteric fever is extremely unstable—to use the words of Dr. Cayley, the temperature is *labile*. It is quickly depressed or raised by causes that would in health have little or no effect. Thus, slight exertion, changes in diet, mental emotion, will often cause considerable transient alteration, not only in the convalescence, but also during the course of the attack. The action of remedies still further illustrates this point. Large doses of quinine scarcely affect the temperature in health, while in enteric fever 1.3–2 grammes (20 or 30 grains) given at once, or in the course of as many minutes, will reduce the temperature three or four degrees, and keep it down for several hours. Marked deviations from the typical course of the temperature are always due to special causes (Fig. 10). These causes in many cases cannot be discovered by the most searching investigation. On the other hand, upon inquiry, clinical facts of importance are often discovered, and it is therefore the duty of the physician, in every case where marked deviations occur, to make diligent search for their cause.

The fact that we have a primary and a secondary fever to deal with, in the course of an ordinary attack, is of considerable importance, both with reference to our knowledge of the pathology of the disease and the treatment. In this respect, as I have already pointed out, enteric fever resembles small-pox, in which we have, first, the primary fever due to the direct action of the poison, and lasting usually about three days; this is then followed by a period of remission, to which there finally succeeds a secondary septic fever due to suppuration. In simple cases of scarlet fever the primary pyrexia lasts five or six days, and terminates commonly in lysis; but, where ulceration of the throat or implication of the glands

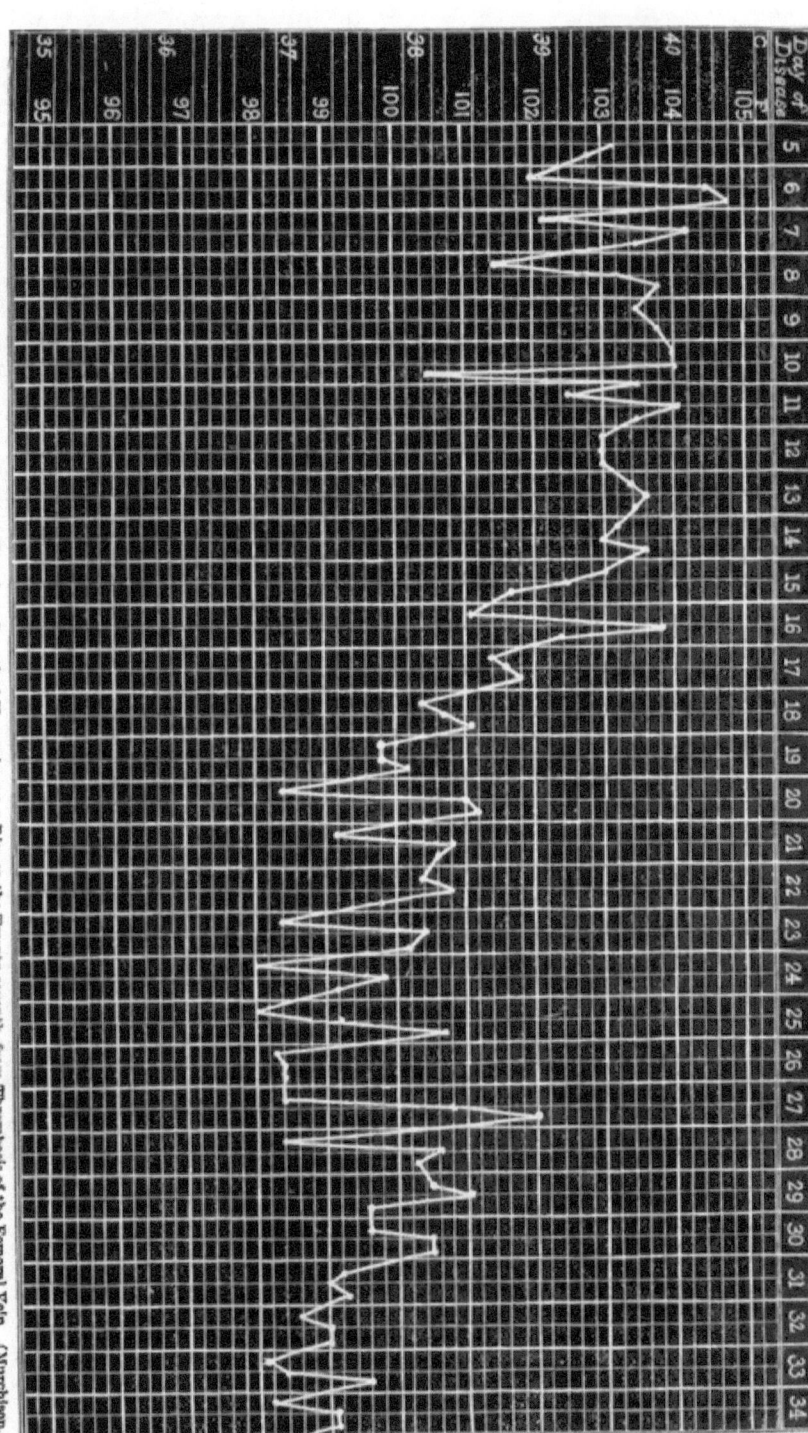

FIG. 10.—Temperature in Enteric Fever; Sudden Fall on the Tenth Day, from Intestinal Hemorrhage; a Rise on the Twenty-seventh, from Thrombosis of the Femoral Vein. (Murchison.)

occurs, we have a secondary septic fever coming on, either with or without a period of remission. On the other hand, in typhus fever, which is, as a rule, unattended by suppurative or ulcerative processes, there is no secondary fever, and the long pyrexia, due to the specific cause, comes to an end by crisis on or about the fourteenth day.

Between typhus on the one hand, and scarlet fever on the other, enteric fever stands midway. It resembles typhus in the long duration of the primary fever; it resembles the protracted cases of scarlet fever in that the secondary fever arises before the subsidence of the primary, so that there is no intervening period of remission. The change is indicated, however, by the alteration in the type of the fever, which commonly takes place during the third period, and not unfrequently as early as the fourteenth day. This change is, as a rule, gradual. It is sometimes, however, sudden, and is marked by a distinct perturbation of temperature, consisting often of an evening rise in excess of that of the previous days. This rise resembles the precritical rise of typhus and relapsing fevers, and is followed by a considerable morning fall, which is analogous to the crisis of the diseases just named. The analogy between enteric and typhus fevers, in respect of the duration of the primary pyrexia, is made more apparent by the fact that it is at this period, namely, about the middle of the third week, when the type of the fever becomes distinctly remittent, that copious perspirations take place, together with the eruption of sudamina, and that the rose-colored spots now cease to appear. This resemblance becomes still more apparent from the consideration of the abortive forms of enteric fever, which are characterized by sudden onset, rapid augmentation of the temperature to a considerable height, continued intense febrile movement until about the fourteenth day, when defervescence takes place by rapid lysis, altogether unlike the lingering decline of fever that is characteristic of ordinary cases. Such cases are analogous to modified small-pox, in which we have the primary fever well marked, but, in consequence of the slight local lesions of the skin and the absence of suppuration, there is no secondary fever. It is probable that they are to be explained upon the same ground, namely, that, while the constitutional disturbance due to the primary action of the typhoid poison is very great, the intestinal lesion, for some unknown reason—doubtless dependent upon the constitutional peculiarities of the patient—is moderate, and the glandular deposit undergoes resolution without ulceration or sloughing. Dr. Cayley suggests that the cases of enteric fever that are from time to time described as having been cut short by special remedies or plans of treatment, are really of this character, the observer having ascribed to the remedy changes which are, in fact, natural phenomena of particular cases of the disease.

SYMPTOMS REFERABLE TO THE CIRCULATORY SYSTEM.

The pulse is increased in frequency. This increase is directly and chiefly, in enteric fever as in other febrile diseases, dependent upon the rise in temperature. In general terms the frequency of the pulse corresponds to the temperature. It rises during the first week, continues high during the second and third, and gradually diminishes in frequency during the fourth. It is further true that the daily variations in the pulse run parallel with those of the temperature. The pulse is less frequent in the morning than in the evening. The absolute frequency of the pulse is, however, less in enteric than in other fevers. There are, in fact, some cases in which, although high fever is present, the frequency of the pulse does not, for some part of the time, exceed that of health; and in the mildest cases of enteric fever, and in cases of intestinal catarrh without fever, due to the cause of enteric fever, the pulse is sometimes less frequent than in health. These facts, as Liebermeister has pointed out, render it probable that infection by the poison has a depressing influence upon the pulse. During the period of the primary pyrexia the pulse does not usually rise above 120, and in many cases it does not exceed 100 during the whole course of the disease. In 100 cases Murchison ascertained that it exceeded the normal standard, at some time of the fever, in all but one; in 97 cases it exceeded 90; in 85 cases it exceeded 100; in 70 cases it exceeded 110; in 32 cases it exceeded 120; in 25 cases it exceeded 130; in 10 cases it was above 140; and in 2 above 150. In 6 cases of 100, the same observer found the pulse fall to 60; to 56 in 2; and in a single case to 52; in one case under his observation the pulse fell to 37, and never, throughout the whole course of the disease, exceeded 56, but rose with convalescence to 66. In severe cases the pulse is apt to be frequent; and where in an adult it continues steadily above 120, the prognosis becomes *pro tanto* unfavorable. Cases, however, occasionally prove fatal, in which the pulse does not exceed 100.

The frequency of the pulse in enteric fever is, like the temperature, readily modified by slight causes. Simply lifting the patient into the upright position may temporarily accelerate the pulse from 20 to 30 beats per minute. During the first week or ten days of the attack the pulse often retains to a moderate degree the force of health; but after this, or sometimes earlier, it becomes soft, compressible, and dicrotic. In the advanced stages of severe cases it may be small, undulating, irregular, or uncountable. These alterations are dependent upon changes in the heart, and where death takes place without complication at the height of the disease, it is commonly due to heart-failure. The following series of sphygmographic tracings show the progressive changes of the pulse during the course of the attack (Figs. 11 to 15).

The enfeeblement of the heart, characteristic of enteric fever in its later periods, and which is a direct result of the continued high temperature, is manifested also by changes which take place in the impulse, and the quality of the systolic sound. These in severe cases become progress-

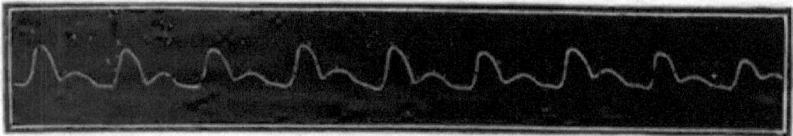

Fig. 11.—End of First Week. Strong Heart Action; Moderate Dicrotism. Frequency, 104.

Fig. 12.—Third Week. Action of Heart Strong; Marked Dicrotism. Frequency, 108.

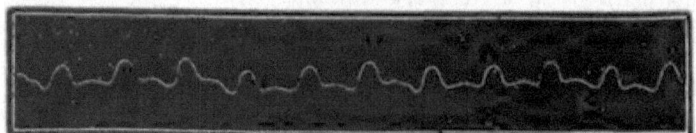

Fig. 13.—Third Week. Action of Heart Weak. Frequency, 128.

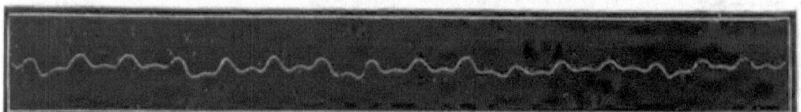

Fig 14.—Beginning Heart Failure. Frequency of Pulse, 144.

Fig. 15.—Heart-Failure after Profuse Intestinal Hemorrhage.

ively feebler, until the former is imperceptible and the latter almost or even quite inaudible.

To the enfeeblement of the circulation are also referable a certain amount of venous stasis showing itself in duskiness of the surface, and a lowering of the arterial pressure which shows itself in diminished secretion of the urine. Hypostatic congestion of the lungs and many other complications arise from the same cause.

It is to the diminished power of the circulation also, that is due the marked coldness of the hands and feet often occurring in severe cases, while the internal temperature still remains high; this condition is, therefore, an important sign of impending danger from failure of the heart. To the same cause we must refer the common danger of collapse in enteric fever. The greater the weakness of the heart, the greater does this danger become. Collapse may result from various accidents, such as intestinal hemorrhage, the shock following perforation, or even a sudden copious diarrhœa or violent vomiting. A sudden fall of temperature, either spontaneous, or in consequence of the administration of remedies, may also occasion collapse. Still more frequently collapse occurs as a result of the sudden change from the recumbent to the erect posture. Whatever its cause, collapse must be looked upon, under all circumstances, as an extremely dangerous accident of the disease; for the transient weakness of the heart may quickly pass into complete paralysis, and so cause death. Liebermeister states that the collapse which occurs in consequence of a sudden fall of temperature is generally devoid of danger, and may even be a favorable sign.

SYMPTOMS REFERABLE TO THE NERVOUS SYSTEM.

Headache is one of the earlier and more constant symptoms. The proportion of cases in which it is absent is extremely small. Louis found it to be absent in but 7 out of 133 cases, and Murchison in 5 out of 82. It is probably not less common in children than in adults. It is most severe in the first week, and by the end of the second week, or earlier, it has usually ceased. According to Sir William Jenner, it usually ceases spontaneously about the tenth day.[1] It is commonly confined to the forehead or temples, sometimes it extends over the whole head, and more rarely it is referred to the occipital region alone. Its intensity, usually moderate, commonly increases toward evening. It is described by patients as dull rather than shooting or darting, although in some instances I have known it to be sharp, piercing, or agonizing.

Slight vertigo is often associated with the headache in the early days of the disease. As a rule, it comes to an end at the same time as, or before the headache; exceptionally it remains till the close of the attack.

Pains in the back and extremities are commonly present from the onset, in this, as in most of the other acute infectious diseases. These pains are sometimes vague; at others they are fixed, and aggravated by movement, like the soreness which follows bruises. Sometimes the pa-

---

[1] On the Treatment of Typhoid Fever. Lancet, November 15, 1879.

tients describe them as aching or boring. Occasionally they assume a distinctly neuralgic character, and sometimes they are confined to the joints, and are attended by tenderness, slight swelling and redness, so that they simulate acute rheumatism. They usually subside some time during the second period of the disease.

Delirium occurs in a majority of all the cases. Many cases, however, pass through the whole course of the attack without delirium or distinct impairment of the mental faculties. Thus, Louis found that in 32 out of 134 cases there was neither somnolence nor delirium; and Murchison states that, out of 100 cases in which this matter was noted, 33 passed through the attack without impairment of the intelligence. These cases do not necessarily belong to the lightest forms of the disease; of Murchison's 33 patients, 3 died—2 from perforation of the bowel and 1 from epistaxis; and of Louis' 32 cases, in which there was no delirium, 8 were fatal—6 from perforation. These statistics are of interest as showing that no direct ratio exists between the local intestinal lesions and the intensity of the primary febrile movement. For, although there is good reason to believe that the disturbance of the nervous system, in the early days of typhoid fever, is in a measure directly due to the action of the poison, it is certain that the graver disturbances of the nervous system, among which are to be classed somnolence and delirium, and which in their complete development constitute the "typhoid state," are largely due to the prolonged high temperature. Here, however, we see a considerable proportion of cases, in which the graver nervous symptoms are absent, perishing in consequence of the extent and intensity of the local intestinal lesions.

The character of the delirium varies greatly; it is often slight and occasional, occurring chiefly in the night-time, or upon waking from sleep, in patients who are otherwise entirely rational. This form of delirium may become active and noisy, and then, as the patient becomes more prostrate, may pass into the low, muttering delirium, to which the name of typhomania has been given, or into a wandering, fatuous state, with trembling like that of alcoholism. Sometimes the delirium is active and noisy from the first, the patient talking in a loud voice, screaming or shouting, and being restrained with difficulty. This form of delirium may suddenly supervene upon either of the others; it is therefore of the utmost importance that the patient should at no time, after the appearance of delirium, be left to himself, even for brief intervals. Exceptionally, maniacal delirium occurs early in the disease, and sometimes it is the first symptom which attracts the attention of the friends of the patient. As a rule, however, delirium does not commence before the middle or end of the second week, upon the subsidence of the headache. In a small proportion of the cases it does not appear till late in the course of the disease, and lasts only a few days.

In children it occurs somewhat earlier than in adults.

In many instances delirium, if mild, occurs only at night, and in all cases it is more marked during the night-time.

During the first and second periods of the disease the patient is often disturbed by *wakefulness*. This symptom is, however, much less marked in enteric than in typhus fever.

Somnolence usually supervenes some time during the course of the second week. It is at first slight, but becomes, especially in severe cases, gradually more profound. It usually precedes delirium, and, after it is established, alternates with periods of wakefulness and spells of delirium. The patient is often dull and drowsy by day, and wakeful, restless, and delirious during the night. In cases of great severity the somnolence becomes more constant and deepens into complete unconsciousness, which lasts to the termination of the case. Somnolence is met with also in children.

Muscular weakness is present, to some extent, in all cases from the beginning of the attack, and increases with its progress. A large proportion of the patients are, nevertheless, able to assist themselves, to sit up in bed, and even to rise to stool throughout the whole course of the attack. In the mild forms of the disease, patients, although very weak, are often able to go about, and it is not rare to encounter walking cases as hospital outpatients, in the second or third week of the attack. In grave cases muscular debility is very often complete.

Muscular tremulousness is present in a considerable proportion of the severer cases. The tongue trembles as it is protruded, the lips quiver, and movements of the hands are trembling and uncertain. This phenomenon is most common in those addicted to the use of alcohol, and in old and very feeble persons. More rarely it occurs in young and temperate persons, and it is occasionally observed where there is no impairment of the mental faculties.

Retention of urine and *involuntary evacuations* occasionally occur. They are apt to take place in those cases in which the prostration is extreme.

Rigid contractions of groups of muscles in the trunk, neck, or extremities, are met with in a few cases. They are most frequent in females. In the advanced stages of severe cases, *subsultus tendinum*, picking at the bedclothes, and vague graspings in the air, are observed. In such cases protracted *hiccough* may also occur. *General convulsions* are rare. They occur with greater frequency in children than in adults. It would appear that, although occasionally associated with albuminous urine, they also occur independently of that condition, but are, in all instances, of the gravest prognostic import.

Liebermeister distinguishes four different degrees of febrile disturbance of the nervous system, which occur successively in severe cases. In the first degree there is general malaise, restlessness, headache, and disturbed sleep. These symptoms correspond to the first half of the first week. They are not associated with disturbance of the intellect, and cannot be distinguished from the symptoms of the prodromic period, which are due to the action of the poison upon the nervous system, without increase of temperature. In the second degree the patient is apathetic, dull, his memory is blunted. There is temporary disturbance of the intellect, amounting to transient delirium. These symptoms correspond to the second half of the first week and the beginning of the second. In the third degree there is marked somnolence, from which the patient, however, can be temporarily aroused. This alternates with delirium, sometimes muttering, sometimes violent and associated with restlessness and excitement. This group of systems begins, in severe cases unmodified by treatment, some time in the second week, and continues into the fourth. In the fourth degree of the disturbance of the nervous system there is loss of consciousness, out of which the patients can no longer be aroused. This degree is gradually developed from the third degree, and commonly begins some time in the third or fourth week. With the defervescence, the mental condition slowly improves; it is long, however, before the patient regains his old sharpness of memory and ability for continued mental effort.

The organs of special sense present certain symptoms which are sufficiently common to have a certain amount of diagnostic value in obscure cases. Thus, *epistaxis* is common. It may occur at any period of the disease, but is apt to occur early in its course. It is often slight, not exceeding a few drops, and is for this reason frequently overlooked. To this fact is doubtless to be ascribed the varying statements of the books as to its frequency. I am satisfied that slight epistaxis occurs in a considerable proportion of the cases of enteric fever in Philadelphia, at some period of the course of the disease, and often repeatedly. The quantity of blood lost is seldom great; yet Murchison states that it may amount to several pounds, or even be so profuse as to occasion death. If epistaxis be considerable, it is sometimes followed by a transient fall of temperature; but, with this exception, it is never followed by any relief to the general symptoms or to those of the nervous system. Da Costa states that epistaxis is not often absent in grave cases.

Subjective auditory sensations, ringing and humming, often annoy patients during the early days of the disease. They are said to be most marked, and to last longest, in the cases that are most severe.

Deafness is very common. It usually affects both ears, but may be confined to one. It is sometimes very marked. It commonly appears toward the end of the second week, and, in most instances, is in part due to catarrhal processes implicating the Eustachian passage, and in part to the blunted sense-perceptions incident to the action of the poison. One-sided deafness has been ascribed to local inflammation of the ear. Deafness of both ears was at one time looked upon as a favorable symptom, but the closer investigations of more recent observers show that this opinion is no longer tenable. Among the symptoms connected with the organs of special sense, the condition of the pupil demands attention. In a large proportion of the cases the pupils are abnormally dilated at some period of the disease. Murchison found the pupils dilated in at least three-fourths of his cases. This symptom commonly coexists with delirium, and comes on, like delirium, upon the cessation of the headache. It may, however, be present after the middle of the second week, in cases where delirium is absent. In respect to the condition of the pupils, most cases of enteric fever are in strong contrast with typhus, in which the pupils are, for the most part, contracted. But, in certain grave cases of the former fever, after great stupor or unconsciousness has occurred, the pupils are contracted, and may be as small as they are in typhus.

Conjunctival injection is very rare in enteric fever, which differs in this respect from typhus and relapsing fever. If present at all, it appears later than in typhus, and is usually much less intense. It was noted by Murchison in 8 out of 100 cases, by Louis in 38 out of 60, and only three times in 13 cases observed by Jenner. It is not a symptom of importance.

Cutaneous hyperæsthesia was observed by Murchison in about five per cent. of the cases under his care. It is most common in children and females, and appears both during the course of the disease and in the convalescence. It is usually restricted to the abdomen and lower extremities, and is commonly associated with symptoms of spinal origin, such as rhachialgia, tenderness over the spinous processes of the cervical and dorsal vertebræ, and the like. The tenderness of the abdomen due to this cause is sometimes exquisite, and is to be carefully distinguished from that of peritonitis.

Cutaneous and *muscular anæsthesia*, with numbness of the extremities, also occur in rare instances.

This group of symptoms is more common in severe epidemics than in the sporadic forms of the disease, and is to be regarded with apprehension. Murchison, however, states that hyperæsthesia alone is not a formidable symptom.

THE SKIN.

The eruption of enteric fever appears, as a rule, between the seventh and twelfth days. Exceptionally it is met with as early as the fourth, or not discovered until as late as the fourteenth day. In children it appears a little earlier than in adults. It is not invariably present. Out of 5,988 cases admitted into the London Fever Hospital during twenty-three years, it was noted in 4,606, or in 76.92 per cent. Dr. Murchison's suggestion that, in some of the remaining 1,382 cases, the fact of the spots not being observed was perhaps due to their not having been looked for with sufficient care, is probably correct. The same observer states that the spots are more frequently absent in patients under ten and over thirty years of age, than between ten and thirty, and illustrates his remark by the following statistics: of 1,413 cases between ten and thirty, the eruption was absent in 142, or 10 per cent.; of 252 patients over thirty, it was noted as absent in 40, or nearly 16 per cent.; out of 107 cases under ten, it was not noted in 37, or 34.5 per cent. From the same series of statistics we learn that no eruption was discovered in 127 of 905 males, and in 97 of 916 females.

There is no relation between the abundance of the eruption and the severity of the symptoms.

The typhoid eruption is characteristic of the disease, and, when found, clearly establishes the diagnosis. It consists of small, slightly elevated, rounded or oval, isolated spots of a rose-pink color. They are from half a line to two lines in diameter, indistinctly marginate, and alike to the eye and the touch, faintly rounded and convex, but not acuminate, although some observers state that a minute vesicle may in rare cases be discovered at their centre. They are frequently compared to flea-bites, from which, however, they differ in the absence of the central mark and in their paler color. They disappear wholly on strong pressure, and return immediately when the pressure is removed. They may be made to disappear and reappear under the eye by placing a finger upon each side of the spot and making traction: as the skin becomes tense they disappear; when it is relaxed they return. They are developed in successive crops, each spot lasting three or four days, and, as it fades, being replaced by a new one at no great distance, which runs the same course, fading in its turn, and so on, till about the middle of the third week. They are not found during convalescence, but reappear, along with the other characteristic symptoms of the disease, in true relapses. They are never present on the dead body.

Their most common situation is the abdomen and the lower part of the chest, anteriorly. They are occasionally present upon the upper part of the thigh, and are sometimes to be met with between the scapulæ. In some instances, they are present upon the back alone, and in doubtful cases should be sought for in this situation. They have been met with, in very rare instances, upon the arms and legs, and Murchison mentions a single case in which they were found upon the face. The duration of the eruption, in cases that are not unduly protracted, is eight or ten days. The spots are usually few in number, and discrete; hence, they may be readily overlooked. It often happens that not more than six or eight can be discovered, and in most cases the number present at one time does not much exceed a score. They are, however, sometimes very numerous, but are never confluent as in typhus.

Each spot runs its course without change, and usually disappears, leaving no trace upon the skin; although in some instances a faint pigmentation, which does not disappear upon pressure, persists.

Tabular arrangement of the chief points of distinction between the eruption of enteric fever and that of typhus:

Enteric Fever.	Typhus.
The spots are pink or rose-colored until they fade, leaving no trace.	The spots are pink or dirty pink at first, subsiding into reddish brown stains.
Undergo little or no change.	Become darker, and often show a minute extravasation of blood at the centre.
The spots are neither converted into petechiæ, nor do petechiæ appear interspersed with them.	Petechiæ very often appear.
Circular or slightly oval in outline.	Less regular in outline.
Usually restricted to the abdomen, thorax, and upper part of the thighs, and the interscapular space.	Commonly distributed over the greater part of the body and extremities, with the exception of the neck, face, head, and palmar and plantar surfaces.
Few in number and discrete.	Copious and confluent.
Elevated throughout.	Raised at first, but persisting as stains after the elevation disappears.
Momentarily disappearing on pressure.	Disappearing upon pressure only during the first day or two.
Rarely appear before the seventh day.	Commonly on the fourth or fifth day.
Appear in successive crops.	Appear at once, and arise in successive crops, although their efflorescence may occupy several hours, or a day or two.
Each crop lasts three or four days, and fades as others appear.	Most of the spots last until the defervescence.
No subcutaneous marbling or mottling.	Skin often indistinctly marbled or mottled between the spots.
The abundance of the eruption not at all proportionate to the general gravity of the case.	In many instances the severity of the general symptoms is in direct ratio to the copiousness of the eruption and the darkness of its color.
Not seen after death.	Often seen after death.

The eruption is occasionally preceded by a faint scarlet rash seen in patients whose skin is fair and delicate. This rash is not very common; it is not peculiar to enteric fever; but it is met with in other diseases attended by pyrexia. If well-marked, and particularly if it be associated with slight sore throat, as has sometimes happened, the disease may be mistaken for scarlet fever.

True petechiæ are rare.

Sudamina appear at a later period in the disease. They consist of minute, transparent vesicles, scattered plentifully over the body, and are often, but not invariably, attended with profuse sweating. They are very common in typhoid fever, but are without specific character, and occur with perhaps equal frequency in other febrile affections.

Slight desquamation occasionally occurs during convalescence; the hair falls out; and changes occur in the nails indicating the arrest of nutrition which has attended the course of the attack.

Emaciation is usually great; often extreme.

The physiognomy of persons ill of typhoid fever is peculiar, though less characteristic than that of typhus. Some patients, especially if the attack be mild, show but little alteration of expression during its whole course. Much more commonly the expression is dull, weary; the face pale, with circumscribed flushing over one or both cheek-bones. This comes and goes, and is sometimes called forth or intensified by the administration of food or stimulants. The dilatation of the pupils adds to the peculiarity of the expression; and, in the later stages of the attack, the wasted tissues, the sunken eyes, the circumscribed flushing and hurried breathing, suggest the appearance of patients in advanced pulmonary phthisis.

SYMPTOMS REFERABLE TO THE DIGESTIVE TRACT.

The tongue at first has a somewhat swollen and flabby appearance; it is at this period also moist and covered with fur, commonly thin and whitish, or yellowish white, sometimes thick and creamy or pasty. Its edges and tip are unusually red. It may remain moist and furred during the whole course of the attack, or, during the second week, the coating may break up into flaky patches of a whitish color, while the surface of the tongue remains bright red. This redness is in peculiar contrast to the pallor of the lips in the advanced stages of the disease, and has given rise to the name of "red-tongue fever," by which enteric fever is known in some sections of the West. It is more common, after the middle of the second week, to find the tongue dry, red, glazed, and slightly or even deeply fissured, or it is dry, with a brownish streak along the middle, or a triangular brownish patch at the tip. In cases in which the typhoid state is well-developed, the tongue is usually covered with a more or less thick, brownish crust.

It is rare to find the tongue firmly retracted into a globular mass, as is sometimes seen in typhus, and inability to protrude it is less common than in that disease.

The lips often crack and bleed, and in children, by reason of picking, they frequently become very sore and painful. In grave cases sordes collect upon the teeth. Hemorrhage from the gums is a rare occurrence in enteric fever.

In the first week there is usually slight catarrhal inflammation of the faucial mucous membrane, with enlargement of the tonsils. The accumulation of the altered secretion in the naso-pharyngeal space occasions in some patients considerable discomfort. Later the throat becomes **dry**, and there is, as a result, more or less difficulty in swallowing.

The appetite is, as a rule, greatly impaired; it is wholly lost **when the** tongue becomes dry. In mild cases, when the tongue **retains its** moisture, some appetite may be present throughout the attack. **I have seen a** case in which, with a very red but moist tongue, evenly coated **at first,** but later showing only flaky patches of whitish fur, the appetite was good during the whole course of the disease. In this patient, **a lad aged nine**teen, the highest evening temperature was 40° C. (104 ° F.). **The secre**tion of saliva is in most cases greatly impaired, and there is good ground for believing that a like impairment of the secretion of the pancreatic juice takes place. For this reason starchy articles of food are not so well digested as albuminous foods; and it is probable that, while small amounts of arrow-root or gruel may be advantageous in certain cases, because they augment the food-volume, in the majority of instances any considerable quantity of starch is injurious and likely to **add to** the intestinal irritation.

Thirst is commonly present in the early stages; in many instances it is urgent.

Nausea and *vomiting* occur in the early stages of a small proportion of the cases. These are sometimes among **the earlier symptoms, and being** associated with headache and general malaise, **lead the patient to suppose** that he is suffering from "biliousness." **More** frequently **these symptoms** appear during the second week. The vomiting in most cases **is only** occasional; it is sometimes persistent and distressing. In **the latter case** it is apt to be associated with epigastric soreness and pain. Some observers look upon vomiting at the beginning of the attack as **a favorable** symptom; by **others it is** regarded **in the** opposite light. There are no statistics by which to settle this question, **but so** far as my own observation goes, early vomiting, which **is not** common in enteric fever as it occurs in Philadelphia, has been followed by the severest forms of the disease. Vomiting after the end of the second week is of grave import; it is often the first sign of peritonitis. The matters vomited usually consist of food, sometimes in a partially digested state, **or they** consist simply of gastric mucus stained green with bile.

Abdominal tenderness and pain are present in the majority of the cases. They are not, however, **necessary symptoms**, and are sometimes absent throughout the attack. **Palpation is to be made with circumspec**tion in the later periods of the attack, lest mechanical violence give rise to peritonitis, or even perforation of the ilium at a point of deep ulceration. Bartholow[1] mentions a case in which fatal peritonitis, due to rup-

---
[1] Practice of Medicine, 1880.

ture of the spleen, was caused during convalescence by a not violent blow. The tenderness is elicited by light pressure in the right iliac fossa, but it is not necessarily limited to that region. In many cases it is also experienced in the umbilical region and even in the left side of the abdomen. Spontaneous pain is also often complained of. Pain and tenderness in the abdomen are largely, if not wholly, due to local morbid processes, and are to be looked upon as to some extent the measure of the extent and intensity of the intestinal lesions. This statement must, however, be qualified by adding that a most serious, or even fatal lesion of the gut may sometimes occur without previous marked pain or tenderness.

Meteorism is present in most cases; according to Sir William Jenner[1] it is observed to some extent in all cases. Murchison states, on the other hand, that out of 100 cases he found meteorism in 79, and that the abdomen remained flat throughout in 21; and Louis noted meteorism in only 89 out of 134 cases. The amount of distention varies from slight fulness to a tympany so great as to interfere with the contraction of the diaphragm and impede respiration. In this way meteorism increases the danger of congestion of the lungs. It does not usually appear until after the first week, and is most developed in severe cases. Thus, Murchison noted it in 20 out of 21 fatal cases; Jenner in 18 out of 19 fatal cases, and Louis in one-half of his fatal cases. Furthermore, the first-named of these observers found that, out of 17 cases in which extreme tympany arose, 7 died; while of 62 in which it was moderate or slight, 14 died, and of 21 where it was absent, none died, and Louis noted great meteorism in only 7 cases among 88 in which recovery took place. These statistics are of great interest as indicating the importance of the intestinal lesions in regard to the prognosis. Tympany, like abdominal tenderness and pain, is in part a measure of the extent of the mischief wrought in the intestines. It is due to excessive development of gas and to deficient expulsive power in the bowels. The first of these factors has its pathological genesis in the impaired quality of the digestive fluids, and a tendency to the rapid decomposition of imperfectly digested food; the second in general lowering of nerve-tone, or in local injury to the bowel. Jenner calls attention to the fact that a single deep ulcer will paralyze the action of the bowel and lead to such an accumulation of flatus as produces enormous distention of the abdomen. Weakness of the abdominal muscles contributes also to the accumulation of flatus. The conditions which underlie abdominal distention in this disease attain their maximum during the latter half of the third and in the fourth period of the fever, and it is at this time that meteorism may, in the more severe cases, become both a troublesome and an alarming symptom.

The spleen is, as a rule, enlarged.—Augmentation in the bulk of this

---

[1] On the Treatment of Typhoid Fever. Lancet, November 15, 1879.

organ is a prominent and characteristic symptom; it occurs early, and may often be demonstrated before the close of the first week. It increases during the second week, and diminishes again during the fourth. The amount of enlargement is usually considerable; at the height of the disease the organ sometimes attains three times its natural bulk, or more, and can be felt through the abdominal wall. The enlargement is greatest in persons under thirty years of age. In a few cases it is absent altogether, and this is more common in old than in young persons.

Diarrhœa is one of the most common symptoms in enteric fever. Although cases occur in which this symptom is absent throughout the whole course of the attack, they are to be looked upon as exceptional. Out of 100 cases in which diarrhœa was made the subject of special observation by Murchison, it occurred in 93. The time at which the diarrhœa first appears is very variable. It is sometimes present in the prodromic period, or if not present at this time, it is often induced by purgatives taken by the patient under the impression that he is suffering from a bilious attack. Diarrhœa sometimes appears early in the course of the disease, but ceases after a few days, and does not return. More commonly it is a prominent symptom during the whole course of the illness. It sometimes happens that diarrhœa is absent until the third or fourth week of the disease, and is then profuse. The movements are not often attended with pain, and never by tenesmus. Their frequency varies. In the greater number of cases they do not exceed three or four a day: not infrequently, on the other hand, they may amount to twelve or fifteen in the course of twenty-four hours. There is no constant relation between the urgency of the diarrhœa and the extent of the intestinal lesions. Diarrhœa may be altogether absent in cases in which, after death, extensive and deep ulceration is found. Profuse hemorrhage or perforation sometimes occurs in cases unattended by diarrhœa or any other previous abdominal symptom. Prolonged constipation is attended with the danger of the formation of firm scybala which are liable to do harm by increasing the extent and depth of the ulceration, or by directly leading to perforation.

When diarrhœa occurs during the first week, the stools are thin and brownish; but toward the end of the second week they assume the appearance peculiar to the disease. They are then liquid and of an ochrous color. On standing, the stool separates into two layers; a supernatant fluid and a flaky sediment. The former has a yellowish or pale brown color; its specific gravity is 1015, and it contains about 40 parts in 1,000 of solid matter, which consists chiefly of albumen and soluble salts, particularly chloride of sodium. The deposit is made up of particles of undigested food, disintegrating intestinal epithelium and blood-corpuscles, shreds of sloughs, which are separated from the intestinal ulcers, and multitudes of crystals of triple phosphate (Murchison). The reaction of the typhoid stools is alkaline. Sometimes, instead of being watery, they are frothy or pultaceous, or they may be mixed with blood.

Gurgling in the right iliac fossa is often elicited upon palpation. Associated with tenderness, this symptom undoubtedly has diagnostic value; but it occurs so constantly in other affections, attended by diarrhœa, that it cannot be looked upon as a characteristic phenomenon of enteric fever.

Hemorrhage from the bowels is of frequent occurrence and constitutes a symptom of the gravest importance. Sometimes it amounts merely to a few streaks of blood, or a little bloody mucus; in others it is more abundant; or it may be even copious, amounting to one or more quarts. The color of the blood is often bright red, particularly if it be promptly discharged; it may be of a syrupy consistence, or loosely clotted. If it be retained for some time in the intestine (concealed hemorrhage), it becomes tarry in consistence, and of an olive-green or brown color. Murchison states that hemorrhage, amounting to over six ounces, occurred in 58 of 1,564 cases under his observation, or in 3.77 per cent. It occurred in 8 of 134 cases noted by Louis, or in 5.9 per cent. In this estimate the milder cases appear not to have been included. Liebermeister found that hemorrhage from the bowels occurred among the cases treated in the hospital at Basle in 127 of 1,743 patients, or in 7.3 per cent. The proportion among men was 5 per cent. of all cases; that among women, 10. In this series of cases, the lighter hemorrhages are included, those only being thrown out in which a mere trace of blood was discovered. Griesinger observed 32 cases of hemorrhage in 600 patients, or 5.3 per cent.

It is somewhat less frequent in children than in adults.

The date of the appearance of the hemorrhage shows, according to the statistics of different observers, considerable variation. Liebermeister found that in 81 cases of intestinal hemorrhage in which the chronology was carefully kept, 7 took place during the first week; 33 during the second; 19 during the third; 14 during the fourth; and 8 at a later period. Griesinger, in 32 cases of hemorrhage, found no instance in which this accident occurred during the first week; during the second, and chiefly toward the end of it, there were 10 cases; during the third week, 8; during the fourth, 8; in the fifth, 2; in the sixth, 3. Of 60 cases observed by Murchison, the bleeding commenced during the second week, mostly toward its close, in 8; during the third week in 28; during the fourth in 17; during the fifth in 1; during the sixth in 3; during the seventh in 1, and during the eighth in 1; while in one case the date of its occurrence was not noted. In three of Murchison's cases, where it took place on the sixteenth, eighteenth, and nineteenth days, it recurred respectively on the forty-ninth, thirty-third, and forty-fourth days. The last named author states that he has known slight intestinal hemorrhage to take place as early as the fifth or sixth day, and even copious hemorrhage at a period so early in the disease as to preclude, in all probability, the existence of intestinal ulceration.

The source of the hemorrhage varies with the period of its occurrence. During the early period of the disease, prior to the latter part of the second **week**, the hemorrhages arise from the rupture of minute **vessels** within the relaxed and highly vascular tissues of the infiltrated **patches**; in the third and fourth weeks, they are due to the separation of sloughs; and at this period, or later, to the **destructive action** of progressive ulceration. Hemorrhage from the bowels is occasionally associated with nosebleeding, the spitting or vomiting of blood, or with hæmaturia, and with petechiæ, as evidences of profound alteration of the state of the blood, in consequence of the action of the typhoid poison; or hemorrhages from various mucous tracts may occur during the course of the disease, as evidences of the existence of the hemorrhagic diathesis.

Extensive hemorrhage may take place into the bowel, and death ensue in consequence, without the escape of the blood externally. If a large amount of blood escape into the intestine, whether it be voided externally or not, symptoms of collapse speedily ensue. The patient suddenly becomes extremely prostrate; his face grows pale; his pulse weak and frequent; his extremities cold; while **the temperature falls, with great** abruptness, several **degrees**. If the blood be not discharged, an area of the abdomen, previously tympanitic, becomes dull. The temperature sometimes falls to a point below the normal. This fall is followed by the same general amelioration in **the condition of the patient that results from** a decided remission of fever under other circumstances; in particular, by **the diminution** or disappearance of **serious** nervous symptoms due to the prolonged high **temperature**. The change is, however, usually transitory; **within twenty-four** hours the temperature rapidly regains its **former height,** or rises beyond it, and the disease resumes its course.

It is to this transient amelioration of the general symptoms of the disease **that is doubtless** due the opinion entertained by some observers that the occurrence of hemorrhage is productive of benefit to the patient. Among those who hold this view are Graves and Trousseau. The great majority of observers, however, concur in the opinion that it is a dangerous symptom. Of Murchison's 60 cases, 32, or 53.3 per cent., terminated fatally; in 11 of these the immediate cause of death was peritonitis; of the remaining 21 cases, 14 died within three days of the bleeding; and of these **14 cases,** 8 within a few hours. **Of Liebermeister's** 127 cases in which **hemorrhage** occurred, 49, or 38.6 **per cent., died. Of Griesinger's** 32 cases, 10, or 31.2 per cent., **died, 7 of them within four days.**

Most of **the cases** in which **copious** intestinal hemorrhage occurs, have been previously severe, and attended by considerable diarrhœa; in a small proportion of **them,** however, the previous symptoms have been mild; and in a few of them, diarrhœa has been absent. It would appear then that hemorrhage occurs most frequently in the severer cases of the disease, where the mortality would be high without the occurrence of this accident. Furthermore, the fall of temperature attendant upon hemorrhage and the consequent amelioration of the general symptoms of the disease, if they occur in the later periods of the fever, may usher in a permanent improvement. It is probable, therefore, that intestinal hemorrhage, although unquestionably influencing the prognosis unfavorably, is less dangerous than some observers have been led to suppose. A slight hemorrhage probably affects the result but slightly, if at all; even in a grave case, it is of little importance, except in so far as it excites the fear of a profuse recurrence. Copious hemorrhages at any period of the disease are to be regarded with apprehension, because of the increased debility arising from the actual loss of blood. If they occur early, they render the patient less able to bear the prolonged fever; if late, death may ensue from collapse.

It has been thought that the danger of intestinal hemorrhage is increased by the treatment by means of cold baths. And it would appear that the application of cold to the entire surface of the body, by inducing contraction of the superficial blood-vessels, must drive the blood to the internal organs, and thus favor hemorrhage. Liebermeister, however, found that of 861 cases treated before the introduction of the cold bathing, 72, or 8.4 per cent., had intestinal hemorrhage; but that of 882 cases treated after the introduction of the cold baths, hemorrhage occurred in 55, or 6.2 per cent. He concludes, therefore, that "the frequency of intestinal hemorrhage has materially diminished under the cold-water treatment." This point cannot at present be looked upon as settled.

SYMPTOMS REFERABLE TO THE ORGANS OF RESPIRATION.

The frequency of respiration varies with the intensity of the febrile movement, in the absence of pulmonary complications. It rises with the pulse; but in cases characterized by an unusually slow pulse, there is no corresponding slowness of the breathing. At times the respiration is shallow, noisy, or irregular, but these symptoms arise for the most part in the gravest cases.

A certain amount of *bronchial catarrh* is so frequent in enteric fever that it merits consideration as a symptom of the disease rather than as a complication. In a majority of the cases this does not manifest itself by cough; and the cough, when present, is often by no means proportionate to the intensity of the bronchial congestion. Upon auscultation we detect râles which are often loud and ringing.

Hypostasis gives rise to notable enfeeblement of the respiratory murmur at the most dependent portions of the lungs, and to impairment of resonance upon percussion.

THE URINE.

The urine is diminished in quantity during the first and second weeks. Notwithstanding the increased amount of fluid consumed by the patient, the urine excreted may not exceed one-half or even one-fourth the normal quantity. In most cases it is diminished from the commencement of the attack until convalescence, when it becomes, as a general rule, copious and of low specific gravity. Sometimes, however, a considerable increase in quantity takes place about the end of the second week. Its color is at first darker than in health, in consequence of the rapid destruction of the pigmented tissues of the body, and particularly of the red blood corpuscles. In the advanced stages of the disease, and during convalescence, it is pale. As a general rule the urine is acid throughout the disease; toward the end of the attack, however, the acid reaction becomes less intense, and in some instances the urine is at this time even feebly alkaline. The specific gravity varies in proportion to the amount. The scanty urine of the early periods ranges from 1020 to 1030; after the close of the second week, in some instances, and almost invariably during convalescence, the specific gravity falls to a point considerably below the normal. The abundant limpid urine of early convalescence is often as low as 1008 or 1005.

The daily excretion of urea is invariably increased at some period of the attack, and in almost all instances throughout the whole course of the disease. This increase is greatest during the first week; after that, the quantity usually falls off somewhat until convalescence, when it may remain for several days lower than normal. According to Parkes, the average increase is about one-fifth, but occasionally this amount is far exceeded. The quantity of urea excreted is usually greatest, when the temperature is highest, and, as the temperature subsides, the urea diminishes to the standard of health or below it. The amount of urea does not appear to be dependent upon the frequency or intensity of the diarrhœa. It would appear from the observations of Dr. Parkes, that the urea may be reduced during the occurrence of inflammatory complications. In one case, this observer ascertained that the amount of urea, during an intercurrent attack of pleurisy, was one-third less than the average before the occurrence of this complication. The uric acid is always increased. During the latter period of the disease the amount falls to the normal, and during convalescence, it is less than in health. Copious deposits of the urates may occur at any time in the course of the disease. They are not necessarily critical, and are therefore without prognostic value. The chlorides are greatly diminished. Sometimes they do not exceed a mere trace. This diminution in the chlorides cannot be wholly explained either by the diminished amount ingested or by the increased amount voided with the stools. It would appear that they are temporarily stored up in the tissues. With the advent of convalescence, the chlorides are greatly increased.

In many cases the urine contains albumen. Of 549 cases, collected by Murchison from various sources, albumen was discovered in 157, or in 28.6 per cent. It rarely appears earlier than the middle of the third week; the amount is small, and in most cases the albuminuria is transient, disappearing shortly after the abatement of the fever. The appearance of the albumen in the urine is due to the parenchymatous degeneration of the kidneys, and is a direct consequence of prolonged high temperature. It coincides in the chronology of the disease with the appearance of cerebral symptoms of gravity and the other phenomena of the typhoid state (third degree of disturbance of the nervous system—Liebermeister). Acute parenchymatous nephritis occasionally occurs; it will be spoken of under the head of complications.

Hæmaturia is occasionally encountered; it is commonly associated with other hemorrhages. Blood-corpuscles may be found, in connection with albumen and renal epithelium, in the urine of severe cases. Tube-casts are commonly discovered along with the albumen; they are also occasionally met with where albumen is absent.

Leucine and tyrosine, creatinine, and the urinary indigo are occasionally met with. In the later stages of the disease, when the urine is feebly acid in reaction, it often contains large amounts of the phosphates.

COMPLICATIONS AND SEQUELS.

Enteric fever is conspicuous among the acute diseases for the number and variety of its complications and sequels. The prolonged high temperature, the serious impairment of nutrition which affects the tissues of the body in the most general manner, and the enfeeblement of the circulation characteristic of the developed disease, contribute directly and indirectly to the excessive development of certain of the lesions. Thus, on the one hand, phenomena of the disease itself attain the importance of secondary affections, while on the other, the length of time during which the powers of resistance to evil influences from without are lowered, renders the patient especially liable to the development of intercurrent affections, not essentially dependent upon the primary disease, but of an accidental kind.

Hence, the complications and sequels of this disease fall of themselves into two general classes. Of these, the first comprises those which are closely connected with the pathological processes of the particular form of fever, and which are to be regarded as due to an unusual development of the same, either in extent or in intensity. Here are to be considered those complications which we must look upon as the *accidents* of the intestinal lesions, such as hemorrhage, which is so common that it has already been treated of as a symptom, perforation, and peritonitis, with or without perforation. The general lesions, which are of the nature of a widespread impairment of nutrition, leading to parenchymatous degeneration of the muscular system, the glands, and the tissues of the nervous system, may be followed by ruptures of muscles, abscesses, parotitis, nephritis, and various affections of the nervous system; and finally, the enfeeblement of the circulation leads to various venous congestions, hypostasis, œdema, thrombosis, embolism, infarction, and secondary pathological processes dependent upon these occurrences. The second group comprises occurrences not necessarily dependent upon the malady, but to which the condition of the patient renders him peculiarly liable. These are mainly acute inflammatory attacks, such as pneumonia or pleurisy, and the development of intercurrent diseases of an infectious character, as erysipelas or diphtheria.

The impairment of nutrition characteristic of the developed disease manifests itself in a peculiar tendency on the part of the tissues to break down under the influence of slight causes. Hence, trifling injuries may give rise to serious destruction of tissue. The pressure of the teeth, or the sharp point of a tooth, may cause an ulcer upon the tongue which spreads, becomes gangrenous, and refuses to heal until the defervescence. In a like manner, bed-sores are not only intractable while the fever lasts, but they tend to become deep and extensive, despite the most careful efforts to guard the parts from pressure. Venereal ulcers tend to become gangrenous, and sometimes result in the extensive destruction of parts, and Liebermeister has seen old fistulous tracts, dependent upon former disease of the bone, reopen, and necrosis of the bone supervene.

It is a matter of common observation that wounds do not heal well, if the patient develop enteric fever.

Diseases of the respiratory tract constitute an important group of the complications of enteric fever.

Laryngitis occasionally occurs. It is a serious complication, and is not infrequently the cause of death. The laryngeal inflammations, which occur as complications of fevers, may be grouped under these headings: 1, Œdematous Laryngitis; 2, Ulcerative Laryngitis; 3, Laryngeal Perichondritis.[1]

"Practically it is often exceedingly difficult to separate these various forms even at the post-mortem, so far do they overlap each other. Œdema may exist alone, or it may result from either of the others; ulceration may march steadily deeper until the cartilages are involved; or the perichondritis may produce an abscess which will burst, and so form an ulcer. How much more difficult, nay often impossible, then is it to diagnosticate precisely the form of the disease, when, happily, the patient recovers. Dyspnœa, suffocation—this is the one great overshadowing clinical fact

---

[1] See the Fifth Toner Lecture (On the Sanguine Complications and Sequels of the Continued Fevers), by W. W. Keen, M.D., Washington, 1877.

which groups them all together, whatever the form of the disease, or of the preceding fever."—(Keen.)

This is a rare complication in this country; it is rare also in England. Of 13,000 cases treated in the London Fever Hospital, Murchison records only 21 of laryngitis; 8 of these proved fatal. Laryngitis occurred in but 3 or 4 cases of enteric fever. On the other hand, it appears to be very common in Germany. Griesinger met with laryngeal ulceration in 31 out of 118, and Hoffmann in 28 out of 250 cases examined after death. These ulcers are sometimes found in the dead body, in cases where there had been no symptoms referable to the larynx during life. They were at one time regarded as specific in their character, and as due to "typhoid" infiltration of the laryngeal glands. According to Liebermeister they are due to secondary changes, resulting from circumscribed "diphtheritic" infiltration of the mucous membrane. Others do not believe that they are of specific origin, but that they are to be referred to the depraved nutrition of fever-patients, in consequence of which a low grade of inflammation readily follows slight irritation, and tends to rise rapidly into ulceration, and even into local gangrene. Dr. Keen suggests that local stasis of the blood, or clots in the vessels, are not unimportant factors in the production of the laryngeal lesions. The areas of ulceration are usually small; they may, however, become extensive, and the ulcers may extend in depth, implicating the cartilages. They may be few in number, or numerous and confluent. Their most common seat is the posterior wall of the larynx, which is most abundantly supplied with blood-vessels. Hence, they frequently involve the insertion of the vocal chords. They are not uncommon in the epiglottis, particularly at its margins. Hoarseness, even aphonia, difficulty in swallowing, and a troublesome, tickling cough are among the symptoms to which they give rise.

The laryngeal complications of the continued fevers are far more rare in children than in adults, and somewhat less common in women than in men. Cases in which cough has been prominent during the course of the attack, or where the patient in his delirium has used his voice excessively, are especially disposed to these troubles. Laryngitis, during convalescence, may occur from various causes and thus constitute a sequel of the primary fever.

Acute œdema glottidis may arise in consequence of laryngeal ulcers of small extent. It is more commonly due to erysipelas or parotitis, and it is thought by some observers to occasionally occur as a simple œdema in consequence of asthenia.

Perichondritis may occur without previous ulceration, as is shown by the fact that in some instances submucous abscesses are found in connection with local necrosis of cartilage, where no opening in the overlying mucous membrane exists; in other cases the necrosis of cartilage is secondary to the ulcerative processes.

Necrosis of the nasal cartilages has been observed, in rare instances, as a result of fever.

Bronchial catarrh is of sufficiently common occurrence to acquire a certain amount of diagnostic significance. It is sometimes unattended by cough, or by subjective symptoms, and is discovered only upon auscultation. As a general rule the cough is slight, and expectoration scanty, or altogether absent. Exceptionally, there is spasmodic cough, with paroxysms of dyspnœa.

When the catarrhal processes affect the smaller bronchial tubes, they often give rise to lobular collapse and lobular pneumonia. In a considerable proportion of the cases which terminate fatally prior to the end of the second week, death is due to pulmonary complications. Bronchitis, usually associated with hypostasis, is often a troublesome condition in the fourth week, when it may contribute to the fatal termination, or to the indefinite retardation of the convalescence.

Lobular pneumonia was noted by Hoffmann[1] as present in 38 out of 250 cases examined after death. Of these 38 patients, 2 had died in the second week, 8 in the third, 7 in the fourth, 8 in the fifth, and 14 at a later period.

Hypostatic congestion of the lungs and *pulmonary œdema* arise in consequence of the failure of the circulation. Hypostasis develops itself as soon as the force of the heart is notably reduced. This may occur in the course of the second, but is common in the third week. The patient lies quietly upon his back, and the influence of gravity upon the blood in the vessels of the lungs is added to that of the enfeeblement of the circulation. The blood stagnates in the most dependent portions of the lung, in which regions the air is gradually forced out of the alveoli, and there results an airless condition of the pulmonary tissue, not due to inflammation, which is termed splenization. If a sluggish inflammatory process arise in this tissue, hepatization results in consequence of hypostatic pneumonia. These conditions are to be recognized by the enfeeblement of the respiratory sounds to which they give rise, by dulness at the bases posteriorly, a little more marked upon one side than upon the other, and by the well-marked weakness of the heart with which they are associated. They are chiefly to be diagnosticated from lobar pneumonia, by the slight degree of difference in the two sides, by the absence of rigors or increase of fever, and by the gradual manner in which the physical signs of consolidation are developed. Pulmonary hypostasis aggravates the condition of the patient by cutting off extensive areas of alveolar surface, and thus curtailing the function of respiration; it is also to be viewed with apprehension, as an indication of cardiac weakness.

Pulmonary œdema is very common in connection with other affec-

---

[1] See Ziemssen's Cyclopædia, vol. i., Article Typhoid Fever.

tions of the lungs. When death takes place by gradual failure of the heart, it is associated with the development of extensive œdema of the lungs, and the patient is drowned in the serum of his own blood.

Hemorrhagic infarcts occur. They are difficult of diagnosis during life. If due to heart-clot, of which fragments are swept into branches of the pulmonary artery, they may be absorbed. They may, however, undergo purulent changes, resulting in the formation of abscesses, or they may result in circumscribed gangrene of the lung. They are apt to occasion pneumonic infiltration and pleurisy, and in all cases increase the dangers of the patient's condition.

Lobar pneumonia is a common complication. Occurring in the course of the disease, it has the character of secondary pneumonia; the cough is not increased, the chest-pain is absent or slight, and rusty sputa do not occur. It is to be recognized by the signs with which its onset is attended, by the sudden increase in fever, and by the evidences of infiltration discovered upon physical examination of the chest. In rare cases it occurs early, but it is much more common at the height of the disease, that is to say, in the last part of the second or in the third week, and it may not arise till after convalescence is fairly established. In the last case the ordinary characters of primary pneumonia are apt to be present. When it occurs early, or before the patient has come under observation, this complication may be mistaken for the primary disease. The term "typhoid pneumonia" has been applied alike to idiopathic pneumonia with "typhoid" symptoms, and to cases of typhoid fever in which the pulmonary complications have been prominent. It is an unfortunate term, leading to no little confusion, and richly deserves to be discarded from medical writings.

Gangrene of the lung occasionally occurs. It may be diffuse or circumscribed. The former may result from the breaking-down of a lobar infiltration; it manifests itself by the ordinary symptoms, and is, therefore, recognized during life. Circumscribed gangrene frequently follows hæmorrhagic infarction; it may result from the necrobiotic processes in the tissues of patches of lobular pneumonia. It usually remains circumscribed, and is not recognized during life.

Chronic pneumonia, in consequence of the delayed resolution of inflammatory products, not rarely supervenes upon the various pulmonary complications of enteric fever. It may, after a duration of variable length, terminate favorably; much more frequently, the infiltrated portions of lung ultimately break down with the formation of cavities, and the patient succumbs to rapid phthisis. Although no exact statistics upon which to base the opinion exist, it is generally thought that consumption is a much more common sequel of enteric than of the other fevers. This opinion is probably correct.

Acute miliary tuberculosis is an occasional sequel. It may be developed immediately after the attack, or not until the lapse of some weeks.

Pleurisy with more or less abundant effusion is more frequent after enteric than after typhus fever. It occasionally results in empyema.

As has been already pointed out, to the failure of the power of the circulation is largely due many of the complications of the disease. This failure is a direct result of the degeneration of the muscular tissue of the heart, which is to a greater or less extent present in all severe cases. The general nutrition of the tissues is impaired because they do not receive their usual supply of blood, not less than because the supply is of an inferior quality. The amount of blood being decreased and the force of the circulation diminished, it is apparent that a slight amount of inflammatory infiltration may cut off the local supply altogether, and destructive ulceration and gangrene readily ensue. But the mere slowing of the blood-current within its ordinary channels gives rise to many complications of importance. Those implicating the respiratory tract have been alluded to.

Dilation of the cardiac ventricles, both upon the right and the left side, occasionally occurs in consequence of the degeneration of their muscular walls, other recognized causes being absent. This sometimes reaches an extent that enables one to diagnosticate it during life. It is more common upon the right than upon the left side. In a majority of cases terminating in recovery, the increased area of dulness recedes as the quality of the systolic sound and the force of the impulse improve. Excessive weakness of the heart, combined with dilatation, often lead to the formation of heart-clot. Ante-mortem clots occur on both sides of the heart. If fragments of such clots are swept from the right ventricle into the branches of the pulmonary artery, they result in embolism and the formation of hemorrhagic infarcts; while the detachment of fragments from a clot in the left ventricle produces embolism somewhere in the course of the general circulation, oftenest in the spleen or kidneys.

To the same cause, namely, weakening of the force of the circulation, we must refer the occurrence of *venous thrombosis*, which is most frequently met with in the femoral vein. It is a complication of moderately frequent occurrence. Murchison encountered it in fully one per cent. of his cases. Of 17 instances in which it occurred under his observation, it was restricted to the left leg in 14; to the right in 1; and both limbs were implicated in 2. Of these 17 cases, 3 proved fatal, and it is of interest to note that they were cases in which the evidences of the gravest impairment of nutrition existed; thus one died of intestinal hemorrhage and pleural effusion; one in consequence of extensive bed-sores and sloughing of the nates, and the third proved fatal six months after the commencement of the fever, death being preceded by the occurrence of jaundice, albuminuria, and the signs of a very feeble heart.

In the hospital at Basle, 31 cases of thrombosis of the veins of the lower extremity occurred among 1,743 enteric fever patients, the majority being among men. This complication made its appearance commonly during convalesence, but in a few instances in the third or fourth week of the fever. In 24 cases, 16 occurred in men, 8 in women; 18 implicated the femoral vein, 3 the saphena, and 1 the popliteal. Thrombosis of the femoral vein on both sides occurred twice, four times on the right side alone, and twelve times on the left alone. The greater frequency of the occurrence of this accident upon the left side has been explained by the fact that the left common iliac vein, being crossed by the right common iliac artery, does not admit of so ready a flow of blood as the vessel of the other side (Liebermeister). Of the 31 cases referred to, only two proved fatal. The foregoing statistics show that the mortality to be attributed to this complication is low, if due regard be paid to the fact that it occurs late in the disease, and is of itself an evidence of grave impairment of the heart-power, and of the general nutrition of the body. Spontaneous gangrene, in consequence of arterial thrombosis, is much less common in enteric than in typhus fever.

Endo- and *pericarditis* are very rare as complications or sequels of enteric fever.

The complications and sequels arising in consequence of affections of the intestinal tract are numerous; some of them are among the most serious connected with the disease.

Ulceration of the tongue and of the *buccal mucous membrane* are noted as of common occurrence by systematic writers on enteric fever. It would appear that these complications are more common in Europe than in this country. They frequently lead to gangrene, which is usually superficial, but may be deep, extensive, and destructive.

Catarrh of the *mucous membrane* of the *pharynx* and *naso-pharynx* is of sufficiently common occurrence to merit consideration as a symptom rather than as a complication, and has already been spoken of as such. *Diphtheritic processes* occasionally involve the tonsils, half arches, the lower part of the pharynx, and the upper air-passages. All these processes may, by extension, implicate the Eustachian tube and middle ear, and give rise to serious lesions of the organ of hearing, and more or less permanent deafness.

Difficulty in swallowing may arise from mere dryness of the throat, from any of the inflammatory infections of the pharynx, which have just been spoken of, or in consequence of more or less perfectly developed palsy of the muscles of deglutition. In children it appears to be occasionally due to pharyngeal hyperæsthesia, attempts to swallow occasioning spasmodic cough, with the rejection of fluids through the nostril.

Swelling of the *parotid glands* occasionally occurs. It is much less common in this country than in Europe. The enlargement occasionally undergoes resolution without suppuration. More commonly it terminates in the formation of pus, at various points in the gland itself, and in the connective tisssue overlying it, and is then very often fatal. It usually implicates one, much less commonly both sides. Suppuration of the other salivary glands does not occur.

Jaundice occurs in a small proportion of the cases, and is a symptom of great gravity. Murchison met with it in three cases, all of which proved fatal, although, in one of them, the jaundice had disappeared before death. Louis, Frerichs and Jenner have also recorded cases of jaundice, all of which proved fatal. Out of 600 cases, Griesinger observed jaundice in 10, in several of whom recovery occurred. Jaundice is of much less common occurrence in typhoid fever than in any other acute febrile affection. It is sometimes due to an extension of the catarrhal processes from the intestine to the biliary passages. In a certain proportion of the cases, however, it is to be attributed to the parenchymatous degeneration of the liver, incident to the prolonged fever. This degeneration may reach an intensity so great as to amount to a distinct complication, presenting the group of symptoms characteristic of *icterus gravis* or acute yellow atrophy of the liver (Liebermeister). In two of Murchison's cases the liver is noted as having been small and its secreting cells loaded with oil. Other observers speak of the occurrence of a high grade of fatty degeneration of the liver, in fatal cases attended with jaundice. *Abscess of the liver* belongs to the rarer of the complications of enteric fever.

The consideration of intestinal hemorrhage belongs properly to the discussion of the symptoms of the disease rather than its complications, and has already received attention in a foregoing division of our subject.

Closely related to this subject, however, as it is one of the accidents of the ulceration, is perforation of the intestine. This is the most important and dangerous complication of the disease, and is met with, in the course of no other acute disease, with the exception of rare cases of dysentery. Of 1,721 autopsies recorded by various observers in Britain and on the continent, Murchison found 196 instances of perforation, that is to say, 11.38 per cent. of the fatal cases. The same observer states that it occurred in 38 of 1,580, or 3.04 per cent. of the cases under his care; it occurred in 14 out of 600 cases, or 2.3 per cent. of the cases observed by Griesinger.

Perforation is much more common in males than in females. In general terms, it may be stated that age does not especially influence the liability to this accident, although some authorities entertain the opinion that it occurs less frequently in children than in adults, and that it is much more rare in persons over forty years of age than in the earlier periods of life.

Intestinal perforation occurs by far most frequently in the severest cases of the disease, and particularly in those in which diarrhœa, tympany and abdominal pains have been prominent symptoms. In many instances intestinal hemorrhage has preceded the occurrence of perforation. On the other hand, it is of the utmost importance to bear in mind that this accident may occur in cases of the mildest description, and in those in which the bowels have been constipated or confined throughout. It has even occurred where the intestinal ulceration has been limited to a few points.

Perforation is most liable to occur during the third, fourth, or fifth week of the disease, although it sometimes occurs at a later period. Out of 58 cases observed by Murchison, four occurred in the second week; 13 in the third; 16 in the fourth; 13 in the fifth; 8 in the sixth; one in the eighth; one in the ninth, and one as late as the tenth week. Of 22 cases noted by Liebermeister, perforation took place in 2 at the end of the second week; in 6 in the third week; in 2 in the fourth; in 6 in the fifth; twice each in the sixth and seventh weeks, and twice at a later period. According to Nücke,[1] of 185 cases, 84 occurred during the course of the first three weeks, and 99 at a later period.

One of the more important lessons conveyed by the foregoing statistics relates to the danger of perforation not only after the termination of the fever, but even long after convalescence has been fairly established. Instances are not rare in which perforation has occurred after the patient has been allowed to leave his room, or even to go about, and was in every respect apparently almost well.

The earlier perforations take place about the time of the separation of the sloughs from the ulcerated areas of the intestine. The later perforations are due to the extension of ulcerations that show no disposition to heal. Among the immediate causes of perforation may be enumerated indigestible food, hardened fecal masses, ascarides, over-distension of the gut with gas or fæces, vomiting, straining at stool, and sudden changes of posture. When the ulceration has extended to, or has implicated the serous membrane, the most insignificant causes may produce this accident. The vermicular movement following the injudicious administration of a purgative, or excited by an enema, is sufficient to rupture the thinned wall of the bowel. The most frequent seat of the opening is at the lower portion of the ileum. It may occur higher up in the small intestine, or in the caput coli, particularly at the appendix vermiformis. From the statistics of Nücke, we find that of 133 cases, perforation occurred in the ileum 106 times; in the colon 12 times, and in the appendix 15 times. Of 20 cases observed by Hoffmann, the perforation was located in the colon once; in the appendix twice; in the small intestine 18 times. In one case, the perforation being double, was counted twice. The position of the 18 perforations of the small intestine was as follows: once immediately above the ileo-cæcal valve; four times at from four to six inches above it; nine times at from eight to twenty inches; twice at from four-and-a-half to six feet; once ten feet; and in one remarkable case there was from 25 to 30 perforations in the jejunum.

---

[1] Ueber Darmperforation im Typhus Abdominalis. Würzburg, 1873. This work is referred to by Liebermeister.

The perforation is usually a small opening in the serous coat, varying in size from a pin's head to a split pea: it forms the apex of a funnel-shaped ulceration at some point in a Peyer's patch, and is then surrounded by more superficial ulceration; or, and this is less frequently the case, it occurs in a solitary follicle. The margins of the opening are rarely torn or ragged, but usually present a 'punched-out' appearance, and are often surrounded on the peritoneal surface by a narrow ring of recent lymph.

The immediate result of perforation is acute peritonitis, which is, in by far the greatest number of cases, diffuse, although in rare instances the extension of the inflammation has been discovered to have been limited by rapidly formed adhesions resulting in the formation of a circumscribed peritoneal abscess; or a minute opening has been blocked by adhesions formed with the abdominal wall, some adjacent coil of intestine, or a fold of mesentery.

The patient experiences, at the moment of perforation, an intense sudden pain, which rapidly extends over the whole abdomen, but of which the focus is at first in the right iliac fossa. This pain may be accompanied by rigors of greater or less intensity, or it may occur without them. Tympany, if present, usually increases, or if it have previously subsided, recurs. The abdomen becomes exquisitely tender; the patient lies upon his back with his legs drawn up, his face drawn and pinched. Vomiting often occurs. The pulse is small, rapid, or uncountable; the breathing shallow and thoracic; there is tormenting thirst and mostly suppression of urine. Shock commonly occurs, and the patient falls into a state of collapse, with cold extremities, sweating, and a more or less decided fall of temperature. With this fall the mental state of the patient improves, and he may even pass from stupor into a state of mental clearness. In sudden and severe cases death sometimes takes place in the course of a few hours, the mind remaining clear until the end. Much more commonly the patient survives the shock and the temperature rises again; but the symptoms of peritonitis overshadow those of the primary disease and he perishes in the course of from two to four days. In a considerable proportion of the cases perforation takes place without the occurrence of distinct symptoms of peritonitis, and death may result from this cause in cases where it has not been suspected. Its advent may be announced by no other symptoms than a sudden deepening of the prostration, an increase in the pulse frequency and an abrupt temperature-rise; or sudden vomiting, and coldness of the extremities, may be the only changes observed. Death may in some few cases be delayed for several days or weeks, and there is abundant evidence to prove that in rare cases recovery from this accident has taken place. If this statement rested upon no other basis than that of the occasional sudden occurrence of the symptoms of peritonitis in the advanced stages of enteric fever, in patients in whom permanent recovery ultimately took place, it would be open to the criticism that the peritonitis might be due to other causes than perforation. But it is supported by more direct evidence derived from cases where, after the subsidence of the symptoms following perforation, death has resulted from other causes, and the perforation has been found closed by adhesion to some adjacent structure.

Buhl [1] relates the case of a patient who had symptoms of perforation on the twenty-fifth day of enteric fever and was recovering, but died twenty days later, of profuse hemorrhage; a perforation was found completely closed by adhesions to the mesentery.

Analogous cases have been reported by many observers. Recovery has in many instances taken place after the formation of a circumscribed peritoneal abscess, the contents of which have after a time been evacuated either by the bowel, or by an external opening.

Affections of the genito-urinary tract occur as complications of enteric fever.

Transitory albuminuria occurs in nearly one-third the cases. It is, therefore, under ordinary circumstances to be looked upon as a symptom rather than a complication. Acute Bright's disease occasionally occurs, but is far less frequent after enteric fever than after scarlatina. According to Liebermeister it is even less frequent after enteric fever than after pneumonia, facial erysipelas, or measles.

Hæmaturia occurs in connection with hemorrhages from the other mucous tracts, and is not unfrequently one of several evidences of the hemorrhagic diathesis.

Catarrh of the bladder not rarely occurs during convalescence. It is commonly slight and speedily passes away; sometimes it is acute and troublesome. It is chiefly to be attributed to over-distention of the bladder during the course of the fever. But this is not always the case; at this time there is under my care a gentleman convalescent from a light attack of typhoid, who still suffers from mild vesical catarrh, although convalescence is, in other respects, complete. There was not the slightest undue retention during the whole course of his sickness. *Orchitis* and *epididymitis* may occur without previous gonorrhœa.

Menstruation often occurs prematurely during the course of the attack, and is, as a rule, more profuse than is habitual with the patient.

Pregnancy affords a relative, but by no means complete, immunity from the attack. It undoubtedly adds to the danger of the patient, but is not to be looked upon as a formidable complication. Of fourteen cases observed by Murchison, ten recovered; of these ten, two carried the child throughout the attack; the four fatal cases aborted.

Herpes labialis is very rare.

---

[1] Quoted by Murchison.

Facial erysipelas occasionally occurs at the height of the attack or during convalescence. It is a serious complication.

Hemorrhages into the skin, true petechiæ, vibices and the like, occur in persons subject to the hemorrhagic diathesis, or who develop it in the course of the disease. They may also occur in others, but are rare.

Boils and *abscesses* in the integuments, the muscles, or the intermuscular connective tissue are met with in a small proportion of the cases during convalescence. Much more rarely, suppuration of the lymphatic glands of the axilla, and in other regions, takes place.

Bed-sores constitute a common and troublesome complication in severe cases. They are far more frequent in enteric fever than in any other *acute* disease, a fact that is to be explained by the long duration of this fever, the great emaciation, the feebleness of the circulation and the grave general impairment of nutrition. They occur not only over the sacrum and trochanters, but also at the elbows, heels, and occiput.

The hair falls during convalescence. The new hair is often lacking in lustre, but gradually acquires a normal appearance.

The nails both of the hands and the feet show markings that indicate the impaired nutrition of the tissues during the attack. These markings consist of bands or furrows across the whole width of the nail. The portion of the nail developed during the attack is duller than the rest, rough, white, and more or less thinned. Similar changes occur during the course of other severe febrile diseases. They have been described by Vogel,[1] Longstreth,[2] and others.

Among the more important of the complications and sequels of enteric fever are those referable to the nervous system. The importance of this group of secondary affections arises from their gravity rather than from the frequency of their occurrence.

Effusions of blood are noted as of rare occurrence. They may take place into the meninges, or into the substance of the brain itself, and usually occur at the height of the disease. A previous condition of degeneration of the walls of the vessels is a necessary predisposing cause of this accident. Liebermeister states that slight effusions into the meninges give rise to no symptoms, but that considerable effusions occasion symptoms of compression; while effusion into the substance of the brain is followed by the symptoms of apoplexy.

Meningitis occurs but rarely in the course of enteric fever. The cerebral symptoms attendant upon ordinary cases of the disease are in no way dependent upon inflammatory processes affecting any part of the nervous system. A number of cases are recorded in which meningitis has oc-

---

[1] Die Nägel nach fieberhaften Krankheiten. By A. Vogel: Deutsches Archiv für klin. Med., viij. 1870.

[2] Trans. College of Physicians of Philadelphia. 1877.

curred in the course of the disease, or during convalescence, in consequence of disease of the internal ear, or of the development of acute tuberculosis. Murchison states that meningitis may occur, in rare instances, independently of such causes.

Feebleness of intellect and *attacks of mania* show themselves in a small proportion of the cases during convalescence, or at a considerable time after apparent recovery. They are most apt to appear in persons who have a hereditary tendency to mental disorders. These affections are not peculiar to enteric fever, but they occasionally occur after other acute febrile disorders. They are commonly transient, lasting a few days or weeks, less often several months; but all authorities agree in stating that they result in ultimate recovery.

Palsy is an occasional sequel of enteric fever. It presents all the varieties met with after the other acute diseases, and may occur during the course of the attack, or not until several weeks after the commencement of convalescence. Trousseau mentions a case of typhoid fever, in which the beginning of the disease announced itself by a violent pain in the lumbar region, and a true paraplegia such as is occasionally seen in variola. The most common form is paraplegia; but hemiplegia, paralysis of the portio dura, strabismus, and paralysis of individual spinal nerves, may also occur.

Laudouzy[1] has collected cases illustrating the more common forms. Among these is one case of enteric fever in a soldier, where paraplegia began gradually during convalescence about the seventh week after admission to the hospital. There was also squint (paralysis of the left external oblique muscle), which lasted six or eight days, and retention of urine, which made the use of the catheter necessary. The urine was albuminous. This patient recovered. A second patient, a woman twenty-nine years of age, suffered from paraplegia nearly three months after the defervescence; there was vesical and rectal palsy, and paralysis of the velum palati; recovery took place. Other cases are detailed in which hemiplegias, paralyses of the dilator muscles of the glottis, necessitating tracheotomy, aphasia, etc., occurred. The greater frequency of aphasia among children than among adults has attracted the attention of all observers. More frequently the paralytic symptoms are developed at the period of decline of the fever, or in the early days of convalescence. This group of paralyses has a natural tendency to recovery. Paraplegias, hemiplegias, aphasia, the various local and limited paralyses, due to lesions of the nervous system incident to typhoid fever, disappear generally in the course of some weeks or months.

There is, however, another group of paralyses encountered as com-

---

[1] Des paralysies dans les maladies aiguës. Par Dr. Louis Laudouzy. Paris, 1880. See also Bailly: Paralysies consécutives à quelques maladies aiguës. Paris, 1872.

plications or sequels of enteric fever, **of which the** foregoing statement **is not true.** The fever is not the primary cause of the affections **of the nervous system;** it merely calls forth an individual **predisposition already** existing, and the future of the case depends upon **the pathological conditions** underlying the paralysis, that is, upon **the** individual peculiarities of the patient. Palsies in such **patients may result from attacks of** little severity.

Paralysis of the bladder is not uncommon. In **this respect enteric fever** differs from diphtheria, which **is rarely followed by vesical paralysis**—a difference that is remarkable in view of the **fact that in other respects** the palsies following these two diseases closely resemble each other in kind, though not in frequency.

Finally, we must include among the paralyses the sudden death that occasionally takes place in the advanced stages of the disease, from arrest **of the** heart in diastole, and the paralysis of accommodation, which is often present in the early days of convalescence.

Neuralgias and *disturbances of sensation* are less frequent after enteric fever than after some other acute affections.

The organs of **special sense are occasionally the seat of affections that** result directly or **indirectly from enteric fever.** *Otorrhœa* is by no means rare, especially in children. **Inflammatory affections of the internal ear** occasionally result in meningitis.

Deafness, independently of destructive inflammation of the ear, occasionally persists.

Paralysis of accommodation, amblyopic conditions, and even *sloughening of the cornea*, occur in rare instances, and are to be referred to lowered nutrition.

It is often a long time before the patient, emerging from a severe attack of enteric fever, regains his previous health. He may gain rapidly in flesh and present all the appearances of vigorous health, yet lack the ability to sustain any but the most moderate physical or mental effort. As a rule, in such cases, the normal standard of health is gradually regained. It is a remarkable fact that the personal habit of the individual occasionally undergoes marked changes after a severe attack of enteric fever, that is to say, a lean person may exhibit a tendency to corpulence, or a fat person become lean; and it is even more remarkable that changes in disposition also sometimes occur.

The patient may, however, remain permanently weak and anæmic, and continue to emaciate without obvious cause, or the existence of any distinct, local, or constitutional affection. Cases occasionally prove fatal in this way, months after the cessation of the fever, and after death no lesion is discovered, except an abnormally smooth appearance of the mucous membrane of the ileum, and a shrivelled condition of the mesenteric glands (Murchison).

VARIETIES.

The numerous forms attributed to typhoid fever are, for the most part, merely differences in the mode of onset, or in the prominences of certain symptoms or groups of symptoms. The form called "*bilious*" is only a typhoid, which begins with gastro-duodenal catarrh, implicating the biliary passages, and which, therefore, presents among the number of its initial symptoms, *catarrhal jaundice*, and all the accidents which are associated with that condition, notably nausea and vomiting. After several days, rarely more than seven, these epiphenomena disappear, and the typhoid fever runs its ordinary course, sometimes mild, sometimes severe, but in such a manner that no constant relation can be established between this mode of beginning and the ulterior evolution of the sickness (Jaccoud).

The form called "*mucous*" and the form called "*nervous*" are separable from the disease, as it is met with in general, by no more warrantable principle of division; and to distinguish an ataxic from an adynamic form, or other varieties based upon the prominence of particular symptoms, such as an abdominal variety, a thoracic variety, or a cerebro-spinal variety, is neither scientific nor convenient, but only serves, both at the bedside and for purposes of description, to darken counsel. Such methods of classification are to be discarded.

A great variety of forms of enteric fever is, however, met with. Many of these are clearly to be referred to the varying degree of intensity with which the specific poison of the disease acts upon different individuals; others are to be referred to the relative intensity of its action in producing local or constitutional effects, and still others to individual peculiarities on the part of the patient.

Hence, we find, upon the first principle of division, a series of cases ranging from the mildest affections attributable to the especial cause of enteric fever, to the gravest forms of the typical disease; upon the second, a series in which the cases vary according to the relative prominence of the intestinal disease, or the constitutional disturbance (zymosis), the former predominating in some instances, the latter in others; and again, we observe that enteric fever presents notable differences in its course and evolutions at different points of life.

Without attempting a closer analysis of the forms, we may divide the cases into typical and atypical, or, with Liebermeister, into perfect and imperfect.

The typical or perfectly developed cases present the complexus of symptoms already described as constituting the clinical history of the disease, and further illustrated in the analysis of the symptoms, and in the consideration of the complications and sequels.

The atypical or imperfect forms constitute, in most epidemics, a large proportion of the cases, and, when the attention of physicians is more closely turned to the study of enteric fever from an etiological as well as from a clinical standpoint, they will be found, I believe, to be much more common where the disease is endemic than has usually been thought. The cases are partly due to mild **infection**, or, to use an expression already employed in this work, in speaking of other fevers, the smallness of the dose of the fever-producing principle; partly to an imperfect susceptibility on the part of the patient.

Those cases **which approach most nearly to** the typical form of the disease are to be grouped as the mild cases. A second group is constituted by the abortive cases, and following the lightest forms are, first, the cases of intestinal catarrh with fever, and finally those of a febrile intestinal catarrh.

The mild cases present the symptoms of the typical disease modified as respects intensity, and in particular is this true of **the** febrile movement, which is of lower grade. The commencement of the attack is usually gradual; there are prodromes, which pass step **by step into the** declared disease. Chilly sensations may occur; a decided chill is unusual. There is headache, **diarrhœa; the nose** may bleed, and **the eruption appears or not, as the case may be.** Upon the **fourth or fifth day the temperature may reach** 40° C. (104° F.), but it rarely **exceeds that point, and** much more commonly does not attain it. The temperature-range corresponds to that of the typical form, save that upon corresponding days it is about a degree lower. The duration **of this form may be four full weeks;** it is perhaps oftener less than this, each of the four periods not exceeding four or five days. The febrile movement corresponds to the primary and the secondary fever of the fully developed disease. The intestinal lesions do not undergo resolution, but go on to sloughing. According to Jürgensen,[1] the spleen is enlarged in the mildest cases.

The latent, or ambulatory form (*walking typhoid*) belongs to this group. Jürgensen is of the opinion that walking typhoid (typhus ambulatorius) is nothing more than mild typhoid (typhus levissimus) prolonged by repeated errors in diet. In this form all the symptoms are mild, the fever shows itself only in general malaise, prostration, and elevation of temperature, yet the sickness extends over three or four weeks, and the intestinal lesion proceeds to sloughing and ulceration. Herein lies the danger of this form of the disease. The patient regards himself as suffering from some slight indisposition, a "cold," or a "bilious attack," **and** continues **to go about** in a **wretched way, or even, if he be a person of determined will,** to attend to his ordinary occupations, and to eat such

---

[1] Ueber die leichteren Formen des Abdominaltyphus. Sammlung klinischer Vorträge, No. 61. **Leipsic.**

food as his appetite permits, until sudden delirium reveals to his friends the serious character of his illness, a profuse hemorrhage occurs, or, and this is still more common, symptoms of perforation supervene, and are followed, after a few hours, by death. Occasionally more fortunate patients of this class come under the observation of the physician, and the thermometer reveals a temperature of 40° C. (104° F.) or higher, and the history of the case and *ensemble* of symptoms show the disease to be in its third or fourth period.

The abortive form appears to be not uncommon in Europe. In this country it is certainly rare. The attack begins abruptly; prodromes are usually of short duration, or they may be absent altogether. The temperature-range is that of the typical disease, save that it in some instances more rapidly attains its maximum. By the evening of the third or fourth day the temperature may reach 40—40.5° C. (104°—104.9° F.). The invasion is often accompanied by rigors, sometimes by a decided chill. In some instances of abortive typhoid the absolute temperature is very high. Liebermeister has observed in such cases an axillary temperature of 41.1° C. (106° F.) or even higher. There is usually moderate diarrhœa, tympany, enlargement of the spleen, sometimes epistaxis, and often more or less bronchial catarrh. The characteristic eruption is frequently observed, and transient albuminuria is met with. Somewhere between the seventh and the fourteenth day "the sickness takes a sudden turn, and runs a course similar, as regards ordinary enteric fever, to that which varioloid runs as regards variola" (Jaccoud). Cases have been observed where the duration did not exceed five days (Griesinger).

The defervescence is rapid, often being completed in from 24 to 72 hours, and is often attended by profuse sweating. Convalescence is rapid. It is in the highest degree probable that in these cases the intestinal lesions undergo resolution, their evolution being arrested short of the ordinary necrotic processes. We, therefore, have to do with the primary fever due to the action of the special poison, and not with the secondary or septic fever due to ulceration and the formation of sloughs. The parallelism between these cases as compared with typical enteric fever, and varioloid as compared with variola, is complete.

The imperfect cases are to be recognized by the occurrence of the eruption, enlargement of the spleen, and their occurrence in the same house with, or otherwise in such relation to well-developed cases, as warrants the supposition that they are due to a common infection. In 100 cases of this class Liebermeister found that enlargement of the spleen occurred in 71, diarrhœa in 41, and roseola in 21.

A still slighter disturbance of the functions of the body may result from the infection, and give rise to cases of abdominal catarrh with elevation of temperature so slight and so irregular that it scarcely deserves the name of fever, 38° C. (100. 4° F.). And finally, cases of intestinal catarrh occasionally occur, in consequence of typhoid infection, in which there is no elevation of temperature at all. Liebermeister found among such cases many with evident enlargement of the spleen, and a few with an unmistakable eruption. The action of the bowels was irregular; in some instances there was diarrhœa, in others, obstinate constipation; but all the cases manifested a decided impairment of the general health, lassitude, depression, vague pains, often headache and loss of appetite, and a furred tongue. The duration of an apparently trifling indisposition was particularly noticeable, and he calls especial attention to the fact that there was marked diminution in the frequency of the pulse without appreciable alteration in its character, and that the pulse increased in frequency with convalescence, before the patient had quitted his bed.

Dr. Cayley states that many cases and even epidemics of typhoid have been met with in which the temperature has been subnormal throughout the whole course of the disease. He cites the following instance of such an outbreak, which was observed by Dr. Strube:

During the siege of Paris by the Germans in 1870, an epidemic of typhoid fever broke out among the troops, beginning to show itself during the march to Paris, and attaining its greatest height in October. In November a decline took place, which was followed by a fresh outbreak in December. These two outbreaks differed greatly in their characters; the later one resembled in all respects the ordinary form of typhoid; the earlier one presented very different features. In many of the cases the temperature throughout was subnormal, and in others never exceeded the normal point. The roseola was usually profuse; the nerve-symptoms were of marked severity, and were in inverse ratio to the temperature, consisting of violent delirium alternating with stupor; the duration of the fever was very short, defervescence usually taking place at the end of a fortnight. Of the twenty-three fatal cases, in twenty death took place during the first fourteen days. The abdominal symptoms were slight, but the characteristic lesions were found on post-mortem examination. All the cases were characterized by great prostration. These cases presented some features which were probably due to this peculiarity of the temperature: thus, the pulse was but little accelerated, seldom exceeding a hundred; the tongue did not become dry and brown, and the enlargement of the spleen was either absent or much less marked than usual. Dr. Strube attributed the peculiar features of this epidemic to the depressed condition of the troops; they had been exposed to great hardships on the way to Paris, over-fatigued by forced marches, and very insufficiently supplied with food, and the supply continued deficient for some time after their arrival, owing to difficulties of transport. In the later outbreak these conditions were no longer present.

Infantile remittent fever.—This term has been applied to enteric fever as it occurs in children, for the reason that the pyrexia often assumes in them a distinctly remittent type throughout the whole course of the attack. The symptoms and complications are modified by the age of the patient. Children are very susceptible to enteric fever, and Murchison calls attention to the fact that they are often attacked in houses where adults escape.

In the advanced periods of life enteric fever runs a modified course. Its onset is insidious, the febrile movement is less intense than at earlier periods of life, and during convalescence the temperature often falls to markedly subnormal ranges. There is especial danger of collapse. Acute delirium is not so common, and diarrhœa is less apt to be urgent. The characteristic eruption is rarely observed. Perforation is less frequent in early and in advanced life than in the middle periods. Murchison encountered it twice in patients over forty, and I saw it once in the body of a gentleman aged fifty-three.

Typho-malarial fever.—This term, introduced by Dr. Woodward, has been applied to two essentially different conditions: first, typhoid fever occurring either in persons recently subject to malarial influences, or in malarious districts, and modified to a greater or less extent in its course and duration by malaria. The second is remittent fever in one or another of its forms, where the symptoms are of grave character, and where, sometimes, in the course of the disease, the patient passes into that condition which is called the "typhoid state." The term is an unfortunate one, and has given rise to no little confusion concerning the nosological position of the various forms of disease to which it has been applied. It is needless to state, after what has already been said of the etiology of enteric fever, that a *hybrid* disease to which this term is applicable does not exist.

Typho-malarial fever is not a specific or distinct type of disease, but the term may be conveniently applied to the compound forms of fever which result from the combined influences of the causes of the malarious fevers and of typhoid fever (Woodward [1]).

RELAPSES.

By a relapse of enteric fever is understood a second evolution of the specific febrile process after convalescence from the first attack is fairly established (Murchison). Relapses take place after this disease with much greater frequency than was formerly supposed. Murchison observed them in 3 per cent. of 2,591 cases in the London Fever Hospital; Griesinger in 6 per cent. of 463 cases at Zurich; Liebermeister in 8.6 per cent. of 1,743 cases at Basle, and other observers place the frequency of their occurrence at from 1.4 to 11 per cent. This discrepancy is to be explained in part by the difference of opinion as to what really constitutes a relapse, and in part by the fact that, among recent continental observers, the cold-water treatment is generally employed, and relapses are much more apt to occur where the temperature is systematically kept down by cold baths

---

[1] Transactions of the International Medical Congress held at Philadelphia in 1876: Article Typho-Malarial Fever.

than where the fever is allowed to run its course unchecked. Relapses are to be distinguished from the *recrudescences* of *fever*, which are liable to occur during the period of defervescence, or in the early convalescence, and which last from one to several days. These recrudescences may arise from very slight causes, such as errors of diet, solid food, mental emotion, or even moderate exertion. They are, in most instances, dependent upon some local lesion, and, in particular, upon unhealed intestinal ulcers. Their occurrence is to be looked upon, then, as evidence of the persistence of ulceration and is of no little importance as determining the

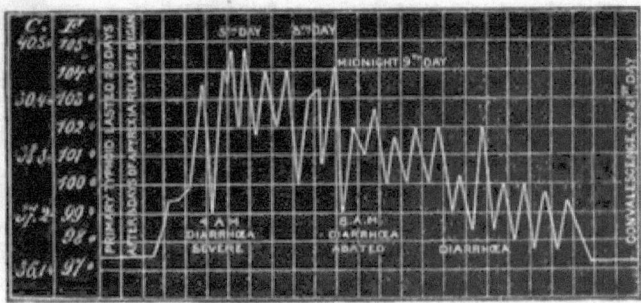

FIG. 16.—Relapse in Enteric Fever. (Irvine.)

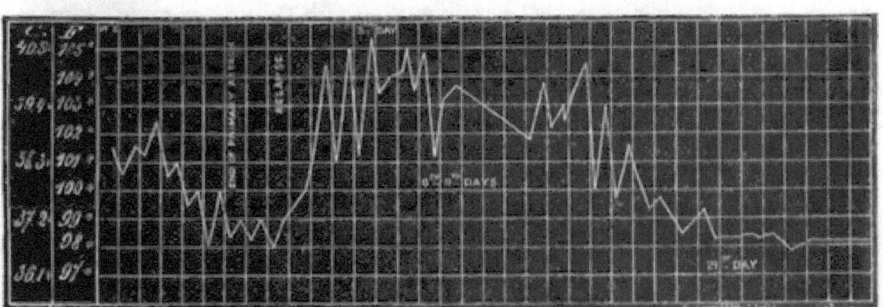

FIG. 17.—Relapse in Enteric Fever. (Irvine.)

treatment during convalescence. They show that the temperature remains *labile*, and that the vaso-motor system, to use the words of Dr. Cayley, is still very unstable. This instability of the vaso-motor system in enteric fever is shown by the readiness with which the *tache cérébrale* is produced, and the last-named observer looks upon the persistence of this phenomenon as a valuable guide, and as indicating that the intestinal ulcers have not yet healed, and that the patient is therefore still liable to the sequels of the disease.

True relapses are attended with a fresh infection of the blood by the specific cause of enteric fever, fresh glandular infiltration, enlargement of the spleen, and a new eruption of rose spots. They are often ushered in by chilliness or distinct rigors, and are attended by the symptoms common in the primary attack.

After a period of apyrexia, varying from twenty-four hours, or less, to several days, the relapse begins with a sudden rise in temperature (Figs. 16 and 17). This interval is marked by normal, or, as is often the case after the defervescence of the acute febrile diseases, subnormal temperatures. Of twenty-nine relapses occurring among twenty-three patients, and analyzed by the late Dr. Irvine,[1] the average duration of the interval was a fraction over five days; in three instances the duration was over ten days, and in four there was no appreciable interval, "at least no interval extending over twenty-four hours." Periods of normal or subnormal temperature of several hours' duration are common toward the close of enteric fever; the relapse may arise at the close of such a period and be, so to say, welded upon the primary attack.

Da Costa,[2] in a valuable contribution to the knowledge of relapses in typhoid, has emphasized the fact that the prodromic stage, so common in ordinary attacks, is absent. The eruption appears earlier, often upon the fourth day, and is apt to be coarser and redder.

The ordinary complications of the primary disease occur in relapse, just as do its ordinary symptoms. Da Costa has called attention to the importance of the transverse markings upon the nails in the diagnosis of doubtful relapse in cases that have not come under observation until after the close of the primary attack. The second ridge of altered nail-growth shows how completely the nutrition suffers during the relapse, and the first ridge is the visible sign of the character of the previous sickness.

The following case is extracted from Da Costa's paper:

"A boy, thirteen years of age, was sent to Bed 12 of my ward, December 6, 1876. The history was most unsatisfactory. He had received a blow on the back of the head about a month before admission, and had had a cough for several months prior to this accident. He had been confined latterly to bed for three weeks; had bled from the nose once; and during the first week had had a diarrhœa, which was readily checked by medicine. The boy, though he was stated to have been delirious for a week, and showed some hebetude, answered questions intelligently. He complained of great weakness, pain in the bowels, and tenderness in the muscles of the lower extremities. He had sordes on the teeth; shallow, frequent respiration; râles in the chest, some fine; a pulse of 120, readily compressible; and an evening temperature of 105°. He soon became extremely delirious; he was very weak; his feet were cold; indeed, his condition was so grave that recovery was regarded as very doubtful, and, notwithstanding the obscurity of the symptoms, wine, chloric ether, and other stimulants were

---

[1] Relapse of Typhoid Fever, especially with Reference to the Temperature. By J. Pearson Irvine, M.D., B.S., F.R.C.P. London, 1880.

[2] Remarks on Relapses in Typhoid Fever: Transactions Philadelphia College of Physicians. Third Series. Vol. iii.

freely resorted to. Some of the symptoms and the history pointed to a brain-trouble, but the case was regarded as one of typhoid fever; and its progress, the tympany, the occasional diarrhœa, the tenderness at the lower part of the abdomen, the look of the tongue, the course the chest-symptoms took, and the markedly typhoid aspect of the face, rendered this opinion more and more certain, although no eruption except sudamina could be found. But what gave us the most information, and told us that we were really dealing with a second attack, thus explaining the extraordinary length of the malady—a length with difficulty reconcilable to the view of ordinary typhoid fever—was that on the 17th it was noted that the nails about half way up showed a white line of impaired nutrition, evidently the result of the first attack of illness, and that near the root another white line was developing, due to the relapse. The patient made a very slow recovery, and was not free from fever until the 26th. When quite himself, we learned from him that he had had unmistakable signs of typhoid fever for several weeks before his admission; that he had been treated for typhoid fever by a physician; and that he was rapidly getting well when the untoward symptoms arose which sent him to the hospital."

Typhoid relapse is usually single; it is less frequently repeated, and in rare instances, of which Irvine had the opportunity of studying five, a third relapse takes place.

The patient whose temperature is represented in the accompanying chart was a male, aged twenty-four, and was under the care of Dr. Pollock, in Charing Cross Hospital. He was admitted in October, 1877, and eight days afterward his primary attack of typhoid, which had lasted four weeks, seemed to be at an end. The temperature became subnormal, but did not remain so, for within twenty-four hours it rose suddenly and quickly, and, by the third afternoon of what proved to be relapse, was 104.6° Fahr. It had risen from 97° Fahr.—nearly eight degrees—without remission. The pulse was never above 120, and usually about 100; the bowels were constipated. There was a remission of the temperature on the fifth morning, but on the fifth evening it was again 104.6° Fahr., and it fluctuated for the next five days between 104° Fahr. and 103° Fahr., with a slight tendency to daily defervescence. It did not fall on the eighth and ninth days, as in more favorable cases, and on the tenth evening was as high as 104.7° Fahr. The patient during these days was ill as he could be; he had a weak and frequent pulse, was troubled with vomiting, and was only kept alive by stimulants. He passed two or three typhoid stools daily, and had a typhoid eruption. One could not but contrast this dangerous case with others of favorable omen, in which the temperature fell at the very time when here it remained persistently high. On the twelfth evening the temperature had a favorable fall, and on the morning of the fourteenth day was only 101° Fahr. For the two following days it exacerbated, but on the sixteenth evening fell decidedly, and gradually descended to subnormal on the twenty-first day of relapse. On the sixteenth day the patient was exceedingly low, and his stimulants were increased, with the best results. Constipation was marked at this time, and simple enemata were given. On the twenty-first day convalescence began, and for about seven days temperature was subnormal. Then came a *second relapse*. During the apyrexial interval constipation persisted. Relapse set in suddenly, and the only warning was given by the thermometer. On the first day temperature was 98° Fahr., and on the fifth day 104° Fahr. The patient looked ill, but not seriously ill. He had no diarrhœa; on the contrary, constipation was still obstinate. From the fifth to the seventh the temperature fell decidedly, and on the morning of the seventh day was only 100.2° Fahr. But the evening temperature remained high to the tenth day (103.6° Fahr.), and the patient, judging from general symptoms, was not free from danger. On the tenth evening a "critical" fall began, and went on to the following morning, when the temperature was but 98° Fahr., a fall of nearly *six degrees* having occurred in less than twelve hours. A slight rise followed, but only to 101° Fahr., and day by day the temperature fell, and became subnormal on the eighteenth day of relapse. For the succeeding eight days convalescence seemed established, and the patient craved for food, the temperature continuing subnormal and constipation being decided. Then came a *third relapse*, as well-marked as its predecessors, for from 97.4° Fahr. on the first day of relapse the temperature rose with little remission to 103.8° Fahr. on the fifth day—that is, nearly six and a half degrees. It may be said in this relapse that from the fifth morning there was a distinct downward

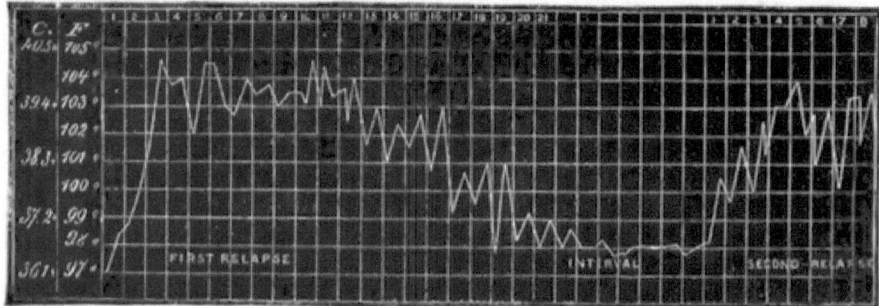

FIG. 18.—Multiple Relapse in Enteric Fever. (Irvine.)

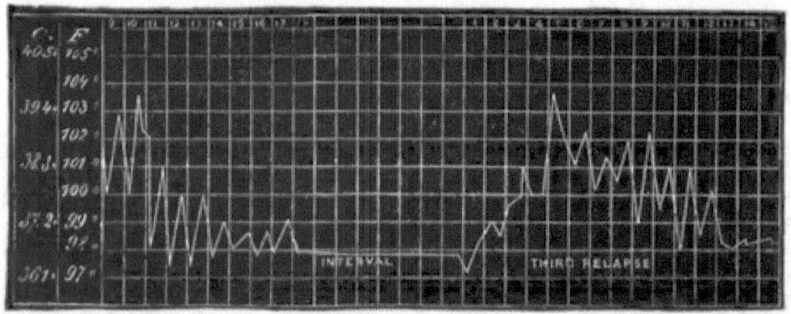

Continuation of Fig. 18.

tendency to the end of the attack, but the relapse had a marked similarity with the first and second, and with those met with in other cases. As in many favorable instances, there was no tendency, at any hour from the fifth to the tenth day, to the elevation of temperature reached on the fifth day, and the patient day by day seemed to improve, the general symptoms being comparatively insignificant. On the ninth morning there was a considerable fall to 99° Fahr., and though the temperature on the same evening rose to 102.3° Fahr., the daily fall afterward showed permanent tendencies. The third stage of the relapse began, but cut short; from the tenth day the temperature went down, and on the *fourteenth* day became subnormal, where it remained for many days, during which an uninterrupted convalescence was entered upon.

The relapse declares itself by an **unexpected** and prolonged rise in temperature. Sometimes the rise is almost continuous; at others, morning remissions occur, but usually the maximum is attained on the fifth evening. The fever remains subcontinuous until the eighth or ninth day, when a **marked** and critical fall takes place. This fall, in the **absence** of hemorrhages or other accidents, is of favorable **prognostic significance;** but it indicates the end of the *primary* fever of the relapse, not the end of the relapse. Upon the tenth day a decided rise takes place, the fever almost reaching its previous height, but from this time the morning remissions are marked.

The duration of the relapse is usually, but not invariably shorter than that of the primary attack. Of fifty-three cases noted by Murchison, the mean duration of the first attack was 26.58 days, of the interval, 11.27 days, and of the relapse 15 days; while the mean total duration of the sickness was 52.86 days. Dr. Cayley mentions a case in which the temperature was maintained at febrile heights, without distinct remissions, for eighty days. As a rule, the relapse is milder than the primary attack, but exceptions occur. In one-third of Murchison's cases the relapse was more severe than the first attack, and of the fifty-three cases, seven died.

There is danger **of the patient's dying** of exhaustion when the illness is protracted **by relapse or repeated relapses; but death may occur in consequence of any of the events which bring it about in ordinary attacks.** Thus, Irvine records a **case in which death** suddenly took place on the fifteenth morning of relapse, in consequence of failure of the **heart;** another, where it occurred on the twenty-fifth day of relapse, and there was found post-mortem, suppuration of mesenteric glands, and recent general peritonitis, without perforation; and a third in which the temperature fell upon the twenty-eighth day of the illness to 98° F., but rose again and the patient died upon the thirty-eighth day; the necropsy revealed perforation and general peritonitis. Da Costa mentions a case in which a second relapse was much protracted by intense pulmonary congestion, and in which several violent intestinal hemorrhages happened after the sixty-sixth day of the original seizure. Fortunately, recovery took place.

Post-mortem examination of the bodies **of those who have** died in the relapse of enteric fever discloses the **lesions of the** primary disease. The individual intestinal lesions are less numerous, for the reason that only those glands and patches of Peyer are involved that escaped during the first attack. The ulceration is, therefore, higher up in the ileum, and co-exists with the recent cicatrices of the former attack, which are most numerous and most extensive near the ileo-cæcal valve.

The cause of relapse in enteric fever is involved in no little obscurity. It is certain that it is in no case to be attributed to errors in diet, overexertion, or any other non-specific cause. It is here that the distinction between *recrudescences* and *relapses* becomes practically most important.

Unfavorable non-specific influences constantly cause recrudescences; they never cause relapse. The latter is due to reinfection by the specific cause of the disease; it is, in fact, a repetition of the primary attack. It is held by many observers that the second infection takes place from the source of the original poison. When relapses occur in patients who have been removed to hospital early in the attack, this explanation is inadequate. It is most probable that relapse is due to resorption of the poison from the lymph-follicles of the ileum and from the mesenteric glands. Some portion of the poison does not undergo those changes in the body, which are necessary to its destruction or elimination, until a later period than that to which the primary attack is due. In a majority of instances the patients, being protected by the illness just passed through, do not suffer; exceptionally they are not protected, and the relapse occurs. Relapse is certainly not more common than second attacks of enteric fever, due to an independent infection at a remote period.

Anatomical Lesions.

Enteric fever differs from the other continued fevers, with the exception of cerebro-spinal fever, in the invariable presence of specific anatomical lesions. These lesions are so characteristic that an examination of the body after death will in all cases make known the nature of the disease, even when the symptoms have been obscure or are unknown. It is important, however, to bear in mind that the lesions of the intestines and of the mesenteric glands do not constitute the disease, but that the poison, which is its specific cause, is taken up by the fluids of the body, and gives rise to general disturbances, which are an essential element of all fully developed cases, and that this constitutional disturbance manifests itself at a very early period in the disease. The more important symptoms of enteric fever are directly attributable to the general disease, and not to the special lesions. The lesions, therefore, fall naturally into two groups. The first embraces those arising from the local action of the specific poison, and includes changes in the lymphatic system of the intestinal canal. These changes consist of an intense inflammation, with new-growth, producing an increase in the size of the lymph-follicles which constitute Peyer's patches and the solitary glands, and subsequent necrotic processes resulting in the partial destruction of these tissues. Secondary changes in the mesenteric glands, and enlargement of the spleen, are also to be referred to this group. These are the lesions which are to be regarded as characteristic of enteric fever. They are present in the mild and abortive, as well as in the fully developed cases.

The second group includes lesions which are not the direct result of the local action of the special poison, but are due to the constitutional infection. They consist of degenerative changes involving the tissues of various organs, and are to be found generally manifested throughout the body, and particularly in the liver, the kidneys, **the voluntary muscles,** the heart, the salivary glands, and the pancreas. Analogous **changes** probably take place in the structure of the nervous system, particularly the brain. These changes are not peculiar to enteric fever; they occur in other acute febrile diseases, and are dependent upon the intensity and duration of the pyrexia. They **attain** their fullest development, however, in enteric fever for the reason **that this disease** is characterized by an unusually long continuance of the febrile movement.

Cadaveric rigidity is usually marked and of long duration. Emaciation is often extreme. The integuments of the dependent part of the body are apt to be more or less discolored, but the deep livid discoloration of typhus is rare. Except where death has taken place in consequence of pulmonary complications, the face is not often livid. The characteristic rash of enteric fever is never observed on the dead body, even in those cases where the spots have been numerous **immediately before** death. Sudamina, and other accidental eruptions, **persist.**

The digestive tract.—The pharyngeal mucous membrane is in most instances healthy, but it occasionally exhibits signs of **recent inflammation,** and **sometimes distinct** points of **ulceration.** The most common seat of the ulcers is at the lower part of the pharynx. They are usually superficial, but may extend to the muscular coat. The pharyngeal mucous membrane may be the seat of a diphtheritic exudation. Occasionally the œsophagus shows the evidences of ulcerative processes similar to those met with in the pharynx.

When present in the œsophagus, the ulcers are most numerous at the lower or cardiac extremity of this viscus, and vary from simple excoriations to deep lesions implicating the muscular coat. The foregoing changes are never found when death occurs earlier than the third week of the disease. **They are not met with after death from** typhus, or other acute diseases, but are not on this account to be regarded as analogous to the specific lesions of the intestine (Murchison).

The *stomach* presents no changes peculiar to this disease, and is in many cases healthy. It occasionally is the seat of morbid appearances consisting of hyperæmia, softening and superficial erosions of the mucous membrane.

The *duodenum* usually presents no anatomical **changes.** Sometimes it exhibits the evidences of increased vascularity, or some enlargement of the mucous follicles. Ulceration does not occur.

The *small intestine* is mostly healthy at its upper part; it does not, as a rule, contain much gas. The *jejunum* and the upper part of the ileum may be moderately distended, the lower portion of the ileum is commonly collapsed. If peritonitis has preceded death, the intestines are commonly more or less distended. The tympany which belongs to the disease is due to the presence of gas in the colon. Invagination of the intestine, unaccompanied by the evidences of inflammation, is occasionally met with at one or more points. This is probably due, as Murchison suggests, to the death-struggle, and is found after other diseases, in which death is preceded by a high degree of torpor of the cerebro-spinal system. Round or tape-worms are occasionally voided during enteric fever, and are sometimes also found in the intestines after death. Small masses of fecal matter of an ochrous yellow color, intestinal mucus, sloughs, and, if there have been intestinal hemorrhage during life, more or less blood are met with in the intestines. The mucous membrane of the ileum is usually hyperæmic; this redness may be uniformly distributed, or it may occur in patches. It is most intense in the neighborhood of the ulcerated glands, and in particular in the region of the ileo-cæcal valve. When death takes place in the later stages of the disease, the mucous membrane of the ileum often presents a grayish appearance. It is frequently softened in consequence of post-mortem changes.

The foregoing lesions are not peculiar to enteric fever, nor are they constant in it.

Those which are about to be described, and which involve the agminate and solitary glands of the ileum, are characteristic of enteric fever, and are constantly met with in the bodies of those who have died of that disease. They constitute the specific or primary local lesions, and present different appearances, according to the period of the illness in which death has taken place. The progress of the pathological processes which result in these lesions, may be divided into four periods, namely: the stage of medullary infiltration, the stage of softening and necrosis, the stage of ulceration, and the stage of cicatrization. Two or more of these stages are often represented by the lesions found in the same body, as the morbid process always commences at the lower extremity of the ileum, near the ileo-cæcal valve, and at a later period involves the patches higher up. The periods occupied by these stages usually consist, in severe uncomplicated cases, of about a week each, but, as with the periods of the febrile movement, we must reckon sometimes four or five days to each, sometimes as many as eight or nine.

The first stage.—The mucous membrane of the intestine, particularly that surrounding Peyer's patches in the lower part of the ileum, is hyperæmic and swollen. The agminate and solitary glands are infiltrated by an excessive proliferation of cellular elements; the follicles are swollen and distended. The neighboring mucous membrane is also infiltrated with cells. This change has been observed in cases where death has taken place as early as the second day. The Peyer's patches are thickened, hardened, and elevated from half a line to two lines above the surrounding mucous membrane. Their surface is usually uneven and of a reddish or a reddish gray color. The number implicated is very variable; sometimes it does not exceed three or four, at other times more or less infiltration is to be observed in nearly all the patches. As the long axis of Peyer's patches corresponds to that of the intestine, it follows that, where a number of adjoining patches are infiltrated, very extensive linear lesions sometimes result from their confluence. The solitary glands in the neighborhood of the diseased patches are commonly involved in a similar process. Usually a comparatively small number of them are implicated, but in exceptional cases the solitary glands principally are diseased, and very rarely there is disease of the solitary follicles without any implication whatever of the Peyer's patches. The solitary follicles of the large intestine, especially those in the neighborhood of the ileocæcal valve, are likewise involved, and they sometimes form, by the extension of the infiltration to the neighboring mucous membrane, patches of considerable size.

The second stage.—The hyperæmia of the mucous membrane now decreases, but the infiltration of the solitary and agminate glands goes on. Some of the swollen patches undergo partial or complete necrosis, which may be superficial, or may proceed to various depths, implicating the muscular or even the serous coat of the intestine. This process is usually nearly completed by the end of the second week, and at this time, or in abortive cases even earlier, reparative processes begin, and the affected patches that have not been the seat of sloughing, undergo resolution. The sloughs are of a yellowish brown or greenish color, from staining with the intestinal contents, and particularly the bile; sometimes, being infiltrated with blood, they are dark in color.

The third stage.—The sloughs are now gradually detached. Ulcers of varying depth and of a size and form corresponding to the area of the necrosed tissue, are formed. If an entire patch be involved, the ulcer is elliptical, and of considerable size. Ulcers resulting from necrosis of infiltrated solitary glands are usually small and round. They may, however, enlarge by secondary implication of the surrounding tissue. The edges of the intestinal ulcers are usually abrupt, the surrounding tissue being thickened and overhanging. At the close of the third stage the sloughs are for the most part detached.

The fourth stage is that of cicatrization. The swelling at the edges of the ulcers gradually diminishes, and the surface becomes covered with a delicate layer of granulations, which is transformed into connective tissue, and ultimately covered with epithelium. Where the ulceration has extended into the muscular coat, neither the mucous membrane nor the villi, according to Hoffmann,[1] are reproduced.

The cicatricial surface-tissue is at first adherent to the underlying coat.

---

[1] Ziemssen's Encyclopædia of Medicine: Liebermeister's article on Typhoid Fever.

It usually becomes, after a time, movable, and may, if the ulceration have been superficial, even be coated with villi. The gland-structure is, however, not regenerated. The resulting scar is slightly depressed, firm, smoother and less vascular than the surrounding mucous membrane. It is never surrounded by puckering, and never gives rise to diminution in the calibre of the bowel. Where cicatrization takes place in this simple way, the time required to heal each single ulcer is probably about a fortnight. Not infrequently the process of healing is much more complex. While one part of the ulcer is undergoing cicatrization, the sloughing in another part may continue, so that the ulcer may be said to have become serpiginous. Such ulcers often persist for a long time, prolonging convalescence, and occasionally causing death by perforation at a comparatively remote period from the commencement of the attack. The scars are often more or less strongly pigmented.

Not all the patches necessarily slough. In a certain proportion of them the morbid processes are arrested prior to the stage of necrosis, and in the abortive cases, it is probable that all the patches of infiltration undergo resolution, without sloughing, the swelling gradually diminishing until the patches and follicles at length resume their normal condition.

The thickening of the patches is due not only to the infiltration of the follicles, but also to an increase in the interstitial connective tissue. If, in the period of resorption, the follicles undergo resolution more rapidly than the interstitial net-work, a reticulated surface remains; or the follicles may break down and be discharged, while the hypertrophied connective tissue remains unaffected, thus giving rise to a similar reticulated surface. At the same time, numerous minute points of pigmentation are formed in the seat of the softened follicles, and these persist, presenting a peculiar appearance which has been thought to resemble that of the newly shaven beard. This appearance is not characteristic of enteric fever, as was formerly thought. It is met with after death from other diseases, and in the bodies of those who have never been the subjects of enteric fever.

Analogous changes take place at the same time in the mesenteric glands. These bodies become more or less swollen in consequence of cellular hyperplasia and increase of their connective tissue. The enlargement of the glands is greatest in those portions of the mesentery which correspond to the diseased portions of intestine. In some cases all the mesenteric glands are more or less swollen. The swollen glands are hyperæmic, bluish, and tense. They may attain the size of a small bean or chestnut; sometimes they are as large as a pullet's egg. Later in the course of the disease they become pale, gray, or reddish gray. After the detachment of the sloughs from the intestinal ulcers, the enlarged mesenteric glands shrink, and gradually regain their normal appearance. Some of them, however, may undergo partial softening. If this be not extensive, complete resolution may ultimately occur; if, however, the softening be considerable, resorption does not take place, but the **softening material** undergoes cheesy **metamorphosis and ultimately becomes calcareous.** Sometimes, however, the **softening results in the formation of pseudo-abscesses,** which may burst into the peritoneal cavity and give rise to general peritonitis.

Other lymphatic glands, particularly those in the fissure of the liver, the retro-peritoneal, and the bronchial glands, are occasionally found enlarged. The lymphatic follicles at the root of the tongue and in the tonsils undergo changes analogous to those just described, giving rise to enlargements which appear early in the course of the disease, and usually disappear without further change; although in some cases softening, rupture, and subsequent ulceration result.

The changes in the spleen are to be regarded as analogous to the changes which take place in the lymph-follicles of the intestine and in the mesenteric glands. The spleen is tense and hyperæmic. On section, it is of a brownish red color; in the early periods of the disease it is of moderate consistence; later its tissue is soft, pulpy, or even diffluent. The enlargement is almost always present when death occurs before the thirtieth day. At a later period the capsule is wrinkled, the tissue firmer, the stroma more prominent, and the color paler. Hemorrhagic infarctions are often met with. The enlarged and softened spleen of the later stages of typhoid fever is liable to be ruptured by mechanical force, as in the instance referred to by Bartholow, and in some instances it is said to have undergone spontaneous rupture. In a small proportion of the cases enlargement of the spleen does not take place. This is more frequent in elderly persons than in the young. Sometimes thickening of the capsule has apparently prevented the occurrence of the enlargement, and in other instances it has been thought that the spleen was abnormally small before the occurrence of the disease.

Gerhardt[1] states that, in many cases in which a relapse takes place, the enlarged spleen is not diminished during the non-febrile interval between the original attack and the relapse.

The second group of anatomical changes comprises, as has been said, parenchymatous degenerations of the various organs of the body. These changes are the result of the long duration of the febrile movement. They are therefore not confined to typhoid fever, nor are they characteristic of it.

The liver is occasionally hyperæmic, but in most cases it is normal in appearance, or it may be pale. In many cases it is softened, and upon microscopical examination the cells are found to be granular, loaded with fat, the nuclei indistinct, or no longer to be seen. These microscopical

---

[1] See Leibermeister's article on Typhoid Fever, in Ziemssen's Cyclopædia.

appearances are sometimes observed where the organ does not appear to be softer than normal. The changes in the liver-cells are proportionate to the intensity and duration of the febrile movement. In cases where this has been slight these changes are little marked, or even wholly absent. The amount of bile is usually markedly diminished, and in the later periods of the disease it is thin and almost colorless.

The kidneys also show parenchymatous degeneration. The epithelium is granular, the contour of the cells indistinct, and the nuclei disappear. These changes affect first the cortex, later the pyramids. In many cases they are but little marked. They are usually associated with albuminuria, although Liebermeister states that he has repeatedly noted absence of albuminuria throughout the whole course of the disease, where at the autopsy advanced degeneration of the kidneys was discovered.

Softening of the *muscular tissue of the heart* is very common. This softening is due to the parenchymatous degeneration which occurs in the severer cases of the disease, and is, like similar changes in other organs, proportionate to the intensity and duration of the febrile movements. In its higher degrees the degeneration gives rise to changes in the muscular tissue that are easily recognizable. The heart is soft and of a pale, gray, yellowish or "faded-leaf" color, the muscular tissue is easily torn, and the organ thrown upon the table settles down into a formless mass. The changes consist in the deposit in the muscular tissue of numerous minute granules, which are often arranged in long rows. If they be slight, the striations are still visible; but in the higher grades the muscular fibres are filled with granules, and the striations disappear altogether. The waxy change referred to farther on is less frequent in the heart than in the voluntary muscles. The feebleness of the heart, which is characteristic of typhoid fever, and particularly of the advanced stages of severe cases, is proportionate to the degree of degeneration found after death.

Evidences of recent endocarditis with thickening of the aortic or mitral valves, are sometimes met with. The pericardium is usually healthy. Recent pericarditis belongs to the rarest of the anatomical changes observed after death in enteric fever.

Fatty degeneration of the minute arteries of the brain, kidneys and other organs was demonstrated by Hoffmann, who also called attention to the frequency with which thickening and opacity of the inner coat of the larger vessels, and particularly of the pulmonary arteries, occur.

The blood is dark-colored, with small, soft coagula. Pale, fibrinous clots are frequently found in the heart. If death takes place in the latest stage of the disease, or during convalescence, the vessels are often nearly empty, the blood thin and watery, and the tissues œdematous.

Changes in the voluntary muscles, similar to those already described as occurring in the muscular tissue of the heart, are of very frequent occurrence in enteric fever. They were first described by Zenker,[1] who distinguishes two forms. The first is a granular degeneration, which in its highest degree does not differ from ordinary fatty degeneration. Less fully developed, it consists in the appearance in the fibres of minute granules, tending to form themselves into rows and obscuring the striations. The second form is a waxy degeneration, by which the muscle-substance is converted into a glistening, colorless mass, in which the striations are no longer to be seen. The granular degeneration is more frequent, but the two forms are often associated, sometimes one, sometimes the other predominating. These changes are not peculiar to typhoid fever, but occur in other severe febrile diseases, and are probably in all cases the result of the long-continued high temperature. They are most marked, usually in the second, third, and fourth weeks. If death takes place at a later period, the degeneration of the muscles disappears, or is only to be observed in its results, namely, softening, hemorrhages, and pseudo-abscesses in the substance of the muscular masses. The rectus abdominalis, the adductors of the thigh, the pectorales major and minor, the diaphragm and the tongue are more frequently implicated in these changes, though all the voluntary muscles share in them to some extent. The excessive loss of power, which appears both at the height of the disease and during convalescence, is due in part to impairment of the functions of the nervous system, and in part to these degenerations of muscular tissue.

The central nervous system presents, in most instances, no gross anatomical changes sufficient to account for the symptoms during life. More or less extensive adhesion of the dura mater to the inner surface of the cranium is occasionally found, even in the early periods of the disease. There is occasionally increased injection of the pia mater, and of the vessels of the brain-tissue itself. Later in the course of the affection the pia mater is often œdematous, and sometimes opaque, while in most cases there is moderate distention of the ventricles, with œdema of the brain-substance itself. Some observers have thought that the amount of cerebral œdema found after death was in direct relation to the prominence of the mental disturbance during life. When death takes place late in the course of the disease, the convolutions are often flattened, and spots of softening, due to imbibition of serum, appear.

According to Hoffmann, frequent changes are found in the *salivary glands*. Early in the disease they are firmer in consistence than normal; the acini are found, upon microscopical examination, to be filled with large, multinuclear, granular cells; later, the cells lose their sharp outline, become turbid, and are filled with granules. The glands gradually resume

---

[1] Ueber die Veranderungen der willkurlichen Muskeln im Typhus abdominalis. Leipzig, 1864.

their normal appearance. These changes are regarded by Liebermeister as analagous to the parenchymatous degeneration which occurs in other organs of the body. The parotid, submaxillary and sublingual glands are implicated. *The pancreas* is the seat of similar changes.

The organs of respiration show no anatomical changes peculiar to enteric fever. The epiglottis is congested—sometimes ulcerated or œdematous; or, if diphtheria complicates the case, often the seat of false membrane. These changes are not met with except in the advanced stages of the disease. The larynx, as has already been pointed out, may be also the seat of more or less extensive ulceration. The trachea is usually normal in appearance, or somewhat congested. It is rarely ulcerated. In the bronchial tubes, those changes are met with which underlie the various forms of bronchial catarrh occurring in other diseases.

The lungs almost constantly present changes referable to the enfeeblement of the circulation. Hypostasis is very frequent ; it is limited to the most dependent portions of the lungs. When hypostasis is incomplete, the cut surface of the congested lung-tissue discharges upon pressure reddish serum with bubbles of air ; when complete, the pulmonary tissue is deprived of air, and we have the condition to which the term *splenization* has been applied.

Pulmonary œdema is common.

The evidences of lobular and lobar pneumonia, and of acute miliary tuberculosis, occur in a certain proportion of cases. These complications have already been considered in a foregoing division of this article. Recent pleural adhesions, and serous and purulent pleural effusions, are sometimes met with. The bronchial glands are occasionally enlarged.

Diagnosis.

The diagnosis of well-developed, typical cases of enteric fever, after the first week, is unattended with difficulty. During the first week, however, it is often impossible to form a positive diagnosis; but even then the nature of the disease may be suspected, if there be febrile movement with nocturnal exacerbations, each night attaining a higher temperature, and especially if there be bleeding at the nose, diarrhœa, either spontaneous or readily produced by laxatives, and appreciable enlargement of the spleen.

The direct diagnosis of the developed disease rests upon the continuance of the febrile movement and the appearance of abdominal symptoms, namely, diarrhœa, abdominal pain, enlarged spleen, and tympany. If, in addition to these symptoms, lenticular rose-spots appear, the diagnosis becomes certain.

If neither the eruption nor the abdominal symptoms occur in the course of the second week of the disease, the diagnosis can be established only by a careful differentiation from the other febrile disorders, which more or less closely resemble enteric fever. These diseases are to be divided into two classes: first, those which **resemble enteric fever in** the first week of its course; and secondly, those with **which it** is liable to be confounded in its more advanced stages.

To the first group belong *simple continued fever* and the *exanthematous diseases*. Diarrhœa is not, however, present in those diseases, **nor** is their onset commonly characterized by the occurrence of marked prodromes. Furthermore, the character of the temperature-range in all these affections differs greatly from that in typhoid fever, being marked by an abrupt rise, which lacks the distinct morning remissions of typhoid, and attains its maximum with greater rapidity. Moreover, in most cases of simple continued **fever,** the attack comes to an end in less time than is **required for the full** development of typhoid. The exanthemata cannot be distinguished from typhoid fever with absolute certainty in their pre-eruptive periods. Nevertheless, the presence of naso-pulmonary catarrh in a doubtful case would lead us to suspect measles; or the presence of a sore throat would lead us to suspect scarlet fever, while the intensity of the febrile movement and the character of the lumbar pains in **small-pox** serve to distinguish it in its early stages from typhoid fever.

After the first week, typhoid fever may in some instances be confounded with the following diseases: *typhus, relapsing fever, remittent fever, small-pox, influenza, enteritis, peritonitis, meningitis, tuberculosis, trichiniasis.*

Typhus, see page 241.

Relapsing fever, see page 302.

Remittent fever.—Enteric and remittent fevers not unfrequently prevail together in malarious countries, and all physicians practising in such countries are familiar with that form **of enteric fever which has already** been alluded to **under** the name of typho-malarial fever, and which is, **in** point of fact, enteric fever modified by malarious influences. On the other hand, severe remittent fever not infrequently presents strong clinical resemblances to enteric, particularly when complicated with marked intestinal symptoms. Thus, in both diseases, vomiting, diarrhœa, enlargement of the spleen, prominent cerebral symptoms, and the condition known as the typhoid state, may occur. The more important points of distinction are the occurrence of the eruption, the subcontinuous or imperfectly remittent character of the temperature-range in the second week, and the long course of enteric fever.

Small-pox.—Murchison states that he has frequently known a copious eruption of lenticular spots to be mistaken for variola. This is an error of diagnosis that should **under** no circumstance occur. The eruptions are essentially unlike. They differ in date of appearance, in character, and in evolution. The rose-rash of typhoid does not appear before the seventh day of the illness; it is absent from the face, it disappears upon pressure, and undergoes but little or no change from the time of its appearance till it fades, leaving no trace; that of variola appears during or after the third febrile exacerbation of the initial stage, that is, upon the third day of the disease; it first shows itself upon the face and hairy scalp. From the beginning it is hard, shot-like, and acuminate; it undergoes characteristic and unmistakable changes with great rapidity, and leaves a more or less persistent conspicuous scar.

Influenza occasionally closely resembles enteric fever. The following symptoms occur in both these affections: fever marked with weakness, sleeplessness, delirium, sweating, and occasionally diarrhœa; more or less pulmonary catarrh, deafness, epistaxis, and a dry, red tongue, are likewise seen in both. The differential diagnosis rests chiefly upon the occurrence of influenza in wide-spread epidemics, the short duration of the attack, the atypical temperature-curve, and the absence of the other abdominal symptoms that are usually associated with the diarrhœa of enteric fever.

Enteritis may be confounded with enteric fever. The former is, however, a local disease, and unattended by the constitutional disturbances which are characteristic of and essential to the latter. If fever be present, it is symptomatic; there is no great prostration, no delirium, the spleen is not commonly enlarged. Rose-spots are absent, the disease is of a relatively short duration, and the abdominal pain is more conspicuous and severe than that of enteric fever.

Peritonitis due to other causes than perforation is to be discriminated from that arising in the course of typhoid fever, by the antecedent history of the case. If the patient, however, does not come under observation until after the appearance of the symptoms, it may be impossible, in the absence of a previous history, to determine whether they be due to perforation or not.

Meningitis, see page 95.

Acute tuberculosis presents many points of resemblance to enteric fever. It is to be remarked that in a considerable number of the cases the formal rules for the discrimination of the two diseases, are, at the bedside, unavailing. Only by a prolonged study of the complexus of symptoms presented by the patient does the diagnosis become possible, and in some cases the most experienced clinicien must be content to leave the decision of the question to the investigations of the post-mortem-room. Hectic, delirium, vomiting, varied cerebral symptoms, even palsies and the *tache cérébrale* occur in both diseases; the absence of the characteristic rash of enteric fever loses its diagnostic value from the fact that it is often absent in the early periods of life when the difficulty in diagnosis usually arises. The chief points of difference are these: in enteric fever the temperature-range is typical, or more or less closely conformed to a definite type, whereas that of tuberculosis is extremely irregular. In enteric fever, diarrhœa and tympany occur; in tuberculosis, diarrhœa is rare, and the abdomen is apt to be flat or even scaphoid. In enteric fever, epistaxis, intestinal hemorrhage, and enlargement of the spleen, occur; in meningitis these symptoms are absent. The headache of enteric fever is dull, while that of meningitis is acute and commonly associated with intolerance of light and sound.

In *trichiniasis*, there is pyrexia with vomiting and diarrhœa, succeeded after a short time, by typhoid symptoms. The resemblance of this disease to enteric fever ceases with those symptoms. The rose-spots do not occur, and epistaxis and enlargement of the spleen are rare; while, on the other hand, the severe muscular pains and local and general œdemas that are almost constant symptoms in trichiniasis, are not encountered in enteric fever.

Prognosis and Mortality.

A knowledge of the duration of the disease is of the utmost importance with reference to prognosis. The mean duration of enteric fever is from three to four weeks. The doctrine of critical days, which is borne out by the course of other continued fevers, is very imperfectly illustrated in enteric fever. This is to be attributed to the fact that the febrile movement consists of two distinct parts: a primary fever, due to the specific infection, and a secondary fever, due to the resorption of septic materials from the intestinal lesions—the latter beginning before the former has terminated—being, so to speak, engrafted upon it in such a manner that the two overlap. In the abortive cases, where it is more than probable that the pathological processes of the ileum do not go on to sloughing, the disease comes to a close with the cessation of the primary fever, the secondary fever being altogether absent. These cases commonly terminate abruptly, with copious sweating, or other evidences of true crisis, and ordinarily have a duration not exceeding fourteen days. In cases which run a more usual course, the secondary fever is superadded to the primary at the close of the second or early in the third week, and always terminates by prolonged lysis. A precritical and critical perturbation of the temperature often occurs at this period, and it is then that the continuous or subcontinuous fever becomes distinctly remittent in type. If a crisis occur at all, it takes place during the course of the attack, and is marked by the secondary septic fever that has already become established before the termination of the primary fever of infection. It is common for the defervescence to be completed by the twenty-first or the twenty-eighth day.

Of 200 cases which terminated in recovery, and in which the data of commencement could be fixed with tolerable certainty, Murchison found the duration to be from ten to fourteen days in 7; from fifteen to twenty-one days in 49; from twenty-two to twenty-eight days in 111; and from twenty-nine to thirty-five days in 33. Thus, in all but 7 cases, the duration of the sickness exceeded two weeks; in nearly three-fourths it exceeded three weeks; and in one-sixth, it was more than four weeks. The average duration of these 200 cases was 24.3 days. The average duration of 112 fatal cases observed by the same author was 27.67 days, and of 215 fatal cases studied by Hoffmann, 28.9 days.

If the fever continue beyond the close of the fourth week, it is almost certain that some complication exists or that a relapse has taken place. Dr. Irvine has shown that relapse may occur without an interval of more than twenty-four hours between it and the termination of the primary attack, certainly after an interval so short as to be readily overlooked. The cases in which fresh spots have appeared daily until the thirty-fifth day, and the cases alluded to by Murchison, where, with mild symptoms, fresh spots appeared almost daily from the fourteenth to the sixtieth day, are to be accounted for only, it seems to me, by the supposition that one or more relapses, following very brief and therefore unnoticed intervals of apyrexia, have thus unduly prolonged the attack. Jenner has expressed the opinion that, except in cases of relapse, fresh spots never appear later than the thirtieth day, and that febrile symptoms later than that date are always the result of some incidental complication. The secondary fever is often prolonged by the non-healing of some of the intestinal ulcers. Whether the prolongation of the fever be due to this or to some other complication, such cases show extreme prostration, wasting, and a tendency to the ready breaking-down of tissue, manifested in the formation of bed-sores, and other similar accidents of a lowered nutrition.

Enteric fever may terminate in recovery, in the abortive cases, as early as the tenth day; many observers have noted the termination of cases of this variety even as early as the seventh day, and in very rare instances, recovery has taken place as early as the fifth day. Recovery at so early a period suggests the probability that such cases are, in fact, not instances of enteric, but of simple continued fever; and, in the absence of the characteristic eruption and of the strongest probability that concomitant circumstances can suggest, it would be better to suffer their nosological position to remain undetermined.

In well-developed cases death is not common earlier than the fourteenth day. Murchison has noted it as occurring as early as the twelfth, or even the sixth day. Several observers have recorded cases fatal as early as the fifth day, and a few instances are to be found in medical literature, where death took place upon the fourth, the second, or even the first day of enteric fever. It thus appears that death may occur at any period of the disease. In uncomplicated cases, it is most common about the close of the third, or the beginning of the fourth week; and in such cases it rarely takes place earlier than the third week. Death may, however, occur later than the fourth week in cases not especially complicated, and in consequence of the direct or indirect results of the ordinary lesions.

Liebermeister regards intestinal hemorrhage, perforation of the intestine, and the like, although representing in a certain measure the results of changes peculiar to enteric fever, as complications, and he formulates the immediate cause of death in cases that are uncomplicated, as the *fever and its consequences*. That is to say, the patients die either of paralysis of the heart or paralysis of the brain, both of which are the results of the temperature-rise.

The character of the fever, in any particular case, is of importance in determining the prognosis.

Four hundred cases observed in the hospital at Basle and studied with reference to the influence of the absolute temperature upon the prognosis, without special antipyretic treatment, gave the following percentages:

Of those in whom the axillary temperature did not attain 40° C. (104° F.), 9.6 per cent. died. Of those in whom the axillary temperature reached or exceeded 40° C. (104° F.), 29.1 per cent. died. Of those in whom the axillary temperature rose to or beyond 41° C. (105.8° F.), over 50.0 per cent. died. Fiedler [1] found that more than half of those patients whose temperature had risen to, or exceeded 41.1° C. (106° F.), died. According to Wunderlich, there is very great danger as soon as the temperature reaches a height of 41.2° C. (106.16° F.), and a very tedious recovery is the best that can be hoped for. A temperature of 41.4° C. (106.52° F.) is followed by nearly twice as many deaths as recoveries, while 41.75° C. (107.15° F.), or higher, is rarely followed by other than the fatal termination. The same observer adds that one of his cases of enteric fever recovered, after a temperature of $42\frac{1}{8}$° C. (107.825° F.) had been reached during a rigor in the course of the disease. It is to this case, doubtless, that Murchison refers in stating that recovery had been known to follow a temperature of nearly 108° F.

Very high temperatures with well-marked remissions are of less ominous prognostic import than nearly continuous high temperatures in which slight or no remissions occur, even though the maxima attained be somewhat lower. Thus, Wunderlich states, that after a temperature exceeding 41° C. (105.8° F.) in the morning hours, death is almost certain. Fiedler saw all the patients whose temperature in the morning rose to or surpassed 41.25° C. (106.25° F.), with a single exception, die; and of those whose morning temperature rose to 40.8° C. (105.44° F.), only upon a single day, more than half died.

It is thus apparent that the daily fluctuations of temperature are of very great importance as determining the prognosis. The greater the

---

[1] Deutsches Archiv für klin. Medicin, Bd. I., quoted in Leibermeister's article, and by Wunderlich.

daily fluctuations, the more favorable the prognosis. A fever in which the morning fall is marked, is much less disastrous to the organism than one that is continuously high, and we may hope for a shorter duration where early in the second week the fever shows a strong tendency to remit in the early hours of the day, even when the evening exacerbation is relatively high.

The beginning of the attack promises a fair indication for the duration of the fever which is to be expected. The more sudden the appearance of the disease, and the more rapid the rise of temperature in the first week, so much the more should we expect in general a short, or even abortive attack (Liebermeister).

A closer study of the temperature, during the first week, than is customary with American physicians, is desirable in every-day practice. In uncomplicated cases the temperature of this period is the key to the temperature of the attack, and is therefore of the greatest importance in regard to prognosis. By reference to the schematic representation of the temperature-range, and to that of a mild case recorded by Wunderlich, (see pp. 154 and 157), it will be seen that the height which the temperature attains by the end of the first week is nearly or about that which is to be expected during the course of the attack. Unless complications occur, the temperature rises but little above that point. Moreover, if the temperature at the end of the first week be of moderate elevation, the fever will probably not only be of moderate intensity, but its duration will be correspondingly short.

The character of the circulation and the action of the heart, both of which are directly modified by the fever, are of great moment in prognosis. The most frequent immediate cause of death is cardiac failure. Hence, a close study of the signs and symptoms referable to the circulation is scarcely less important than the close scrutiny of the temperature-range.

While the impulse of the heart retains to some extent the force of health, and the systolic sound remains distinct, the dangers belonging directly to failure of the circulation, or to its indirect consequences, remain at a minimum, although the temperature-rise may be considerable. If these remain favorable until the time at which decided remissions in temperature occur, the prognosis is correspondingly hopeful. The pulse, therefore, becomes in enteric fever a symptom of the greatest importance. Those clinicians who deride the time-honored habit of pulse-study and pulse-counting, ignore one of the most valuable methods of investigation in febrile disorders. It is further true that the trained finger is an instrument of greater precision, and more useful for all purposes at the bedside, than the most ingeniously constructed sphygmograph.

While the pulse remains strong and of moderate frequency, the heart cannot be weak, nor are the most serious dangers of uncomplicated cases to be immediately feared. When the pulse becomes feeble, or its frequency for any length of time exceeds 120 beats per minute, the prognosis is in a considerable measure rendered unfavorable by this very fact. In children, or exceedingly nervous persons, or where a special tendency to pulse-frequency exists, a rapid pulse is less ominous. The frequency toward the end of the fever is often very variable; a decided fall, therefore, has less direct value in prognosis than in most other febrile affections.

From a collection of histories of cases in the hospital at Basle, in which the death-rate was about sixteen or seventeen per cent., Liebermeister found that, of those in whom the pulse reached or exceeded 120, 40 out of 63 died. Of these 63 there were 37 in whom the pulse did not rise above 140: of this number 19 died; 26 patients had pulse-frequency which rose above 140, and 21 of them died. It rose above 150 in twelve cases, and 11 of them died.

The disorders of the nervous system, which chiefly result from the long-continued high temperature, and particularly somnolence, stupor, and delirium, vary greatly in different individuals, in proportion to their ability to endure fever, rather than invariably with the height of the temperature. In general it is, however, true that those cases in which cerebral symptoms are most prominent and severe, are the most dangerous, and that the prognosis becomes less hopeful as the functions of the brain become more deeply deranged. I turn again to the invaluable statistics of Liebermeister. Of 983 cases in whom the disease ran its course in the Basle Hospital, without any specially noteworthy brain-symptoms, 34, or 3.5 per cent., died. Of 191 cases attended by slight delirium or excitement, appearing only at night or for a brief period, 38, or about 19.8 per cent., died. Of 176 attended with well-marked delirium, 96, or 54 per cent., died. Of 43 cases in which stupor or coma were present, 30, or 70 per cent., died.

Disturbances of the general nervous system, not due to the action of the fever, but which are *accidental*, such as apoplexies, local or general convulsions, or the fixed head-pains, vomiting, and local palsies arising from meningitis, are of great prognostic import, seeing that they point to special lesions of a serious character, which are not of the fever, but superadded to it. Less grave are the spinal palsies which arise as complications in the later stages of the attack or during convalescence, and which have already been described.

Before entering upon the consideration of the influence of individual peculiarities upon the prognosis, it is necessary to turn our attention to the general mortality.

The rate of mortality in private practice is undoubtedly lower than among patients treated in hospitals. This is due to various causes, among which are to be named the better previous condition of those patients who can afford the expense of treatment in private practice; the fact that, among the poorer classes, principally the more severe cases seek admission into the hospitals, and the further fact that hospital patients suffering from typhoid rarely come under treatment until the disease has made considerable progress. It is a matter of common observation that the mortality of cases neglected in the beginning of the disease is very much higher than that of those who early come under treatment.

The statistics of private practice, however, are in most instances unreliable, being commonly based upon insufficient collections of facts, and too often not altogether free from the suspicion of bias on the part of the observers, who tabulate them mostly with a view of illustrating the supposed efficacy of particular methods of treatment.

Hospital statistics are more trustworthy. They are based upon facts observed by many different physicians, and in collections sufficiently large. They are, it is true, open to the objection that for many reasons, of which the chief are given above, they indicate a death-rate somewhat too high.

The following statistics indicate the hospital death-rate of enteric fever.

The London Fever Hospital:

1. A period of twenty-one years, during which all pauper patients were received; 8,000 cases, 1,519 deaths; mortality 18.9 per cent.

2. A period of nine years, during which pauper cases have been excluded, and only patients of a better class, mainly artisans, servants, policemen, clerks, and other persons sufficiently well off to occupy private rooms, have been received; 590 cases, 80 deaths; mortality 15.9 per cent.

Cases are included in the above statistics which were admitted moribund and died within 48 hours.

The Pauper Hospital at Homerton: 1,509 cases, 255 deaths; mortality 16.8 per cent.

The Stockwell Pauper Hospital: 1,223 cases, 301 deaths; mortality 22.6 per cent.

St. George's: 387 cases, 76 deaths; mortality 19.6 per cent.

Guy's: 295 cases, 57 deaths; mortality 19.3 per cent.

University College: 163 cases, 29 deaths; mortality 17.7 per cent.

St. Bartholomew's: 635 cases, 104 deaths; mortality 16.3 per cent.

St. Thomas's: 445 cases, 70 deaths; mortality, 15.7 per cent.

Middlesex: 461 cases, 72 deaths; mortality 15.6 per cent.

King's College: 318 cases, 39 deaths; mortality 12.2 per cent.

Total number of cases, 14,125; deaths, 2,522; rate of mortality 17.8 per cent.

Dr. Murchison found that of 27,051 cases collected from various sources, and many of which have been included in the foregoing list, 4,723 proved fatal, a death-rate of 17.45 per cent.

Jaccoud, with a collection of 60,000 cases, observed a mortality of about 20 per cent.

ENTERIC OR TYPHOID FEVER.

The General Hospital of Vienna, with 17,000 cases, has a mortality of 22.5 per cent.

The Hospital at Basle, with 1,718 cases, shows a mortality of 27.3 per cent. And the principal Continental hospitals give, according to Dr. Cayley, a mortality varying from 16 to 25 per cent.

English army statistics, for six years ending 1877, are as follows: On Home Service, 545 cases, 131 deaths; mortality 24 per cent. On Foreign Service, 1,383 cases, 564 deaths; mortality 40.7 per cent.

Royal Navy: Period of six years ending 1878; 414 cases, 110 deaths; mortality 26.5 per cent.

Massachusetts General Hospital: 303 cases, 42 deaths; mortality 13.5 per cent.

Of 73 cases analyzed by Flint, 18 were fatal, or about 24 per cent.

The foregoing figures show in general the results of the expectant treatment of enteric fever, although they, without doubt, include a small proportion of cases treated upon special plans.

Age exerts an undoubted influence on the mortality of enteric fever. Murchison states that of 1,772 cases, in which the age was known, the average was 21.25 years; of the cases which recovered, 1,444, the average was 20.7 years; of those which died, 328, the average age was 23.54 years.

I have rearranged one of Dr. Murchison's tables in such a manner as to show the number of cases admitted to the London Fever Hospital in each of the decennial periods of life during twenty-three years, together with the corresponding number of deaths and the percentage of mortality.

There were, under 10 years,	616 cases,	70 deaths,	or	11.36 per cent.
Between 10 and 20 "	2,762 "	397 "	14.37 "
" 20 " 30 "	1,764 "	361 "	20.46 "
" 30 " 40 "	498 "	129 "	25.90 "
" 40 " 50 "	188 "	47 "	25.00 "
" 50 " 60 "	56 "	17 "	30.35 "
" 60 " 70 "	25 "	11 "	44.00 "
" 70 " 80 "	2 "	1 "	50.00 "
Age doubtful,	77 "	1 "	1.29 "

This table shows that, although the **death-rate of enteric fever is distinctly influenced by the age of the patient, the** extent of this influence **is much less** than in typhus. Its figures are, however, misleading, unless allowance be made for the fact that, for reasons that are obvious, a relatively smaller number of individuals sick of enteric fever find their way into hospitals in the early periods of life than among adults. In point of fact, with children the prognosis is decidedly more favorable than it is later in life. This is to be explained, in part, by the comparative infrequency of severe cases of enteric fever in childhood, the intestinal lesions being, as a rule, neither so extensive nor so marked, and complications and sequels neither so frequent nor so severe; partly by the fact that the temperature, though often attaining great heights, is less continuous on the one hand, and on the other much better borne, so that an intense febrile movement is less apt to give rise to the degenerative changes in the heart which are so frequently the immediate cause of death. After forty years of age the mortality rapidly increases. Among 1,743 cases treated in the hospital at Basle, there were 130 who were more than forty years old; of these, 39, or 30 per cent., died; while the mortality among the cases under forty amounted to only 11.8 per cent.

Sex influences the mortality of enteric fever to an insignificant extent. Most observers state that the death-rate among females is slightly greater than that among males. Murchison has shown that this excess of the mortality among females is not accounted for by the influence of child-bearing upon the course of the disease, for it is much greater between the ages of five and fifteen than between the ages of fifteen and forty-five. After the age of forty the mortality is greater among men than among women.

The season of the year does not appreciably influence the death-rate in enteric fever.

The personal habit and *the constitution* of patients are of importance. Thus, it is a well-known fact that corpulent persons do not bear the disease so well as those who are lean; they are liable to a more intense febrile movement and are less able to resist its destructive influence upon the tissues. Even debilitated, ill-nourished or anæmic individuals bear the disease better than the corpulent.

Nervous, excitable persons manifest the symptoms of the disease, and particularly those referable to the nervous system, much more decidedly than those of a phlegmatic or torpid disposition.

Those whose habits have been intemperate or who suffer from diseases of the kidney, or are gouty, are especially liable to the gravest accidents of enteric fever.

Pregnancy is regarded by most observers as a most formidable complication of enteric fever; the mother usually aborts or miscarries, and considerable hemorrhage is apt to occur. Murchison, however, looks upon it as a less serious complication than is commonly imagined. The prognosis is undoubtedly unfavorably influenced by the puerperal state.

The occurrence of the fatal accidents to which the patient is liable, by reason of the existence of the intestinal lesions, cannot be foreseen; but the percentage of deaths from these causes is small, and becomes still smaller if all but those cases properly treated from the early days of the attack, are thrown out. In this sense, it may be said that the prognosis is influenced by the treatment, and the time at which the treatment is commenced.

Death may take place by coma at the end of the second, or early in the third week, about the time of the termination of the primary fever. It much more frequently takes place by asthenia in the end of the third or during the fourth week, or at a later period. Finally, it may occur by sudden collapse in consequence of intestinal hemorrhage, perforation, or sudden failure of the heart, at any period later than the end of the second week.

Treatment.

Under this division of the subject are embraced the following topics, each of which demands separate consideration: 1. Prophylaxis. 2. The general management of the patient and dietetics. 3. Special forms of treatment. 4. The expectant or "rational" treatment. 5. The treatment of special symptoms, complications, and sequels. 6. The management of the patient during convalescence.

I. PROPHYLAXIS.

A growing knowledge of the various ways in which enteric fever is propagated, and of the habits of its exciting cause, warrants the confident belief that the disease may not only be greatly restricted in its prevalence, but even that it may be ultimately gotten rid of altogether. An efficient prophylaxis is theoretically within reach; its practical realization in communities in which the disease is endemic, depends upon the energy and steadfastness with which measures for the destruction of the poison and the prevention of its spread are carried out. What these measures are is to be directly deduced from the statements made regarding the causation of enteric fever in a previous section of this article. They belong to the subject of public hygiene, and are of sufficient importance to demand the closest attention of all local and general sanitary organizations, for enteric fever destroys more lives, that could be saved, than any other acute disease whatever. They are, nevertheless, largely within the personal control of the physicians of every community. It is the highest duty of the doctor to see to it that no new case of disease arise by direct or indirect contagion from any patient under his care. In enteric fever we have to do with a disease in which this is wholly possible. Not only may the spread of the contagion be prevented, but the poison may be wholly and absolutely destroyed; and that before it has acquired the power of infection. The remark is certainly true, that "measures of prophylaxis will be efficient in proportion to the strength of our belief in the material nature of the typhoid poison, and in the possibility of destroying it or preventing its spread." It is to be regretted that this belief has but little strength in the minds of many physicians—deeply regretted that the study of disease from an etiological standpoint occupies so little as it does the attention of most practitioners in their every-day work. The danger that a house-epidemic of enteric fever may arise from a single case suffered to become a focus of contagion, is to be constantly borne in mind. It is not house-epidemics alone that are to be prevented, but also the spread of the disease to distant points by means of the pollution of running streams, of the air, or, in cities, by continuous sewers that may convey the poison to comparatively remote localities.

The one efficient measure that includes all others is the proper treatment of the dejections.

The dejections of every case should be promptly and thoroughly disinfected. The destruction of organisms in the stools, and the arrest of their development, may be accomplished by the action of powerful chemical agents.

For this purpose solutions of carbolic acid, in the proportion of one to twenty or one to forty, or of sulphate of iron, or of chloride of zinc, are to be employed. Liebermeister uses a porcelain bed-pan, the bottom of which is strown with sulphate of iron each time before being used. Immediately after the passage, crude muriatic acid is poured over the fecal mass in considerable quantities, as much as one-third or one-half of the amount of the discharges being used. He also advises that, when practicable, the contents of the bed-pan should be emptied into trenches dug anew at short periods, and carefully filled up, care being taken that they are located at a distance from the sources of water-supply. Except in rural districts, this is of course impossible, and the dejections must be emptied into the ordinary water-closets or privy-vaults.

All bedding and articles of clothing soiled with the excreta of the patient must be immediately removed and thrown into water containing carbolic acid or chloride of zinc, and thoroughly boiled within the course of a few hours.

Search must in all instances be made for the original cause of infection, and measures taken to correct faulty arrangements which lead to the pollution of drinking-water or of the air.

II. THE GENERAL MANAGEMENT OF THE PATIENT AND DIETETICS.

The successful treatment of enteric fever is largely dependent on the attention which is given to the general management and nursing of the patient.

In the first place, it is important to see that he is not exposed to the continual action of the poison. If the original source of infection be found, upon inspection, to be connected with faulty sanitary arrangements in the house or neighborhood, it may be necessary to remove the patient to more favorable surroundings.

In hospitals, enteric fever patients are generally treated side by side with patients suffering with other diseases. This practice is unattended with the danger of infection if proper precautionary measures be taken with reference to the disinfection and **removal of the dejections, and the** cleanliness of the patient's person and bedding.

In private practice, patients commonly come under observation during the prodromic stage, or early in the first period of the disease. Not infrequently they complain merely of general malaise and evening headache. Sometimes they are under the impression that they have caught cold, but **much** more commonly they are found to be suffering from diarrhœa induced by purgative medicine taken under the impression that they are suffering from a "bilious attack." If the fever has already declared itself, the use of the thermometer will put the physician upon his guard as to the nature of the sickness, but even in those cases seen during the period of prodromes, the languid expression of the patient, his general lassitude and constitutional disturbance, are **usually greater than commonly attend the ordinary trivial ailments to which the patient attributes his symptoms, and are sufficient of themselves to excite a suspicion as to the** nature of the **disease.** Such a suspicion alone should be sufficient warrant to order **absolute rest in bed.** If the suspected case prove to be in fact a simple ailment, but little **time is lost to the patient;** if, on the other hand, his symptoms prove to have **been those of the forming stage** of enteric fever, the early rest in bed **cannot fail to favorably influence** the subsequent course of his attack. **All observers agree in the state**ment that the course of the disease is more favorable, and the mortality lower, in those cases which are treated from the beginning of the attack, than in those not coming under medical care until after the disease is fully developed. Some of the worst cases of enteric fever occur in those who have struggled against the early symptoms of the disease, and continued to go about and to perform their daily duties, until forced to betake themselves to **bed by the intensity** of the febrile movement, the urgency of diarrhœa, or by sheer weakness. There is reason to believe that the fatigue of long journeys, and particularly of railroad travelling, exerts a most injurious influence **upon the subsequent** course of the attack, in those who, after its **commencement, undertake such journeys in order to** reach home. Of primary importance, then, is absolute rest in bed **from the** beginning of the attack. The patient is not to be allowed to rise for any purpose whatever from the beginning of his sickness until some days after complete defervescence. The use of the urinal and the bed-pan must be rigidly insisted upon. Many patients declare that it is impossible for them to empty the bowel in the recumbent posture; but, after trial, they soon acquire the habit.

The room should be large and well ventilated. The temperature should be maintained uniformly at 15.5°—21° C. (60°—70° F.). When practicable, it is desirable that the patient's apartment should be heated by an open fireplace rather than by hot air, and that communicating rooms be used, between which the doors may be kept open, and from one to the other of which the bed of the patient may be from time to time transferred, a window of the unoccupied room being kept constantly open. This arrangement not only secures abundant and satisfactory ventilation, but the change also, to a considerable extent, favorably influences the mental condition of the patient. Thorough ventilation must in all cases be secured both day and night, and whilst direct drafts are to be avoided, it must be impressed upon the attendants that fever-patients are not likely to take cold.

Mental quietude is no less important than bodily repose. Visitors are not to be admitted to the patient during the course of the fever; all **business affairs and matters of annoyance are** to be withheld from him; **disturbing influences of every kind are to be avoided. Even pictures, ornaments, or articles of furniture, that especially attract his attention,** may become **causes of** disturbance, and should then **be quietly removed.** His attendants should care for his wants quietly, noiselessly, holding no conversation with him except to reply briefly to his questions; even in the early days of convalescence, the visitations of friends should be restricted in number **and length of time.** Among the minor duties of the nurse, which are, however, of not inferior importance, is the frequent changing of the position of the patient's body, moistening his mouth, cleansing his tongue, the prevention of the accumulation of sordes, and the most scrupulous care of his person in other respects. If the evacuation of the urine and fæces in the bed cannot be prevented, the discharges and soiled clothing are to be changed without loss of time. In such cases it is sometimes necessary to use two beds, the patient being rolled or lifted in the horizontal position by the attendants, from one to the other.

Fluid is to be administered without stint. The best drink is **pure** water, either of the temperature of the room, or iced. The kind of bever**age may, however, be** left to the choice of the patient. It **should be changed from time to time.** Apollinaris, seltzer, **or** other similar mineral **waters,** lemonade, iced tea with lemon-juice, wine and water, milk and water, or milk and seltzer, koumiss, thin barley-water, or water commingled with jelly, are all grateful; but the amount of fluid must be as great as, or even greater, than that consumed in health.

It often happens that patients fail to partake of the necessary amount of drink, unless it is proffered them, even when apparently fully conscious. **It is important, therefore,** that the nurse offer the patient drink at short intervals. It is often taken with eagerness, though not asked for. The **amount** at each time should, however, be moderate.

The diet should be rigidly restricted. The directions of the physician as to its kind, quantity, and the intervals at which it is to be given, must be definite and explicit. A record of the amounts given, as well as of the intervals between the administrations of food, is to be kept by the attendant and submitted at each visit. Neither general **directions nor** general reports, are sufficient. The diet throughout should be nutritious, easily digestible, and for the most part liquid. If overfed, the patient suffers from indigestion and an aggravation of the intestinal symptoms, particularly the diarrhœa; if underfed, the disturbances of nutrition increase, and convalescence will be prolonged. It is desirable to give the maximum quantity of proper food that can be assimilated, and not to exceed this amount. How much it may be can only be determined by careful study of individual cases. During the earlier stages of the dis**ease, up** to the end of the first week, it is desirable not only that the diet of the patient should be very digestible, but it is also important that it should be of only moderate amount. During this time the hyperplasia of the intestinal glands is taking place, and every possible intestinal irritant is to be avoided. Up to this time the tissues of **the body retain, in a** measure, the nutrition of health, and an alimentation restricted in quantity is free from **the** dangers that attend it later in the course of **the** attack. **After the** beginning of the second week as much food is to be given as can be properly digested.

Milk occupies the first place among fever-foods, but it is neither **to be** given indiscriminately in all **cases, nor is it** to be given in unlimited quantities. Some persons digest pure milk imperfectly and only in small amounts, while others are able to digest as much as two quarts of rich milk in the course of twenty-four hours. The first step in the digestion of milk is the formation of curd. If large, firm curds form in the stomach, **milk becomes in** fact a solid food. The formation of such curds may be in part avoided by the addition of lime-water to the milk, in the proportion of one part of the former to three or five of the latter, and by administering the milk very slowly and in small quantities at a time. Milk may be given raw or boiled, warm or frozen, or it may be coagulated into soft curds by means of rennet. Buttermilk is often an exceedingly grateful change to patients weary of milk as it is ordinarily administered in the course of fever.

Meat-broths of moderate strength, containing a little **barley and** flavored with vegetable essences, are also to be given. They may be made of beef, mutton, veal, or chicken, and should be varied from day to day. Sometimes the addition of claret **or port serves to overcome the growing** disgust which this kind of food after a time excites in certain patients. Meat-juice may also be given, hot, cold, frozen, or in the form of jelly, as the patient fancies it. Clam-soup, or oyster-soup, made by chopping the oysters and the addition of milk, may be occasionally substituted for the meat-broths. Once or twice a day, coffee or tea, well diluted with milk, may be administered. If the appetite requires it, a moderate quantity of thickened gruel, or arrow-root, or of bread and milk, may be given once a day. Some patients appear to do better with an occasional meal of such semi-solid food, but in general terms it may be said that starchy articles of diet are objectionable.

Food should be administered at intervals of two hours during the day and three during the night, the milk and broths alternating. Where the quantity taken at a time is small, or where the prostration is extreme, the intervals must be shortened. During convalescence, solid food may be gradually resumed, but the diet must for a long time be of a kind readily digested. The patient should be warned of the risk attending the eating of the seedy fruits, olives, nuts, and similar indigestible substances, for many weeks after convalescence is fairly established.

Alcoholic stimulants form no necessary part of the routine treatment of enteric fever. During the primary fever—that is to say, up to the end of the second week of the disease—their use, except to meet special indications, is probably, in most cases, injurious rather than beneficial. During the secondary fever the indications which call for their administration are twofold: of these the first is dependent upon the degree of general prostration, as particularly manifested by weakness of the heart's action, and upon the prominence of nervous symptoms. Thus, a feeble or imperceptible cardiac impulse, and a correspondingly faint, or almost inaudible, systolic sound, call for their administration; while the evidences of nervous prostration usually developed at the same time are only to be successfully combated by the abstraction of heat or the administration of stimulants, or these two combined. The second indication for the use of stimulants at this period of the disease is related to the nature of the intestinal lesion. It is now that the process of sloughing is going on; the fever of this period is due to gangrene and ulceration. In accordance with well-established principles of surgery, external gangrene, attended with septic fever and general depression, is treated by the free administration of alcohol. The fact that in typhoid fever these conditions arise in consequence of internal rather than external gangrene, cannot influence the indications for treatment.

Many cases, however, require no stimulation throughout the whole course of the attack; a smaller number demand it in the last days of convalescence; while still fewer, and these are of the most severe character, call for the alcoholic preparations in greater or less abundance from the middle of the second week to the termination of the disease, and sometimes well on into convalescence. It is useless to give alcohol in the early stage of the disease in the hope of anticipating or of preventing the occurrence of prostration and debility. It is to be prescribed, however, as soon as indicated by the severity of special symptoms or the evidences of general prostration. It is impossible to lay down any general rule as to the amount. The quantity should be only as much as is necessary to modify the symptoms for which it has been prescribed. The character of the systolic heart-sound, the pulse, and the nervous symptoms, are our best guides as to the amount and frequency of its administration. If the systolic sound grows more distinct, the pulse slower and the mind clearer under its administration, it may be continued or even cautiously increased. If, however, the action of the heart becomes more rapid, the delirium increases, or the drowsiness deepens under its use, it is to be diminished or abandoned altogether. The best effects of alcohol are, in most instances, attained by from four to eight ounces of spirits, or by from a pint to a pint and a half of sound claret, light Burgundy, or champagne, in the course of the twenty-four hours. More than this is seldom required. Whiskey or brandy may be given in the form of milk-punch or commingled with water. Under ordinary circumstances an interval of from two to four hours should intervene between each dose. In mild cases, toward the close of defervescence, when the temperature during remissions reaches for a time subnormal ranges, great benefit is often derived from the occasional administration of small amounts of alcohol in some form; sherry, either alone or as wine-whey, is eligible. If the urine be albuminous, alcohol is to be given with the utmost caution, and its effects upon the amount and character of the secretions must be carefully investigated at short intervals.

III. SPECIAL FORMS OF TREATMENT.

No medicine or method of treatment by which enteric fever can be arrested is at present known. Many different methods of treatment have been advocated, and innumerable drugs have been lauded, as exercising a special favorable influence upon the course of the disease. Bloodletting, emetics, laxatives, various astringents, turpentine, have been at different periods regarded as useful or necessary in the treatment of this disease. Most of them have no longer even a historical interest; a few are occasionally employed for special purposes. Quite recently Dr. Pepper[1] has advocated the systematic use of *nitrate of silver* in the treatment of enteric fever.

He administers this drug in doses of 0.010—0.016 gramme (gr. $\frac{1}{6}$—$\frac{1}{4}$), for an adult—usually in pill; or, for children, in solution in mucilage of acacia, three or four times daily soon after food. If constipation exist, extract of belladonna is given in combination; if there be a tendency to looseness, a small amount of opium is added. Dr. Pepper states that, in

---

[1] Remarks on some Points in the Treatment of Typhoid Fever, by William Pepper, M.D.: Philadelphia Medical Times, February 12, 1881.

a long series of cases thus treated, and in which the most scrupulous attention to every detail was observed, there has been a remarkable freedom from grave complications and a most gratifying percentage of recoveries (ninety-seven per cent.).[1] Nitrate of silver has been frequently employed, in the past, in the treatment of enteric fever,[2] but never before in the systematic manner advocated by Dr. Pepper. Its action is primarily and chiefly directed against the intestinal lesions.

The *mineral acids*, namely: muriatic, phosphoric, and sulphuric, are largely used in this country. One or another of these, abundantly diluted, and flavored with lemon-syrup, or orange-peel syrup, makes an agreeable drink; they are generally thought to exercise a favorable influence upon the course of the disease, and they should always enter, as Professor Flint suggests, into the treatment, inasmuch as they in no wise conflict with other therapeutic measures.

Calomel, at one time given because of its supposed antiphlogistic influence upon the intestinal lesion, has recently again come into use in Germany in the treatment of enteric fever. Many observers agree in the statement that, given occasionally during the first week, it not only favorably influences the course of the disease, but that it has also a tendency to shorten its duration. It may be given in 0.5 gramme doses (gr. vijss.) on alternate days during the first week; if the temperature be high, this dose may be repeated on successive days. Liebermeister gives three or four 0.5 gramme doses during the first twenty-four hours of the treatment. The diarrhœa at first increases, but soon subsides, and is afterward less troublesome. In most, but not in all cases, the first doses are reported to have been followed by a distinct but transient lowering of temperature. Moderate ptyalism occurred in some of the cases in which it was deemed necessary to repeat the dose on successive days.

Iodine in the form of potassium iodide, in doses of 1.3—4.0 grammes (gr. xx.—lx.) during the twenty-four hours, or of Lugol's solution, gtt. iij.—v., s. q. s. h., has also been thought to exercise a special favorable influence upon enteric fever. Liebermeister states that, in more than two hundred cases, iodine thus employed produced no marked effect on the course of the fever, the temperature showed no important departure from the ordinary course, the intestinal symptoms were but little modified, although in some instances they seemed to be slightly improved, the iodine eruption did not occur, and coryza appeared in only one light case. The death-rate in these cases was, however, notably lower than that in

---

[1] It is to be regretted that these cases have not been tabulated, and that Dr. Pepper's paper conveys no definite information as to their character or severity, or as to whether they occurred in hospital, or in private practice, or in both.

[2] See Murchison: The Continued Fevers of Great Britain. Second edition. London, 1873. Pp. 652, 653.

other cases **treated at** the same time, and in all respects in the same manner, with the exception of the use of the iodine. The following table shows the percentage of mortality, those cases not being included which proved fatal within six days after their admission to the hospital:

Cases.	No.	Died.	Percentage of Mortality.
Treated non-specifically	335	47	13.2
Treated with calomel	216	19	8.8
Treated with iodine	229	25	10.9

In this country Professor Bartholow has used, apparently with decided success, the following modification of the iodine treatment:

R. Tinct. iodinii.................. 8.00 c.c.   fl. ℨ ij.
     Acid. carbolic.................. 4.00 c.c.   fl. ℨ j.
M. Sig.—1 to 3 drops three times a day.

The antipyretic treatment consists of the systematic employment of measures to reduce the **temperature of the body**. In view of **the fact that by far the greater number of fatal cases of typhoid fever die from** the direct or indirect effects of the prolonged high temperature, this plan of treatment has much to recommend it upon theoretical **grounds**. The concurrent testimony of those observers who have applied it systematically to large numbers of cases points **to substantial practical results**, both in mitigating the severity of the symptoms, and in notably reducing **the** mortality. The principle upon **which** it is based is by no means new. From the earliest days of medicine the reduction of the temperature has **at all times** been looked upon as one of **the most** important indications in the treatment of fever. The main point **in** the management of enteric fever is to control the temperature. The measures by which this can be accomplished are hydrotherapy, quinine, the salicylates, and digitalis. These are capable of depressing the temperature for a more or less extended period; their systematic employment in such a manner as to control the febrile movement throughout the attack constitutes what is technically known as the *antipyretic treatment*.

The cold-water treatment was first systematized by Dr. James Currie, of Liverpool (1797), who used it in febrile affections, according to certain clear indications.

He employed, as a rule, cold affusions, frequently repeated, and occasionally cold baths. His method was adopted by many physicians, and soon came into extensive use both in England and on the continent, in **the treatment** of many febrile **affections, and** especially in the management of typhus and typhoid fever, and scarlatina. It gradually fell into neglect and was for a long time almost forgotten. The cold-water treatment of fevers was revived by Dr. Ernst Brand, of Stettin (1868), and rapidly came into use in Germany, Austria, and Switzerland.

The methods of hydrotherapy are various. Cold water may be so applied as to reduce the temperature, by means of the cold bath, the graduated bath, cold affusions, cold packing, cold compresses, and cold sponging. These methods vary in their effects, and different methods are applicable to special cases; but that one of them is to be preferred by which the desired end is reached with the least inconvenience to the patient.

The cold bath is, for general use, not only the most effective, but it is also the least troublesome to apply. The following is the plan employed by Liebermeister at Basle; it differs but little from that generally in vogue elsewhere upon the continent, and that now practised by Dr. Cayley in London. I am not aware that the treatment of enteric fever by cold bathing has been practised with the same degree of system and vigor, and upon an extended scale, by any observer in America.

"For adult patients the full-length cold bath, of 20° C. (68° F.) is to be prepared. The same water can be used for several successive baths for the same patient; the bath-tub remains standing full, and the water, representing about the temperature of the room, answers the purpose without change. The duration of the bath should be about ten minutes. If prolonged much beyond that, it becomes unpleasant to the patient, and may even prove injurious to him. If feeble persons are much affected by the bath, remaining cold and collapsed for a long time, the duration should be reduced to seven or even to five minutes. A short, cold bath like this, will have a much better effect than a longer one of lukewarm water. Immediately after the bath the patient should have rest; he is, therefore, to be wrapped up in a dry sheet and put to bed. (The bed may with advantage be warmed, especially at the foot.) He should be lightly covered and given a glass of wine. With very feeble patients it is well to begin with baths of a higher temperature, say 24° C. (75° F.); but a less decided effect will follow. In such cases the method of Ziemssen is to be especially recommended, if the surroundings permit. A bath of 35° C. (95° F.) is at first employed; cold water is gradually added until the temperature of the bath is reduced to 22.2° C. (72° F.), or below. These baths should be of longer duration."

In severe cases the temperature is taken every two hours, day and night. As soon as 39.5° C. (103.1° F.) in the rectum, or 39° C. (102.2° F.) in the axilla, is reached, the bath is given. Individual peculiarities are to be regarded. It may be advisable to give the bath before the temperature runs quite up to the heights above mentioned, or to give a bath of shorter duration, or of warmer temperature, or the gradually reduced bath of Ziemssen.

The aim of this plan of treatment is to keep the temperature during the whole course of the disease within the bounds of a moderate fever heat. This cannot be accomplished by one bath or by a few baths. If the treatment be systematically carried out, from four to eight baths in the course of twenty-four hours will in ordinary cases be necessary. In very severe cases Liebermeister has repeated the baths every two hours, so that twelve baths have been given every twenty-four hours, and in some instances the number of baths required by a patient during his **entire** illness has exceeded 200. Each bath ought to cause a reduction of temperature of 1°—1.66° C. (2°—3° F.). If the temperature be not modified to this extent, the following baths should be colder or longer. It is not necessary to take the temperature in the bath, for the reason that it continues to fall for some time afterward; it should be taken about half an hour after the removal of the patient from the bath. In children the baths may be made warmer and of shorter duration. In cases marked by great nervous depression with only a moderate elevation of temperature, cold baths of short duration, or cold affusions, are recommended for their stimulating effect on the nervous system.

The graduated bath is particularly useful in the treatment of children, and where the cold baths are inadmissible, as in aged persons, or those suffering from disease of the heart **or lungs, and in cases of extreme prostration.**

Cold affusion is regarded by **the advocates of the antipyretic treatment** as of inferior **value in reducing temperature.** It is, however, more pleasant to the patient, and may be employed in cases where baths are inadmissible, or where a stimulating **effect upon the nervous system** is desired.

Cold packs are also inferior to **bathing** as a means of reducing temperature. They are usually well borne even by feeble patients, and are particularly applicable in the treatment of children, to whom they may take the place of baths. They are very troublesome to apply. The bed being protected by a gum cloth, the patient is thoroughly wrapped in a sheet wrung out of cold water, the face and feet alone being left free; he is then lightly covered with a blanket. A course of four consecutive packs, of ten to twenty minutes' duration each, is said to be about equivalent in its effect upon the temperature to a single cold bath of **ten** minutes.

Cold compresses give rise to local lowering of temperature, but **have** no great influence on the general heat of the body.

Cold sponging has but little influence upon the internal temperature; it therefore cannot be regarded as entering into the antipyretic treatment, properly so called. It is **useful for purposes of cleanliness, and is** in most instances grateful to the patient. I am in the habit of ordering my patients sponged two or three times daily with water containing aromatic vinegar.

Among the more important contraindications to the antipyretic treatment, and in particular to the cold baths, are hemorrhage from the bowels, great feebleness of the circulation, and coldness of the extremities and surface of the body, with high internal heat. This method of treatment is also inadmissible in subjects advanced in years, and in those suffering from chronic bronchitis, pulmonary emphysema, and organic disease of the heart. Dr. Cayley does not regard albuminuria as a contraindication.

Among the *medicines* capable of reducing the temperature of the body, *quinine* occupies the first place. In order to secure its full effect it must be given in large doses. It is useless to give small doses at considerable intervals. From 1.3 to 2.6 grammes (gr. xx.—xl.) are necessary to produce a decided fall of temperature in an adult. This amount should be administered within the space of an hour, 0.5 gramme (gr. vijss.) being given every ten minutes until the full dose is taken. A decline of 1.6°—2.2° C. (3°—4° F.) usually follows in the course of from six to twelve hours. As the effects of the medicine pass off, the temperature gradually rises again, but does not usually attain its original height until the expiration of twenty-four hours. It is best given at night, some time after the evening exacerbation has reached its height, as the effects are more marked upon a falling than upon a rising temperature. It may be administered in powder or in solution, and should be followed by small amounts of hot broth. If vomiting occur, quinine may be administered in small enemata along with opium. Symptoms of cinchonism usually follow, but they are less marked than after similar doses in afebrile diseases or in health. Among the more constant effects of large doses of quinine is profuse sweating.

The salicylates, given in large doses—4.0—6.9 grammes (gr. lx.—cv.) in the course of twenty-four hours—rapidly and powerfully depress the temperature. *Sodium salicylate* has come largely into use in the treatment of typhoid fever in Germany. Its administration in large doses is sometimes followed by gastric disturbances, increase of diarrhœa, and a tendency to hemorrhage. It also appears to exert an unfavorable influence upon the kidneys, occasionally manifested by an increased tendency to albuminuria. The chief objection to this medicine relates to its depressing effects upon the circulation.

Digitalis, administered in full doses, is also capable of depressing the temperature in typhoid fever. For this purpose 0.666—1.3 grammes (gr. x.—xx.) are recommended to be given in divided doses extended over a period of about thirty-six hours, and followed by a full antipyretic dose of quinine. By this procedure a complete intermission can be produced, even in severe and obstinate cases, where quinine alone has but little effect upon the temperature. Digitalis, both in substance and in the form of the infusion, is often badly borne by the stomach; it is inadmissible where the action of the heart is feeble, the rule for its administration in enteric fever being exactly opposite to that which regulates its use in the treatment of organic diseases of the heart.

The advocates of the antipyretic treatment of enteric fever claim that under its use not only is the mortality greatly reduced, but that, to use the words of Liebermeister, the entire appearance and bearing of patients is such that the old picture of a typhoid fever patient is no longer to be seen, and that the disease has in fact lost a great part of its terrors. This observer informs us that, in the hospital at Basle, there were treated upon the expectant plan, between 1843 and 1864, 1,718 cases of typhoid fever; of these 469, or 27.3 per cent., proved fatal. From 1865 to September, 1866, there were treated, under an incomplete antipyretic plan, 982 patients; of these 159, or 16.2 per cent., died. Between September, 1866, and 1872, there were treated, by the antipyretic plan systematically carried out, 1,121 cases; of these 92, or 8.2 per cent., died. After the elimination of certain errors in these statistics, he concludes that the mortality under the antipyretic treatment is ten or eleven per cent. against a mortality of twenty-five or thirty per cent. under the expectant plan. In the hospital at Kiel, the mortality under the antipyretic plan, as pursued by Jurgensen, was 3.1 per cent.; that under the expectant plan, between the years 1850 and 1861, was 15.4 per cent. In the military hospital at Stettin, the mortality under the antipyretic plan was 4 per cent.; under the expectant plan, 25.6 per cent.

Dr. Brand found that, of 8,141 cases treated antipyretically, 600 died, making a mortality of 7.4 per cent.

In by far the greater number of enteric fever cases, as the disease is known to American physicians, the systematic antipyretic treatment, by means of cold baths, is clearly unnecessary by reason of the mildness of the pyrexia; in many others it is clearly inadmissible, and in all cases it is difficult of application, requiring a **degree of** attention and a number of trained assistants not always available in hospitals, scarcely ever to be secured in private practice. To these causes is doubtless largely due the fact that it has not come into use to any considerable extent in this country. Prejudice in the minds of the people, and perhaps also among medical men, contributes to the opposition to this method of treatment. Even the suggestion of a modified antipyretic treatment, necessary to save life, too often encounters the decided opposition of the friends of the patient, who look upon cold compresses, the pack, or the douche, as adding to the horrors of the situation. Moreover, those physicians who are favorably impressed with the accounts of this treatment and its results, enter into half-way measures at a late period of the disease, without the energy and enthusiasm necessary to the realization of its best effects.

With reference to the reduction of temperature by means of drugs, and especially by means of large doses of quinine, the way is clearer, and this practice is growing in favor in America. For my own part, I look upon large doses of quinine, at intervals of forty-eight to seventy-two hours, as an essential part of the management of all cases in which the evening temperature rises above 40° C. (104° F.).

IV. THE EXPECTANT TREATMENT.

The expectant or rational treatment of enteric fever is that generally employed at the present time. Notwithstanding the diminished mortality following the employment of the antipyretic treatment in Germany, it has never been generally introduced in France, Great Britain, or the United States, and the physicians of these countries for the most part still adhere to the expectant or the modified expectant plan. This method of treatment is based upon the knowledge that enteric fever, like the other acute infectious diseases, is of definite duration and cannot be cut short, that is to say, cured, by therapeutic measures. The patient, once having become the subject of the infection, must pass through the successive stages of the fever before he regains his health. If then life can be maintained for a definite time and no serious complication or sequel remains, recovery will take place. The patient is to be carefully watched, he is to be placed under the most favorable hygienic conditions, disturbing and injurious influences are to be prevented or removed, and efforts are to be made to combat unfavorable symptoms and to avert complications. The successful management of enteric fever upon this plan presupposes on the part of the physician an intimate knowledge of the course of the disease, of the relative importance of the symptoms, of the order of their appearance and their duration, and a familiarity with the anatomical lesions, the connection between the lesions and symptoms, and the complications that are likely to arise.

Absolute rest in bed, intelligent and careful nursing, a restricted diet, cleanliness of the person and the bedding, and ventilation, form the basis of the treatment. "If," in the words of Jenner, the most able, as well as the most recent advocate of this method, "medicinal in addition to hygienic treatment is required, it is because special symptoms by their severity tend directly or indirectly to give an unfavorable course to the disease. . . . . My experience has impressed on me the conviction that that man will be the most successful in treating typhoid fever who watches its progress, not only with the most skilled and intelligent, but also with the most constant care, and gives *unceasing attention to little things*, and who, when prescribing an active remedy, weighs with the greatest accuracy the good intended to be effected against the evil the prescription may inflict, and then, if the possible evil be death, and the probable good short of the saving of life, holds his hand."

The special symptoms that are apt to give an unfavorable course to the disease are to be treated for the most part in accordance with the general principles of therapeutics. Some of the symptoms, complications, and sequels are best managed in accordance with the following rules of practice:

V. THE TREATMENT OF SPECIAL SYMPTOMS, COMPLICATIONS, AND SEQUELS.

Headache occasionally causes the patient considerable distress in the early days of the attack. It generally requires no special treatment, and subsides spontaneously about the middle of the second week of the disease. Absolute quiet, darkening of the room, and local applications, sometimes cold, sometimes warm, are, as a rule, all that is necessary to control it.

Sleeplessness is occasionally an important symptom in the early stages of the disease. Like the headache, it commonly disappears or diminishes, without special treatment, some time during the course of the second week. This, however, is not always the case. Sleeplessness is occasionally persistent and exhausting. It then becomes necessary to treat it. During the primary fever *potassium bromide* and *chloral* yield the most satisfactory results. They may be used either in combination or separately. In the personal experience of the writer, chloral alone, in moderate doses, has proved adequate to overcome this symptom in most cases, and its administration has been unattended by cardiac depression or other unfavorable effects.

If other hypnotics fail, opium in sufficient doses will secure sleep. This drug and its preparations, in doses sufficient to induce sleep, must be regarded as objectionable during the early stages of the disease, on account of its unfavorable influence upon digestion and the secretions—an influence not wholly obviated by the hypodermic use of morphia. After the middle of the second week, that is to say, during the secondary fever, opium becomes at once our most efficient and safest means of controlling prolonged sleeplessness and excitability, and its use in fever dependent upon gangrene and sloughing is in accordance with well-established principles of surgery. In the later stages of the disease, chloral is, by reason of its depressing influence upon the circulation, even more objectionable than is opium in the early stages.

Somnolence, stupor, and *delirium* are to be treated by stimulants and the abstraction of bodily heat. In the treatment of these symptoms, *alcohol* stands first and almost alone among the stimulants; *spirits of chloroform* and *camphor* are of use; *ammonium carbonate* is of inferior value, and has been objected to on theoretical grounds, as being liable to increase the alkalinity of the blood. It is frequently used in the treatment of pulmonary complications. If delirium continue or coma threaten, great benefit is often derived from the local application of cold to the head, by means of either the cold douche, or an *ice-cap*. If the brain-symptoms are specially severe, the head may be shaved, and *blisters* may be applied to the **nape of the neck or to the temples; these measures** are of doubtful value, and are only to be resorted to in desperate cases. The lighter forms of **disturbance of the functions of the brain,** as somnolence and transient delirium, do not call for special measures of treatment. They are often relieved, to some extent, by coffee.

Tremor is an important symptom. It indicates extreme prostration. Sir William Jenner has called attention to the fact that tremor, out of all **proportion** to the other signs of nervous prostration, is to be looked upon as a sign of deep ulceration of the intestines. A small, deep slough, the separation of which is especially liable to give rise to intestinal hemor**rhage or perforation, will** often occasion great tremor. Tremor of this kind is to be treated with full doses of *alcohol* and *opium*, not only for **their general effect upon the nervous system, but also with a view to their** local effects in limiting sloughing and ulceration.

Dryness of the tongue, and the *accumulation of sordes* upon the teeth and gums, are to be obviated by the frequent administration of fluids or by pieces of ice allowed to dissolve in the mouth. The patient, if **able** to do so, should rinse his mouth frequently with pure water, or water containing small quantities of claret, aromatic vinegar, or tincture of myrrh.

Diarrhœa, so long as the stools **are of moderate amount** and do not **exceed in** number three or four in the course of twenty-four hours, does **not call for special treatment.** If, however, the **passages are** copious or **very frequent, the strength of** the patient is endangered, and it becomes **necessary to control them.** Sometimes diarrhœa is due to errors in diet, such as the use of solid food, or of excessive amounts of food, particularly milk and the strong animal broths, and abates upon the correction of such errors. It may arise in consequence of the patient's drinking excessive amounts of fluid, which passes through the bowel without being absorbed, and stimulates excessive secretion from the intestinal mucous **membrane** (Jenner). In the absence of these causes, diarrhœa is to be **attributed to** catarrhal inflammation of the intestinal mucous membrane. It is best treated by *bismuth carbonate* or *subnitrate*, in large doses, 1.3 gramme (gr. xx.) s. q. quartâ vel sextâ horâ. To these powders may be added, if necessary, *opium* in 0.01—0.016 gramme (gr. $\frac{1}{6}$—$\frac{1}{4}$) doses, or *deodorized laudanum* in doses of from three to five drops. Other astringents, such as *alum, sugar of lead, nitrate of silver, tannin, catechu,* and *kino*, either alone or in combination with *opium,* are recommended for the control of the diarrhœa. It is more satisfactory at the bedside to use one or two efficient remedies, than to resort to a number of uncertain drugs; and in bismuth freely given, or in opium in repeated small doses, either by the mouth or by enema, or in these two remedies combined, will be found, in almost all cases, an efficient medication against excessive diarrhœa in enteric fever.

If the stools be fetid or highly ammoniacal, Jenner recommends the occasional administration of a teaspoonful of *charcoal*—animal charcoal being preferred, and care being taken that it is in **impalpable powder**. Creosote and carbolic acid are also of service.

Constipation occasionally occurs. If it be but slight, it is often due to the absence of extensive intestinal lesions and the catarrhal inflammation with which such **lesions are associated**. Hence, in mild cases slight constipation requires no treatment beyond the occasional administration of small doses of *calomel* or *castor-oil*, or the *juice of an orange*. Constipation may, however, be due to torpidity of the large intestine, the **fecal matter being** retained for a long time and the stools being hard and dry. Under these circumstances a sort of secondary diarrhœa, due to irritation of the lower bowel by the retained fecal matter, may arise. This form of diarrhœa is attended with a feeling of local distress and tenesmus, which are unusual in enteric fever, and will be promptly relieved by the removal of its cause. Prolonged constipation is by no **means to be taken** as an indication of moderate intestinal lesions; on the **contrary,** deep ulceration of one **or more of** Peyer's patches is not only frequently associated with constipation, but, by its **paralyzing influence** upon the intestine, it is very often the **cause of constipation**. Aperients administered by the mouth **are therefore to be shunned**, lest by inducing peristalsis they forcibly detach **a deep slough**, or otherwise mechanically give rise to perforation where the sloughing extends to, or implicates the serous coat of the intestine. Large enemata are also attended with danger arising from their liability to set up energetic peristaltic movements, which may extend to the lower part of the ileum. The **constipation of** enteric fever is most safely and satisfactorily treated by the daily administration of small enemata of strong, warm **soap-suds** or of thin gruel.

Tympany is present to a greater **or less extent in** almost all cases. It may be due to deficient power of expulsion, or to an undue generation of gas in the intestine, and reaches its maximum, as a general rule, during the latter part of the third, or in the fourth week of the fever; for at this period the causes that produce it are fully developed. These causes are: **first,** sloughing and ulceration of the intestine, which in itself, if deep, is **sufficient** to cause paralysis; **second,** general prostration leading to deficient contraction alike of the intestinal walls and of the abdominal muscles; **and** third, alteration in **the character of the digestive fluids, which, no longer** possessing the antiseptic properties of health, permit the speedy decomposition of the intestinal contents. Flatus accumulates in part in the small intestine, but chiefly in the colon; it varies from an amount scarcely greater than that of health to enormous abdominal distention, interfering with the play of the diaphragm, and, by the outward pressure of the accumulated gas within the gut, adding to the danger of perforation. The indications for the treatment of this symptom are twofold; the first have reference to the loss of nerve-energy, and call for increased stimulation. The second have reference to the nature of the food, and the arrest of the gas-generating decomposition of the intestinal contents. Thus, *alcohol* is to be given, or, if already employed, the amount is to be increased. *Turpentine, camphor,* and minute doses of *opium* may be added to the treatment; the abdomen should be gently rubbed with the hand alone, or with turpentine, at short intervals, or turpentine stupes may be applied. *Charcoal* is to be administered with a view of preventing decomposition of the intestinal contents, and only such food is to be given as will probably leave little or no residue to undergo decomposition in the intestine. At the same time *pepsin* is to be administered along with the *mineral acids*. If the amount of flatus in the large intestine be excessive, paralysis from over-distention may arise. It may then become necessary to carefully introduce into the bowel an œsophageal tube with a view of mechanically removing a portion at least of the accumulated gas.

If constipation coexist with tympany it is to be relieved by the administration of small enemata, such as have been described above, or with the addition of *turpentine*, once or twice a day. Suddenly developing tympany is sometimes a symptom of peritonitis.

Intestinal hemorrhage, if it be slight, does not call for other measures of treatment than the most absolute rest of the patient, the restriction of his diet to substances capable of being most readily digested and absorbed in the stomach and upper intestine, such as essence of meat in small doses, wine-whey, koumiss, etc., and opium in moderate doses, either by the mouth or by enemata. Food and drink are to be iced, and lumps of ice held in the mouth and swallowed. The action of the bowels is to be as far as possible controlled.

If the loss of blood be profuse, the danger becomes imminent, and more active measures are to be promptly resorted to. In addition to *opium*, the remedies to be mainly relied upon are gallic acid, turpentine, and ergot.

Murchison states that in his practice the following mixture was, during many years, almost invariably successful for arresting the bleeding:

℞. Acid. tannic .... 0.66 grm.   gr. x.
   Tinct. opii.................... 0.66 c.c.   ℳ x.
   Spirit. terebinth.............. 0.99   ℳ xv.
   Mucilag........................ 8.00   ℨ ij.
   Tinct. chloroform.............. 1.33   ℳ xx.
   Aq. menth. pip..........ad.   32   ℥ j.
   M. ft. haust. s. q. s. h.

Ergotine may be injected hypodermically in doses of 0.66 grm. (gr. x.) at intervals of half an hour or an hour, until the evidences of bleeding cease. An ice-bag or bladder, filled with broken ice mixed with bran, is to be applied to the abdomen. It is not to be hoped that any direct local effect upon the intestinal lesions will follow the use of the astringent preparations of iron either by the mouth or by the rectum.

Peritonitis, whether due to perforation of the intestine or to other causes, calls for the **free administration of opium**. To an adult, as much as 0.133 gramme (gr. ij.) may be given at once, followed by half that amount every second or third hour until moderate stupor is produced. For at least a time no nourishment, excepting concentrated meat-juices, a spoonful at a time, and brandy and water in not larger amounts, is to be administered. The abdomen may be smeared with a mixture of equal parts of sweet oil, laudanum, and turpentine, or warm fomentations or turpentine stupes may be applied to it. Better than these, however, is the application of large, thinly spread mush or flaxseed poultices well smeared with lard. The Germans recommend ice-bags and ice-poultices.

If opium be not well borne by the stomach, morphia is to be administered hypodermically. Should the patient's life be prolonged, it is of the utmost importance that the bowels be confined as long as it is possible to keep them so. In most cases a movement will take place at the end of several days, even under the continued use of opium; otherwise, after all evidences of peritonitis have subsided, small, lukewarm enemata may be cautiously employed.

In enteric fever palpation of the abdomen is to be practised with great caution, on account of the danger of exciting peritonitis, of causing perforation, or of rupturing the spleen.

The suprapubic region is to be examined by palpation and percussion twice daily as a matter of routine, and whenever necessary the catheter is to be employed.

Frequent exploration of the chest by the methods of physical diagnosis is necessary; complications capable of determining a fatal result may be arrested by the prompt detection and treatment of pulmonary lesions attended by insignificant subjective symptoms.

Hypostatic congestion is to be prevented by guarding against the heart-failure to which it is chiefly due. The control of temperature and the use of stimulants constitute the most important means to this end. **Digitalis** is a dangerous remedy in the feebleness of the heart due to the acute granular degeneration occurring in the continued fevers, and is to be administered with great caution. The patient's position is to be changed from time to time, with a view of preventing hypostasis, and he is to be instructed to occasionally take three or four deep inspirations. If congestion occur, the occasional application of turpentine stupes to the chest is of great advantage.

Bed-sores are to be prevented by frequent change of position, and the removal of pressure by means of cold-water bags or air-cushions. Scrupulous cleanliness and care with regard to the bed are important. So long as the skin is sound, the parts especially subjected to pressure, and therefore liable to gangrene, are to be frequently bathed with equal parts of alcohol and lead-water. If erosions appear they are to be treated in accordance with general surgical principles. Bartholow regards a mixture of equal parts of copaiba and castor-oil as the best dressing for a bed-sore.

Other complications and sequels are to be treated in accordance with general therapeutic indications.

VI. THE MANAGEMENT OF THE PATIENT DURING CONVALESCENCE.

During the early days of convalescence the temperature remains labile, and abrupt recrudescences of the fever are apt to arise from slight causes. It is therefore important that the patient be cared for assiduously for some time after defervescence is complete. For at least a week, morning and evening temperature observations should be taken; and during this time the diet is to be restricted to milk, eggs, custards, farinaceous foods, light puddings, and animal broths or jellies. The visits of friends are to be limited both in number and duration. Undue exertion, even within the limits of the chamber, is to be carefully guarded against, and all conversation upon business affairs, or other matters liable to give rise to excitement or to depressing emotions, is to be avoided. At the end of a week, solid food, and particularly meat, may be resumed; but the effect of such changes of diet upon the temperature and general condition of the patient is to be carefully watched.

The liability to intestinal hemorrhage, perforation, or a relapse, are to be constantly borne in mind, and for a long time the patient's diet is to be restricted to articles of a readily digestible character. If diarrhœa persist, it is to be treated by bismuth and small doses of opium, either alone or combined with the mineral acids; if there be a tendency to constipation, simple enemata may be employed for its relief. Laxative medicines, with the exception of castor-oil in small doses, are inadmissible. Milk-punch, egg-nogg, and wine, are often of service during convalescence; but, in the case of young persons, or of those not in the habit of using alcoholic beverages previous to their sickness, it is important to wholly dispense with alcohol as early as possible. Quinine, iron, and cod-oil, are to be employed if the convalescence be tardy and anæmia persist. A brief sojourn at the sea-shore is not less agreeable than useful; the patient gladly escapes from the apartment which has been the scene of his tedious illness, and finds change of air and of scene invigorating alike to body and mind.

V.

TYPHUS FEVER.

DEFINITION.—A specific febrile disease of from ten to twenty-one—usually fourteen—days' duration, highly contagious, arising under circumstances of general destitution and overcrowding, and prevailing in more or less extensive epidemics. It is characterized by sudden invasion; great and early prostration; a dull, flushed face; injected conjunctivæ; wakefulness, with mental torpor and confusion, passing at the end of the first week into delirium, which may be active and noisy, but is commonly low and wandering; stupor tending to coma; tremors and involuntary evacuations; a furred tongue, soon becoming dry and brown; in most instances, constipation; a copious rash appearing between the middle and the end of the first week, disappearing upon pressure at first, but speedily becoming persistent, and often associated with petechiæ. After death no specific lesion; the blood is broken down, the heart and voluntary muscles are degenerated and softened, the internal organs hyperæmic.

SYNONYMS.—True Typhus:

Febris pestilens; Parish Infection; Infectious Fever; **Pestilential Fever**; Der ansteckende Typhus; Typhus contagieux; **Tifo contagioso**; Contagious Fever; Contagious Typhus.

Febris epidemica; Epidemic Fever.

Morbus pulicaris; Febris purpurea epidemia; Febris stigmata; Febris petechialis; Typhus exanthematicus; La pourpre; Fleckfieber; Das Fleckenfieber; Das exanthematische Nervenfieber; Febbre petecchiale; Spotted Fever; Petechial Fever; Petechial Typhus; Typho-rubeoloid.

Typhus comatosus; Brain **Fever**.

Febris asthenica; Fièvre **ataxique**; **Fièvre adynamique**; **Adynamic Fever**.

Febris putrida et maligna; Synochus putris; Febris maligna pestilens; Febris continua putrida; Fièvre putride et maligne; Faulfieber; Febbre putrida; Putrid Malignant Fever; Putrid Continual Fever.

Pestis bellica; Typhus bellicus; **Morbus castrensis**; Febris militaris; Typhus castrensis; Typhus des camps et des armées; Die Kreigspest; Camp Fever.

Typhus carcerum; Febris carceraria; Maladie des prisons; Jayle Fever; Jail Distemper.

Fièvre des hôpitaux; Malignant Hospital Fever.

Febris nautica; Ship Fever; Infectious Ship Fever; Ochlotic Fever.

Catarrhal Typhus; Irish Ague.

The foregoing are some of the many names by which the fever under consideration has been known and described. For a more complete list the reader is referred to the pages of Murchison.[1] They are variously derived from the contagious character of the fever, its prevalence in epidemics, the eruptions, the presence of cerebral symptoms, the adynamia which attends it, its supposed putrid character, and its malignancy; or from its prevalence in armies, in camps, in hospitals, in prisons, in ships. Ochlotic ($\ddot{o}\chi\vartheta o\varsigma$, a crowd) is an adjective of modern application, derived from the supposed mode of origin of the fever in overcrowding.

Typhus ($\tau\dot{v}\phi o\varsigma$, smoke), used by Hippocrates to define a confused state of the mind, with a tendency to stupor, expresses a prominent condition of the disease. It was first used to designate certain forms of continued fever by Sauvages in 1760. Within the last forty years it has been employed by English writers in a more restricted sense, to designate the particular specific fever which is the subject of the present article. Among continental writers it is still adapted to a vaguely defined group of the continued fevers.

HISTORICAL SKETCH.

According to Hirsch,[2] it must remain uncertain whether the pestilence prevailing in Athens at the time of the Peloponnesian war, and described by Thucydides, was typhus fever or not. Equally uncertain is the nature of the numerous epidemics of contagious fever which occurred in different parts of Europe during the first fifteen centuries of the Christian era, many of which have been supposed by some authors to have been typhus fever. The descriptions of the historians and of the physicians who chronicled them, are alike wanting in precision. The first satisfactory account of typhus dates from the year 1501, when, according to Fracastorius, it

---

[1] The Continued Fevers of Great Britain.
[2] Handbuch der hist.-geograph. Pathologie.

spread from Cyprus into Italy as a new, unheard-of, and, to the Italian physicians, altogether unknown disease. For more than twenty years it prevailed in Italy. If, says Hirsch, we may place confidence in the physicians and historians of that period in the different countries of Europe, we are compelled to believe that, in the beginning of the sixteenth century, typhus fever (Exanthematische Typhus) had for the first time attained general prevalence over the continent. By the middle of this century it had become, in connection with the plague, the predominant form of epidemic disease. At that time the movements of armies and military enterprises contributed, as in fact they have at all periods, greatly to its development and extension; but during that and the two following centuries it appears also as the abiding form of continued fever in every country of Europe, in all states of society, and as playing, under many different names, the most prominent part among epidemic diseases. As Murchison well says: " A complete history of typhus would be the history of Europe for the last three and a half centuries." Consult his work for a very full and satisfactory account of the epidemics that have been the subjects of general and medical history, and for an exhaustive bibliography of the whole subject. The brief outline of the following pages is based principally upon the works of the authors already named—Hirsch and Murchison.

In the years 1550–54, a petechial fever prevailed in Tuscany and destroyed upwards of 100,000 persons. In 1557, typhus was widely prevalent in France. It again prevailed in that country some years later, in connection with the plague.

In 1566, typhus appeared in Hungary, in the army of Maximilian II., and spread over all Europe.

In 1580, an epidemic of typhus arose in Verona. The historian of this epidemic, Petrus a Castro, states that the fever was called "La pourpre" by the French, "Tabardiglio" by the Spaniards, "Petecchie" by the Italians, and "Fleckfieber" by the Germans. Bleeding, both general and local, was recommended at the beginning of the disease, but in the later stages it was regarded as dangerous. This epidemic spread over Italy.

In 1591, famine prevailed in Italy, and at the same time a contagious fever fell upon the people far and wide. The symptoms were the same as those described as attending the epidemic beginning in Verona eleven years before.

A similar fever prevailed in Holland in the latter part of the sixteenth century.

During the thirty years' war (1609–1638) all Europe was devastated by famine and by a contagious fever, which, from the descriptions of various observers, was beyond doubt typhus.

The plague appeared in Leyden and elsewhere in Holland, in 1635, and again in 1669, and on each of those visits it was preceded and followed by a contagious "spotted fever."

About the year 1700, F. Hoffman, professor of medicine at Halle, published a very accurate description of typhus which he had seen among the German troops in 1683. He described the disease under the name of "*Febris Petechialis Vera.*" He advised acid medicines, nourishing food, and regarded nothing better than wine. Under the name of *febris pestilens*, applied by the authors who preceded him to typhus, he described the plague.

From 1757 to 1759, typhus prevailed in Vienna. It was, for the most part, prevalent in overcrowded localities.

About the same time (1757–58) occurred the first epidemic of typhus in Berlin of which any authentic record exists. Its origin was traced to overcrowding, with deficient ventilation and scarcity of food.

In 1764, a dreadful epidemic of typhus and dysentery prevailed at Naples. There was, at that time, great scarcity of provisions, and the poorer classes suffered from starvation. The people from the surrounding country flocked into the city, and their overcrowding and misery were beyond description. The disease raged principally among the poor.

An epidemic of typhus occurred in 1797–1800, at Genoa, at that time besieged by the French. It broke out when the garrison was half-famished.

With the wars which, during the first fifteen years of this century, swept over almost every part of Europe, typhus anew became generally epidemic upon the Continent. It prevailed in the contending armies and among the inhabitants of the countries that were the seat of war, and, arising invariably under circumstances of want and wretchedness, it was especially frequent and fatal among the inhabitants and garrisons of besieged cities.

In 1816–17, true typhus was epidemic in Italy.

Since the peace of 1815, typhus has frequently occurred, in limited or extended epidemics, in different parts of Europe.

The Baltic provinces of Russia and Poland have often suffered from it; Northern and Middle Germany have been frequently infected. In Silesia, wide-spread epidemics have raged on several occasions, and particularly in 1847–48, 1856–57, 1868–69—the last being likewise typhus years in East and West Prussia, and in the Prussia of Posen (Lebert). Sporadic cases of typhus occur in the large cities of Germany almost every year, and sporadic cases or isolated epidemics have also in late years been observed in Sweden, Denmark, Holland, and Belgium.

Northern Italy, of old and in recent years, has been a typhus-centre. The fever spreads thence to Middle and sometimes even to Southern Italy on the one hand, and on the other it crosses the Alps, following the lines of travel into Switzerland.

The statement that typhus does not occur in France is not true. Murchison has collected evidence to prove that it prevailed at Beaulieu in 1827, at Toulon on many occasions between 1820 and 1856, at Rheims in 1839, at Strasburg in 1854.

In the winter of 1854–55, it made its appearance among the English and French troops in the Crimea; but its prevalence was slight compared with that of the following winter, when it was mainly confined to the French and Russian armies. During the first six months of 1856, it is estimated that, out of a force of 120,000 French, 12,000 were attacked with typhus, of whom one-half died (Murchison).

Turning our attention to the British Islands, we find that in 1522 the first of the "black assizes," hereafter to be described, occurred in Cambridge. Murchison regards this outbreak of fever as typhus.

In 1577, a second "black assize" occurred at Oxford, and in 1586 a third at Exeter. These outbreaks of fever, apparently communicated to the public by prisoners brought from foul jails into open court, appear to be the earliest distinct records of typhus fever in England. But these islands, and in particular Ireland, have been the geographical home of typhus fever. For more than two centuries and a half this disease has been endemic in Ireland.

In the spring of **1643, at the siege of** Reading, a fever **broke out in** the army of the Earl of Essex, **and in** the garrison, which was commanded by Charles I. The soldiers of both armies were greatly overcrowded. The fever presented the symptoms of typhus. It was very contagious, and was communicated to Oxford, and thence spread to the neighboring country, where it proved very fatal.

In 1658, a similar fever spread over England.

The great plague of London (1665) was preceded and followed by a malignant continued fever, the symptoms of which point to typhus.

Sydenham describes an epidemic of fever which began in London in 1685, and extended over the whole of Britain.

In the autumn of 1698, after a great failure of the crops, a fatal spotted fever began to prevail all over England. From a period probably extending as far back as the beginning of the seventeenth century, typhus had been known in Ireland as the "Irish Ague." Gerald Boate (1652) mentioned, among other diseases there prevailing, "a certain sort of malignant feavers, vulgarly in Ireland called Irish agues, because at all **times they** are so common in **Ireland, as well among the** inhabitants and the natives as among those who are newly come thither from other countries."

About the beginning of the last century, medical men in Ireland began to pay great attention to epidemic diseases, of which chronological histories, extending over a long series of years, were published later in the century by Rogers, O'Connell, Short, and Rutty. From these authors we learn that in the winter of 1708–9, after a poor harvest the preceding year, and during extremely cold weather, a fever then prevailing in Cork reached its climax. It then "declined sensibly for a year or two," and disappeared. In 1718, a fever, "in all respects the same" as that of 1708, became epidemic in Ireland, and prevailed until 1721, when "it abated of its severity, dwindling insensibly away, till at length it was rarely to be met with." From the description of this fever there can be no doubt that it was typhus.

A similar fever arose in York and elsewhere, in England, in 1718, and, reaching its maximum the following midsummer, declined rapidly, and ceased about the close of the year 1719.

From 1721 till 1728, there was "scarcely any" fever in Ireland. In the latter year, however, after three successive bad harvests, it reappeared, and continued to prevail for four years, reaching its greatest violence in 1731. This fever "did not bear bleeding." On the contrary, a tonic and stimulant treatment was necessary. This epidemic was general over Ireland, and extended also into various parts of England.

Petechial fever was prevalent in Ireland in the spring of 1735, and in 1736, but no great epidemic arose, after 1731, till 1740. The preceding winter was intensely cold, both in Great Britain and Ireland; numbers of cattle and poultry were frozen to death, while the harvests, and in particular the potatoes, were destroyed. There was great distress among the poorer classes, many of whom died of starvation.

In August, 1740, a fever which may be recognized as typhus swept over the whole of Ireland, raging with greatest violence in the province of Munster, where the poor were worst provided for. This epidemic continued through the following year (1741), abating in fury toward its close. In the winter of 1742, after an abundant harvest, it had almost ended. The poor were first attacked, but the rich did not escape. O'Connell computed the loss of life in Ireland in 1740–41, by famine and fever, at 80,000.

Murchison calls attention to the fact that there are evidences of the association of relapsing fever with typhus in Rutty's description of this outbreak.

A little later, in 1740, a very fatal epidemic appeared in England and Scotland. It spread to London in 1741. This fever "could not bear bleeding." It was best treated with bark and acids.

In 1750, and again in 1751, Sir John Pringle described typhus as the "hospital or jayl fever." He remarked that "the hospitals of an army, when crowded with sick, or at any time when the air is confined, produce a fever of a malignant kind and very mortal. I have observed that the same sort arise in foul and crowded barracks; and in transport-ships, when filled beyond a due number and detained long by contrary winds, or when the men were kept at sea under close hatches in stormy weather."

Of treatment he said: "Many have recovered after bleeding, but few who have lost much blood." He recommended bark and serpentaria, and considered wine of great use.

The writers of this period constantly allude to fevers arising in jails, hospitals, camps, and ships, and attribute them either to the concentrated emanations from living human bodies, or to contagion. On shipboard, **typhus was then a very common** disease, especially on the long voyages to this country.

In 1770–71, typhus again broke out, after a long interval, **in Ireland,** and raged with great violence for about a year.

In 1797, there arose in that ill-fated land another great epidemic of fever, which did not terminate till 1803. It was a period of great calamity; Ireland had been threatened with foreign invasion, and was torn with internal rebellion. Political feeling ran high, and the upper and lower classes were arrayed against each other. A great part of the tenantry of the large estates were deprived of work. There was a series of poor harvests; in the summer of 1797 heavy rains injured the crops. The three following years were no better. This condition of things resulted in a lack of food among **the poor.** The prices of the necessaries of life rose enormously. It was the poor **who chiefly suffered, but in proportion to the number of persons attacked the fever was most fatal in the middle and upper classes.** The harvest of 1801 was abundant, and provisions of all kinds were supplied at moderate prices; the epidemic at once began to decline, **and** by the end of the following year had almost spent its **force.**

It spread to England, but was there **less prevalent than in Ireland.** This epidemic was mainly, but not wholly typhus; in Ireland relapsing **fever** was also observed. It was largely in consequence of the prevalence of fever at this time that separate hospitals, for the reception and treatment of fever-patients, were first established throughout the kingdom.

In 1817–19, there arose a very wide-spread epidemic of **fever in Ireland**; it extended also to England and Scotland, **but prevailed** in both much less extensively. It is probable that this epidemic was chiefly constituted of relapsing fever cases, although a considerable proportion of typhus cases were observed.

It is estimated that in this epidemic 800,000 of the 6,000,000 inhabitants of Ireland fell sick, and of these 45,000 died, partly of fever, partly of famine and dysentery.

The next great epidemic appeared in 1826. It began in Dublin in May, **1826, and** prevailed till March, 1827. Meanwhile it spread to Scotland, where it **reached its acme in** 1828. It prevailed to a limited extent in London. This, like the preceding epidemic, was composed of relapsing and typhus fever cases together.

For a period of eight years typhus fever was endemic rather than epidemic. In 1831-32, "there was a considerable increase" of it in Glasgow and Edinburgh. But it was not until 1836 that it assumed the magnitude of an epidemic. This time, as so often before, it broke out in Ireland, and found its way thence into Scotland and England. The fever of this outbreak was typhus. Hence, the **mortality was far in excess of that of the previous epidemics, which were, in great part, made up of relapsing cases.**

In 1842, fever again became epidemic. This differed from previous outbreaks in neither originating in Ireland, nor in implicating it. The disease was general over Scotland, but was by no means restricted to the large cities. It invaded England, but its ravages there were much less extensive than among the Scotch. It was chiefly prevalent among the poorest and most wretched of the population, who were at the time of its outbreak in a condition of, even for them, unusual distress. The cases were almost exclusively relapsing fever; typhus was rare. The mortality was from two and a half to four per cent. Bleeding was but little resorted to; the treatment was of a supporting kind; many cases were thought to demand stimulants. The distinction between relapsing and typhus was clearly recognized in this epidemic, and the cases were separately entered in the registers of the infirmaries of Glasgow and Edinburgh.

Toward the end of the year 1846, a fever epidemic of great magnitude and severity arose in Ireland after an extensive failure of the potato-crop, and at a time of great consequent hunger and want among the people. It prevailed two years, sweeping also over Scotland and England. In the latter countries the cases were mainly typhus, while in Ireland the predominant form was relapsing fever. The amount of suffering caused by this outbreak was appalling. In Dublin alone, 40,000 cases of fever occurred, and it is estimated that, in the whole of Ireland, the total number exceeded 1,000,000. In England, 300,000 cases occurred, and in Liverpool there were 10,000 deaths from typhus. In Edinburgh, 2,503 persons died of the fever, and it was estimated that not less than 19,254, or one-ninth of the population, were ill of it.

The death-rate of this epidemic was everywhere high, but was always highest when the proportion of cases of typhus to relapsing fever was greatest. In Ireland, 8 per cent. died; in Edinburgh, 13 per cent.; in Glasgow, 14.41 per cent.; but, separating the cases of relapsing fever from those of typhus, we find that in Edinburgh the mortality of relapsing fever was 4 per cent., of typhus 24.7 per cent.; and in Glasgow, that of relapsing fever 6.38 per cent., of typhus 21.2 per cent.

At this period (1847-48) a great epidemic of typhus and relapsing fever prevailed in Upper Silesia and elsewhere in Germany.

In Great Britain, and especially in Ireland, typhus fever has its chief geographical home. Pestilential centres of typhus seem to exist upon the Continent—for example, in Northern Italy, the Baltic provinces, and in Silesia; but nowhere in modern times have typhus epidemics occurred so frequently as, or excelled in magnitude, those of the British Isles. Nowhere in the intervals between epidemics have sporadic cases and isolated outbreaks been so constantly observed as there. In these lands, and particularly in Ireland, typhus fever is peculiarly endemic.

No authentic account exists, according to Murchison, of typhus, as it is known in European countries, in Africa or the tropical parts of America; nor has it been observed in Australia or New Zealand, except, on rare occasions, among the passengers landed from emigrant-ships. The same author concludes, after a review of the somewhat conflicting statements of writers upon the diseases of India, with reference to this subject, that typhus fever must henceforth be regarded as one of the diseases of that country. Hirsch informs us that together, with the plague, it is endemic in Simla. The natives of tropical countries possess no immunity from typhus on visiting localities in which it is prevalent.

Turning our attention to the Western hemisphere, we find that typhus fever has prevailed in the United States and British North America, at various times in restricted epidemics. There is reason to believe that most of the epidemics that prevailed extensively in the United States during the early part of the present century, and were described by medical writers of the period under such names as "*spotted fever*," "*petechial typhus*," and "*typhus syncopalis*," consisted of cerebral spinal fever. In more recent times, typhus fever has not infrequently made its appearance in the cities of the seaboard, as a direct importation from Ireland and other transatlantic countries in which it has prevailed. Hence the popular terms, "*Irish fever*," "*emigrant fever*," "*ship fever*." Its importation has without doubt been more frequent in recent times, in consequence of the facility and rapidity of ocean travel and the enormous immigration hitherward. The instances of its supposed autochthonous origin in the United States are readily explicable upon the theory of a prolonged latent existence of the germs, terminating in their becoming the exciting cause of the disease under favorable circumstances; that is, upon the presence and concurrent action of the predisposing causes. Thus, Hirsch states that he has been able to find but seven instances of the spontaneous origin of typhus in the United States between the years 1817-56.

The first of these broke out in the poor-house in Boston, in 1816, and extended to the inhabitants of the city. The second occurred in Philadelphia in 1820, and was exclusively confined to the poor-house. The third, fourth, and fifth occurred in 1818, 1825, and 1827, in the prison at Bellevue, in New York, and prevailed at the same time in the crowded and destitute portions of the city. The sixth of these outbreaks was observed in Westchester County, among railroad laborers. The last occurred in the year 1836, among the most wretched, filthy, and impoverished portion of the population of Philadelphia, and extended over a great portion of the city. This outbreak was described by Gerhard and Pennock, in a paper that remains to this day the most important contribution to medical literature upon the subject of typhus in the United States.[1] It is only necessary to point to the fact that all these outbreaks occurred in or near seaport cities, in direct communication with those parts of the world in which typhus fever makes its home, and that each of them arose in a locality in which the predisposing causes of typhus fever probably existed to a high degree, to show how unstable is the basis upon which rests the belief that the disease was, in these instances, of spontaneous origin. The inability to trace the contagion in any outbreak is not a sufficient warrant for the supposition that it has spontaneously arisen.

Typhus has repeatedly appeared, in consequence of direct and easily traceable importation, in New York, Philadelphia, Boston, and Baltimore.

There are no records of its occurrence in the Gulf States or upon the Pacific slope; and although Drake, in his "Treatise on the Diseases of the Mississippi Valley," treats at great length of the typhous group of fevers, it is clear from his descriptions that he refers principally to outbreaks of enteric and cerebro-spinal fever, and not to typhus fever as we know it. In fact, typhus is not a disease of North America. It occurs here only in consequence of importation, and prevails only in restricted epidemics. Among the more recent outbreaks are to be mentioned that of 1850-52, in Buffalo, described by Flint; that of 1861-65, in New York, of which we have an account in the writings of Loomis; and that of 1864, in Philadelphia, which forms the basis of the excellent paper by Da Costa.

Notwithstanding the existence of typhus in several of the seaport cities of the North during the years of the American war, it is a remarkable fact that there was perhaps entire immunity from this disease in the armies both of the United States and the Confederates. Dr. Clymer[2] states that, as a result of large personal observation and diligent inquiry among the medical officers of the United States army, he is satisfied that, as an epidemic, however limited, typhus never prevailed, even at the depots for returned prisoners of war. He thinks that there is every reason to believe that the cases, 1,723 in number, with 572 deaths, reported to the office of the Surgeon-General of the United States, were not instances of true typhus.

Students of medicine in American cities rarely have the opportunity of familiarizing themselves with the clinical aspects of typhus. Never-

---

[1] On the Typhus Fever which occurred at Philadelphia in the Spring and Summer of 1836; illustrated by Clinical Observations in the Philadelphia Hospital, etc., etc., by W. W. Gerhard, M.D.: Amer. Jour. Med. Sc., vol. xix., p. 289 et seq., and vol. xx., p. 289 et seq. Philadelphia, 1837.

[2] Aitkin's Practice. Vol. i. Article Typhus. Third American edition. Philadelphia, 1872.

theless, occasional examples of this disease find their way into the hospitals of the seaboard towns. Thus, typhus showed itself in Philadelphia in the spring of 1880, and several cases were at that time treated in the wards of the Philadelphia Hospital. During the summer the disease disappeared, but again made its appearance in the autumn, and is still prevalent. The number of cases, fortunately, is extremely limited.

Typhus has been observed under similar circumstances of direct importation in British North America, where, however, the epidemics have, in most instances, assumed more extensive proportions than with us.

Etiology.

I. Predisposing Causes.

Climate has undoubtedly an influence upon the development and spread of typhus. As indicated in the foregoing historical sketch, its home is Europe and the British Isles. If it be endemic in India, it is so to a very limited extent. In other tropical and subtropical countries the typhus of Ireland is certainly almost unknown. It is essentially a disease of cold and temperate climates.

The season of the year appears to exert very little influence upon typhus. Epidemics arise and pursue their course irrespective of the season. It has sometimes been observed that the number of cases has diminished during the summer, and again increased in the last month or two of the year. The diminution does not begin at once upon the advent of warm weather, nor does the increase follow immediately upon cold. A continuance of these conditions is necessary to produce their respective influence upon epidemics. From this it would appear that the influence is due to the different mode of life incident to the seasons, and that the increase of typhus in the winter and spring months is due, not to the weather, but to the protracted overcrowding and deficient ventilation of the dwellings of the poor, and perhaps also to a greater scarcity of food, particularly in times of scarcity, in the winter.

Meteorological conditions exert little or no influence upon typhus; they are by no means constant for different epidemics. Hirsch regards a low and damp situation as powerfully predisposing to the endemic and epidemic prevalence of typhus, but insists that it is by no means a necessary or important factor in the production of the disease. As the result of importation, typhus may occur at a considerable height above the sea-level. In the spring of 1839, Lebert observed a considerable number of cases on the plain and in the valley of the Salvan, in the lower Valais, at a height of 4,000 feet above the sea. The disease in this case was brought from Piedmont, over the St. Bernard pass, and at least one-third of the monks at the Hospice contracted it.

Age affords no exemption from the attack of typhus. It would appear, from death registers and hospital reports, that it is for the most part a disease of adult life. The mean age of 3,456 cases admitted to the London Fever Hospital during ten years (1848–57) was, according to Murchison, 29.33 years; and of 18,138 cases admitted in twenty-three years (1848–70), more than one-half (9,248) occurred between the ages of ten and thirty; the youngest was one month, the oldest eighty-four years.

The evidence furnished by data of this kind is untrustworthy as indicating the relative liability to typhus at different periods of life. Children, for obvious reasons, contribute a relatively smaller number of cases to hospital statistics than adults, and, for the reason that the disease is much less fatal in the early years of life, they contribute an actually smaller number to mortality statistics. In view of the fact that many adults are protected by previous attack, it is probable that all periods of life are alike susceptible to the exciting cause of typhus.

Sex in itself has no influence. Up to thirty years of age rather more males than females contract typhus; above the age of thirty years the reverse is true. Taking all ages together, the number of cases is about the same for each sex.

Occupation, except as it involves actual exposure to the contagion, as the case of hospital attendants, physicians, clergymen, etc., does not predispose to typhus. Patients admitted to hospital suffering with this fever are almost always in destitute circumstances, and, if possessed of a trade, have usually been out of employment so long that it could not be regarded as exerting any influence whatever. Numerous observers have thought that butchers are less liable to typhus than those engaged in other pursuits. The fact may be accounted for by their always having a good supply of food. The laboring classes are more liable than the well-to-do middle classes, probably for analogous reasons.

Habitual alcoholic excesses predispose to typhus. Murchison states that a single act of intoxication may render the subject liable to it; that he has "known several instances of persons exposed for months to the poison in its most concentrated form, who were not attacked until immediately after a debauch."

Previous illness is thought to predispose to typhus. In general hospitals, the convalescents from other diseases not infrequently contract typhus. Many persons, who during epidemics have long escaped the fever, are seized with it after a slight attack of sickness. In the Crimean war, scurvy was found to be a powerful predisposing cause. Phthisis, quiescent before, has frequently been observed to run a rapid course after an attack of typhus.

Fatigue, both bodily and mental, want of sleep, anxiety and other depressing emotions, particularly a dread of the disease, increase the liability to the attack. These influences have, in very many instances, appeared to determine the seizure among medical students, clinical clerks, nurses, and other attendants upon the sick. Among the predisposing influences, is to be mentioned personal idiosyncrasy. Different individuals possess for typhus, as for other contagious diseases, a varying degree of personal susceptibility independent of other circumstances of predisposition.

The mode of life of the individual exerts a powerful predisposing influence. Typhus is a disease of the poor and underfed of large cities. With the exception of persons of the better classes who contract the disease by direct exposure to the contagion, it is, under the ordinary circumstances of its endemic or mildly epidemic prevalence, confined to the indigent classes and those just above the indigent classes of the community. It is only under unusual circumstances, or in fierce epidemics, that it attacks those who are well-to-do. It was found, upon inquiring into the antecedent history of 18,268 typhus patients admitted into the London Fever Hospital during twenty-three years, that they belonged almost invariably to the lowest classes of the population, 95.76 per cent. being the inmates of workhouses or dependent upon parochial relief, whereas comparatively few of the better class of patients, who were able to pay for admission, were affected with typhus. A large proportion of the typhus patients had been on the verge of starvation for several weeks or months prior to admission (Murchison).

The great epidemics of typhus, not only in Ireland and Great Britain, but also upon the Continent, have invariably occurred in times of scarcity. They have followed failure of the crops, and prevailed generally over lands visited by famine; or they have arisen in consequence of the hardships of war, sieges, commercial distress, or strikes in the manufacturing districts, and have remained to a greater or less extent circumscribed.

Overcrowding, beyond all question, plays the most important part among the predisposing causes of typhus. The conditions which constitute overcrowding are, in the words of Dr. George Buchanan,[1] " scarcely to be separated from each other, but may be enumerated as overcrowding of dwelling-houses upon a too limited area, overcrowding of rooms by too many occupants, bad ventilation of streets and houses, domestic and personal dirtiness." To the combined influence of these conditions is due the proneness of the laboring population of great cities to typhus. Murchison found that in London the cases admitted to the Fever Hospital were for the most part brought from the central and most crowded localities, and that on approaching the suburban districts their proportion gradually diminishes. In Liverpool, which has habitually more cases of typhus than any other town of England, and in which the most serious epidemics occur, the huddling together of houses with insufficient space

---

[1] Reynolds' System. Article Typhus, vol. i.

around them is carried on to a greater degree than in any other town in the kingdom. A large number of the houses are built back to back in unventilated courts, and the population is so dense that in some districts each person only gets eight square yards of superficial space. In these parts it is that fever especially flourishes, and, in epidemic periods, passes by none but those who are protected by previous attack (Buchanan). In Edinburgh, where the overcrowded dwellings of the poor and the houses of the better classes are perhaps more widely separated than in any other city, typhus, even in the midst of the greatest epidemics, is almost restricted to the most crowded and wretched parts of the Old City (Murchison). Glasgow is another of the cities in which the houses occupied by the poor are densely crowded together. Its inhabitants have likewise been great and constant sufferers from typhus, which has been found to prevail most fiercely in the more crowded parts, and to leave the more open districts, inhabited by the opulent, almost or quite unscathed.

The crowding together of many persons in small rooms with deficient ventilation is not a less potent predisposing cause. Hence, in former times, the ill-repute of the lodging-houses frequented by the poor in the large towns of Great Britain and Ireland. Of these there are great numbers, and previous to the enactment of laws regulating their management, in 1857, they were pestilential centres, perennial hot-beds of typhus, where the fever, like a spark under the ashes, forever glimmered, ready to burst forth into the flame that so often swept the land from end to end.

Typhus has very frequently arisen, both in early and recent times, under circumstances of overcrowding, in hospitals, prisons, ships, and armies. Sir John Pringle in 1752 gave it the name of "Hospital Fever." The "Gaol Fever" and "Jayl Distemper" of former times was typhus. Common in the overcrowded, foul, and ill-ventilated prisons, it spread thence to the communities around them. Such is the origin of the "black assizes" already alluded to, three of which occurred in the sixteenth and three in the eighteenth century. They are of interest as showing in a remarkable manner the condition of the prisoners and the intense activity of the typhus contagion in densely crowded and unventilated rooms. They certainly do not prove the independent origin of the specific exciting cause. The accounts are transcribed from the pages of Murchison, by whom they were collected from the writings of Ward, Bancroft, Huxham, and others.

"The first occurred at Cambridge, during the Last Quarter Sessions in 1522, the thirteenth year of the reign of Henry VIII. The justices, gentlemen, bailiffs, and most of the persons present in court, were seized with a fever, which proved mortal to a considerable number. No account is preserved of the symptoms of this fever; but the circumstances were similar to those of subsequent black assizes, in which the disease was undoubtedly typhus.

"The year 1577, or twentieth of the reign of Queen Elizabeth, was notorious for the Oxford "black assize." This assize was held at Oxford Castle, on July 4th and two following days, for the trial of Rowland Jencks, a bookbinder and a Roman Catholic, for treason and profanity of the Protestant religion. Jencks was not the only prisoner brought before the court, but the accounts state that, after judgment was pronounced against him, 'an infectious damp or breath arose among those present. Many seem to have been taken ill on the spot. Above six hundred sickened in one night, and the day after, the infectious air being carried into the next village, sickened there a hundred more.' On the 15th, 16th and 17th of July, three hundred more fell sick; and between the 6th of July and the 12th of August five hundred and ten persons perished. The following are mentioned as the symptoms: loss of appetite, great headache, sleeplessness, loss of memory, deafness, and delirium so that the patients would get up and walk about like madmen. The general impression at the time was that the 'infection arose from the nasty and pestilential smell of the prisoners when they came out of the jail, two or three of whom had died a few days before the assize began;' the only other explanation offered being that it resulted from the 'diabolical machinations of the papists,' or, according to the Catholics, that it was a miraculous judgment on the cruelty of the judge for sentencing the bookbinder to lose his ears.

"In 1586, another 'black assize' occurred at Exeter. Some time before, thirty-eight Portuguese seamen had been cast into 'a deep pit and stinking dungeon' in Exeter Castle. They had no change of raiment, and were left to lie upon the bare ground. A contagious fever broke out among them, which, from Hollingshed's description, was evidently typhus. Many of them were sick during their trial, and by them the disease was communicated to those present in the court. The judge, three knights, and many others died, and the disease spread over the whole county. In this instance very few became ill until fourteen days after the trial. The fever was believed to have proceeded from 'contagion by reason of the close aire and filthie stinke of the gaole.'

"There are accounts of a fourth 'black assize' at Taunton, during Lent, in 1730. A contagious fever was communicated by the prisoners, who had been removed from Ilchester jail, to the judges and many others in the court. The Lord Chief Baron, the Sergeant-at-law, and the High Sheriffs of Somersetshire, all died of the disease, which spread widely at Taunton, and proved fatal to several hundreds.

"Twelve years later there was a fifth 'black assize' at Launceston, an account of which is contained in the writings of Huxham. 'A putrid, contagious and highly pestilential fever, which had been *generated in* the prisons,' was widely disseminated by means of the county assize, and occasioned great mortality. Among the symptoms were great prostration and oppression, a florid rash with petechiæ, watchfulness, delirium, tremors, subsultus, black, dry tongue, and fetid breath. The pulse was weak from the commencement, even in the robust, and 'bleeding killed the patient, and not the disease.'

"The sixth and last 'black assize' was that of the Old Bailey, in 1750. Nearly a hundred prisoners were tried, who were all, during the sitting of the court, either placed at the bar or confined in two small rooms which opened into the court. The court was crowded to excess, and many present were 'sensibly affected with a very noisome smell.' Within a week or ten days many of those present were seized with a 'malignant fever,' among the symptoms of which were a weak pulse, delirium, and petechiæ. Its duration was a fortnight. That this was the jail distemper or typhus appears from a pamphlet published at the time by Sir John Pringle. Neither the prisoners under trial, nor any in the jail, were suffering at the time from typhus."

There are many instances where typhus has attacked individuals and families or isolated bodies of men, as in jails or on shipboard, without traceable contagion. Dr. Murchison has collected a number of such examples, which he adduces in support of the theory of the independent spontaneous origin of typhus by the intense action of its predisposing causes, advocated by himself and others. If the infecting principle or contagion be a minute organism capable of indefinitely reproducing itself in the human body and in other favorable localities, as has been rendered almost certain by the discovery of the parasitic exciting cause of the congener of typhus, relapsing fever, it is more in accordance with modern views to suppose a continuous latent existence of the germs, which are called into activity by overcrowding, destitution, and other predisposing causes, than to accept the theory of its independent generation *de novo*. The examples in question are capable of explanation quite as satisfactorily by the former as by the latter supposition.

This brings us to the consideration of the exciting cause.

II. THE EXCITING CAUSE.

Typhus fever is due to an infecting principle, communicable from the sick to the well by actual contact, by means of the atmosphere, by fomites, and by drinking-water. The nature of this principle is unknown. In the words of Lebert, "when contagion plays a part so important, one is forced to admit a specific cause for a disease so absolutely defined and so well characterized. Such can only be an organic poison or an organized germ. A poison may kill, but cannot infect, still less spontaneously multiply to an enormous degree; while everything in the history of this disease admits a ready explanation through organized germs."

Typhus is pre-eminently contagious. When it appears in a community it spreads rapidly among the susceptible persons. The rapidity and extent of its spread is proportionate to the intensity and diffusion of the conditions known as predisposing causes, but persons exposed to none of the predisposing causes contract the disease when in close attendance upon those ill of it. The prevalence of typhus in restricted localities is in proportion to the degree of intercourse between the healthy and the sick. When it breaks out in a house, those living in the same room with the person first attacked are usually the next in order to develop the disease. In hospitals the nurses and resident physicians are much more commonly attacked than the attending physicians or students. The medical assistants in the British fever hospitals rarely escape the disease. During the year 1827, in the Edinburgh Infirmary, ten clinical clerks and twenty-five nurses or servants, caught typhus; all of them had frequent and close communication with the fever-patients; whereas the clerks and nurses residing in the same building, who had no intercourse with fever-patients, almost uniformly escaped.[1] Instances might be multiplied indefinitely in support of this statement. During twenty-three years (1848–70) 288 cases of typhus originated in the London Fever Hospital. Of these, 193 were nurses and other attendants in the wards, 14 were medical officers, 7 laundresses, and only 3 servants not engaged in the wards; 71 were patients admitted for other diseases (Murchison). In 1814, typhus was brought to the Salpêtrière, in Paris, by some soldiers; of the persons attached to the hospital 120 were attacked and eight physicians died. During two and a half months in 1856, 600 of the attendants in the French military hospitals in Constantinople contracted typhus, which was not then prevalent in the city itself.

Typhus is, in all epidemics, imported into localities previously free from it, by infected persons. It is in this way that the disease has made its way to the seaport towns of this country. Hence its names: Irish fever, emigrant fever, and so on. Very often in general hospitals the admission of a single case of typhus is followed by its spread among the attendants and the other patients. The prompt removal of the first cases from the house or locality in which the disease has made its appearance has often arrested its spread, while the neglect of this measure has converted such house or locality into a focus of contagion.

The disease may be and constantly is communicated from the sick to the well by actual contact. This, however, is by no means necessary. The infecting principle is in all probability borne in the expired air of the patients and in the exhalations from their cutaneous surfaces. It may be thus carried into the surrounding atmosphere, and so reach the blood of those quite near at hand by the channel of the breath or by the

---

[1] W. P. Alison, M.D.: Observations on the Epidemic Fever now prevalent among the Lower Orders in Edinburgh. Edinburgh Med. and Sur. Jour., xxviii., 1827.

saliva which they swallow. But the distance to which it may be thus conveyed cannot be great. If the room occupied by the patient be spacious, airy, and clean, the risk of contagion is very slight. Physicians who visit patients in such apartments with due precaution, and pass at once into the open air, incur but little liability to contract the disease; but those, on the other hand, who heedlessly inhale the atmosphere immediately surrounding the patient, or incautiously perform auscultations, or who tarry in his presence, especially if the apartment be small and imperfectly aired, incur great risk. Typhus is never communicated by means of the atmosphere from fever hospitals to the houses in their immediate neighborhood. Lebert states that in his wards at Breslau, during the epidemic of 1868–69, when the greatest attention was paid to ventilation both in winter and summer, neither typhus nor relapsing fever was propagated.

The breath of typhus-patients conveys, and their bodies emit when the bed-clothes are turned down, a peculiar, strong, somewhat pungent odor, which has been regarded by many observers as characteristic of this disease. It has been thought that those in whom this fever-odor is strongest are most likely to communicate the disease to others. The fact is well attested that many persons, who, upon coming into close proximity with patients, have felt a sickening sense of the intensity of this odor, have within a very short period developed the disease. Articles of all kinds with which the patient comes in contact may become carriers of the contagion. It is probable that the germs of typhus, in a dried state, or, at all events, in a condition of diminished activity, may retain their vitality for an indefinite period, in the absence of conditions favorable to their development or multiplication—in other words, in the absence of the predisposing causes of the disease. Not only the bedding and clothing of the patients, but also the apartment in which they have lain, may act as *fomites*. Particular houses, in this way, become hot-beds for the production of the disease; ships used for the transportation of typhus-patients, become the home of the infection, and vehicles used to convey patients to the hospital may communicate the sickness to their next occupants. Those who wash the undisinfected clothing of typhus-patients are peculiarly liable to take it. Woollen substances are more apt to absorb and retain the contagion than other textures, and garments of a dark than those of a light color.

Not only may the disease be thus conveyed to a distance by articles of the most varied description capable of carrying the contagion, but individuals not themselves sick of the fever may be the means of communicating the disease from the sick, or from infected localities, to the healthy at a distance. Thus, in January, 1867, a patient in a surgical ward of the Middlesex Hospital was seized with typhus. She had been in the hospital four and a half months, and in bed all the time. There were no other cases of typhus in the same ward or on the same floor; but a nurse in close attendance upon a typhus-patient down-stairs, though in good health herself, had been in the habit of visiting this patient daily (Murchison). It is, however, fortunate that the contagion, in order to be conveyed by means of fomites, must be concentrated, and that habits of cleanliness and caution, and free ventilation, reduce this danger to a minimum. Drs. Gregory, Tweedie, and Murchison have separately recorded their belief that, in an attendance upon typhus-patients extending over many years, they have in no case been the means of the communication of the fever.

The disease may be contracted by susceptible persons through contact with the bodies of persons who have died of it. There are no facts to prove that it is disseminated from the dejections, as is the case with enteric fever. *The period of incubation* is placed by Murchison at twelve days or less, rarely longer; by Lebert at from five to seven days. There are no reliable facts in support of the statement that it sometimes exceeds three weeks. A number of cases have been recorded in which the symptoms of the disease appeared immediately upon exposure. In these instances the possibility of previous unsuspected exposure is to be considered.

Lebert holds the opinion that the contagion of typhus and relapsing fever, as of other contagious diseases, must in many instances be disseminated by generally acting local causes. He regards the ground—and drinking-water as together playing an important part. This opinion is based upon the simultaneous or nearly simultaneous infection of several persons in the same house or locality at the beginning of an epidemic, a circumstance which the communication of the disease from the sick to the well cannot explain. Typhus is but little contagious during the first week; the period in which it is most likely to be communicated is from the end of the first week to convalescence. After the disappearance of the fever and the return of appetite and digestion, the danger of contagion is slight. It is, however, to be borne in mind that the clothing of the patient, and articles in the sick-room, may, even at this period and long afterward, unless disinfected and exposed to the air, transmit the specific cause of the disease.

The contagion of typhus is destroyed by prolonged exposure to moderate dry heat ($95.5°$ C. [$204°$ F.]).

Immunity from a second attack is enjoyed by a majority of the persons who have suffered from typhus. Nevertheless, many cases of well-marked second attacks attended by the eruption are recorded. It is probable that an abortive attack is less apt to confer immunity than the fully declared disease.

The lower animals, so far as is known, do not suffer from any disease identical with human typhus, nor is it communicable to them (Murchison). The experiments of Mosler, Obermeier, and Zuelzer, upon dogs, rabbits, and guinea-pigs, have yielded contradictory results. Even where these animals have died, after the intravenous injection of the blood of typhus-patients, with the symptoms of an acute infection, it is impossible to affirm that this disease has been typhus, for the reason that typhus presents no specific lesion. At the autopsy of ten rabbits, Zuelzer found localized pneumonic patches in two; in the eight others, congestion of the lungs, the kidneys, and the liver. But, as Jaccoud[1] has pointed out, this does not warrant the conclusion that they died of typhus.

Clinical History.

The evolution of typhus, clinically considered, is continuous rather than by a succession of distinct stages or periods. From the onset of the attack, which is usually abrupt, to the defervescence, which is, in by far the greatest number of instances, critical, the march of the symptoms is progressive; and if stages can be artificially established for purposes of description, they are not separated in nature, but merge imperceptibly into one another. Even the appearance of the eruption cannot be said to begin a distinct period in the clinical history of typhus fever, for the other symptoms are with that event commonly not modified; they are only deepened.

The attack is occasionally preceded by prodromes of a few days' duration. They consist of a general feeling of weakness and indisposition, with headache, loss of appetite, nausea, and restlessness at night. These prodromic symptoms are not, as a rule, so severe as to compel the patient to abandon at once his usual occupations; in some instances, however, he feels so dispirited and his sense of fatigue is so great, that even in this stage he promptly betakes himself to his bed.

In the greater number of cases, and especially in those cases where the development of the fever is rapid and the symptoms are severe, prodromes are wholly absent.

A chill or chilliness marks the invasion of the disease, which is generally so sudden that the patient or his friends are able to designate the day on which the attack began. The chill or chilly sensations are in many cases repeated at irregular intervals during the first two or three days, and, being followed by perspiration, may present a superficial likeness to intermittent fever. In children not infrequently, but rarely in adults, vomiting, often repeated during the first few days, attends the onset. At the same time there is fever, which rapidly augments; the skin is hot, the face flushed, the eyes injected; headache is constant and severe, and a feeling of dulness and confusion, with vertigo upon assuming the upright posture, and noises in the head, distress the patient. He

---

[1] Traité de pathologie interne. Tome ii. Paris, 1877.

complains also of some pain in the back, and dull, sore pains in his limbs and joints. Catarrhal symptoms are common, such as slightly hurried respiration, a little cough, sore throat, swelling of the edges of the eyelids, and lachrymation.

Muscular weakness and an extreme sense of prostration appear early. The patient's face at first wears an expression of weariness, but soon becomes dull and stupid. He falls into a drowsy state, but passes uncomfortable, restless nights. Wakefulness alternates with brief periods of sleep, disturbed by painful dreams and startings; after three or four days he begins to talk and mutter in his sleep, and between sleep and waking there is slight delirium. When awake, the patient is still conscious and answers questions slowly, but generally with correctness, although there is confusion of mind and memory. Already he requires close watching, especially at night, when in his delirium he may leave his bed and wander from the room. In severe cases muscular movements are early unsteady and tremulous, the tongue trembles as it is protruded, and speech is feeble and hesitating.

From the beginning the tongue is large, pale, and coated at first with a white, later with a thick, yellowish brown fur; it speedily shows a tendency to become brown and dry; appetite is lost, there is thirst, the secretion of saliva is diminished, taste is perverted, and a stale, unpleasant odor loads the breath. Nausea is occasionally present, but vomiting is rare. There is constipation as a rule, but in some instances slight diarrhœa occurs. The abdomen is soft and painless, with the exception of slight tenderness in the region of the liver and the spleen. Enlargement of the spleen may be early detected on percussion.

The pulse is increased in frequency from the beginning of the attack; it soon reaches the neighborhood of 110 in the morning and runs up to 120—130, or even higher in the evening, with a much higher rate in children. It is full at first, but compressible—rarely firm or tense; it soon grows feeble, but dicrotism is uncommon.

As a rule the temperature rises rapidly, attaining 39.4°—40° C. (103°—104° F.) by the morning of the third or fourth day, and 40°—41° C. (104°—105.8° F.) the same evening, and remaining nearly stationary at these points until some time in the second week. The high temperatures of relapsing fever are very rare in typhus. An evening temperature of 42° C. (107.6° F.) is seldom observed. A decided difference between the morning and evening temperature is more favorable, even when the evening increase is considerable, than a continuously high temperature range in which the morning remission fails.

On the fourth or fifth day, as a rule, less often at the end of the first week, the characteristic eruption appears. It consists of numerous roseola-like spots of irregular outline and varying in measurement from a line to three or four lines across, scattered singly, like the spots of enteric fever, or, as is by far more common, arranged in irregular groups, like the rash of measles. At first these spots are of a dirty rose-color, very slightly raised above the surface of the surrounding skin, and upon pressure they momentarily disappear. Within the course of a day or two they become darker from the escape of the coloring matter of the blood into the tissues; they are no longer elevated, but appear as faint, dirty brown stains, without defined margin, and fading, not disappearing, upon pressure. Not infrequently, at a later period of the fever, petechiæ show themselves at the centre of many of these spots, while others, especially in grave cases, are converted into dark red stains; yet they cannot be regarded as being in themselves at any period of their course true petechiæ. They closely resemble the rose-rash of enteric fever, differing principally in their numbers and grouping, and in the fact that they appear once for all, and not in successive crops. Their course is typical. They fade during the first half of the second week, and disappear with or without desquamation toward its close. The true petechiæ appear as such about the time the typical rash begins to fade, that is to say, about the eighth or tenth day; they remain longer and disappear more slowly. A faintly reddish, ill-defined mottling or marbling of the skin between the spots or groups of spots, which form the characteristic rash, also occurs to a greater or less extent. It is this that has been described, from its appearing to lie beneath the surface, as the "subcuticular" eruption of typhus. The appearance of the rash varies greatly, and the variation is determined by the general abundance of the two eruptions, by the relative preponderance of one or the other, and, in certain cases, by the extent of the true petechiæ, which are, however, frequently absent altogether. The spots and the mottlings together constitute the "mulberry rash" of Jenner.

The distribution of the typhus eruption is irregular: appearing usually first on the sides of the chest or abdomen, it spreads in a brief time over the chest, abdomen, back, and limbs. It rarely appears upon the neck or face. It has in some instances been observed to first appear upon the backs of the hands. In some cases the roseola-like rash is absent altogether, the faint, subcuticular mottling alone being present. An entire absence of eruption is very rare.

About the end of the first week the depression becomes profound, headache passes into delirium, and the impairment of the mental powers is extreme. The patient is dull of hearing; he answers questions very slowly and vaguely; drowsiness and stupor are marked, and in severe cases there is a tendency to coma. The character of the delirium is variable. It is commonly low, wandering, muttering; occasionally it is at first acute, severe, boisterous. This violence usually soon passes away, leaving the patient in a state of the most profound exhaustion, or it gradually subsides into dulness with muttering. With both forms of delirium there is sleeplessness. The tongue is now dry, fissured, and crusted, sordes collect upon the teeth and lips, the conjunctivæ are deeply injected, the flushing of the face gives place to a dusky pallor most marked about the nostrils and lips, and emaciation progresses. The breath and the skin exhale a peculiar fœtor, there is annoying cough with mucous expectoration, and, upon auscultation, râles are heard in all parts of the chest. The heart-sounds and the impulse are faint and indistinct. The area of splenic dulness is considerably extended. The state of the bowels varies from constipation to irregular, scanty dejections, or a moderate intestinal catarrh; the **urine is scanty**, opaque, high-colored, and very frequently contains albumen. In **severe cases the discharges are** passed involuntarily, or there is retention of urine.

The **symptoms** deepen. The patient utters no complaint. Neither pain nor headache are felt. Appetite is completely lost, thirst no longer distresses him, although he swallows with difficulty, owing to the dryness of his throat. He lies upon his back, stupid, lost, utterly indifferent to everything around him, sometimes moaning or muttering incoherently, sometimes quiet. The eyelids are partly closed, the pupils contracted. Deafness is often present. **When spoken to loudly, he stares vacantly** without attempting a reply. **If asked to put out his tongue, he opens his** mouth and leaves it open till reminded **to close it.** He is unable to raise himself, or even to turn **from side to side; from muscular weakness he is** continually sliding **down in the bed; his hands tremble, he picks at the** bed-clothes and feebly grasps at unseen objects in **the air; there is subsultus.** The pulse is small and weak, often difficult **to** count, less commonly irregular or intermittent. It ranges from **112** to 140 or over. The portions of the skin subjected to pressure show a tendency to slough. The surface now becomes cooler and is often moist. If petechiæ are present they become more numerous.

Death may take place at any time after a condition such as has been described becomes fully developed. In very severe cases it may occur in the course of a few days or before the end of the first week. More commonly the fatal termination takes place between the tenth and the seventeenth days. Death at a later period is uncommon, except as a consequence of complications. The mode of death is by coma, or by asphyxia in consequence of sudden pulmonary enlargement, or by failure of the heart, the pulse becoming imperceptible, the surfaces cold, livid, and bathed **in sweat.**

In abortive cases a favorable termination may take place by critical defervescence at the end of the first or **the** beginning of **the second week.**

In average cases the fever comes to an end about the fourteenth **day**—sometimes as early as the tenth day, sometimes as late as the middle of the third week. The amendment is more or less sudden. The temperature, which in many cases shows a little abatement for some days before the crisis, falls in a single night, or in the course of twenty-four or forty-eight hours, to the normal or even a little below it; the pulse becomes much slower and its character improves; the stupor and coma immediately disappear after a prolonged, refreshing sleep, out of which the patient awakes as from an oppressive dream—conscious, but at first bewildered and confused. The eruption fades and gradually disappears, the tongue cleans and becomes moist at its edges, the appetite returns. The crisis is often attended by moderate sweating or diarrhœa, or both, and by an increase in the amount of urine, with a copious deposit of urates and the disappearance of albumen. In the course of a few days the tongue is moist, the appetite eager, strength begins to return, and the convalescence progresses rapidly, so that many patients are able to resume their work within a month from the beginning of the attack.

Temporary loss of hair not infrequently occurs during convalescence, and in many cases a considerable length of time ensues before the body-weight and the original vigor of mind are regained. The deafness in almost all cases gradually passes away.

Relapses occur, but they are much less common in typhus than in enteric fever.

Analysis of the Principal Symptoms.

Symptoms referable to the nervous system.

A chill or chilliness is, in many cases, the initial symptom. It is, however, often absent.

Headache is among the earlier and more constant symptoms of typhus. When prodromes are present, headache is usually among the number. In most cases it is present from the onset; it is most severe during the first week; it often lasts only a few days, and, as a very general rule, terminates early in the second week, upon the advent of delirium. Its seat is most frequently in the forehead or temples; it is rarely confined to the vertex or occiput; in a majority of cases it is general. It is usually dull or heavy, often moderately intense, and for a few days the most prominent symptom of the disease, but rarely acute or paroxysmal. Sometimes it is slight.

Vertigo, increased upon assuming the upright posture and becoming more marked with the progress of the disease, is usually associated with the headache.

Pains in the back and limbs are prominent symptoms during the early days of the attack. The pain in the back is dull and heavy, but less distressing than the headache or the pains in the extremities. The last are described as of a sore character, as if from severe bruises; they outlast both the headache and back-pains, and often recur during convalescence. They not infrequently implicate the joints as well as the muscular masses.

Delirium is common. Some impairment of the mental faculties is constant; hence the synonym "brain **fever.**" In the latter part of the first week, as a general rule, sluggishness of mind becomes apparent; the perceptions are blunted, mind and memory are confused; the patient cannot tell how long he has been sick, nor the day of the week; he is indifferent to what goes on around him, and **annoyed at being spoken to or questioned.** This mental dulness alternates with drowsiness, which **lacks** the refreshing attributes of sleep and scarcely deserves the name. He becomes more and more dull, and, **although** this degree of **mental disturbance** may not be exceeded **in mild cases,** in most instances the stupor passes into delirium.

The period at which delirium supervenes is variable. It commonly appears as the headache subsides—about the end of the first or the beginning of the second week. It may occur much earlier. In rare instances it has been observed from the first night of the attack; on the other hand, it may not appear till shortly before the critical defervescence, toward the close of the second, or in the early part of the third week in protracted cases.

At first the **delirium occurs only during some part or the whole of the** night, and is absent during the day, **to return again at nightfall.** After a time, it **becomes present** during the day also, and is then worse by night. It is common for patients to be drowsy and stupid in the daytime, wakeful and delirious at night. It usually ceases after the crisis, but in rare cases continues to occur at night for some time into the convalescence, even after the **general** condition **has begun to show a decided improvement.**

The character of the delirium presents the widest range of variation. This phenomenon, as well as the other symptoms of **mental disturbance,** are greatly modified by the mental temperament of the individual, his intelligence and education, previous habits of intemperance and the like, and by the amount of anxiety and fatigue that have preceded the attack.

In uncomplicated cases the severity of the attack may be measured by the degree of mental aberration and delirium (Murchison).

The delirium is generally quiet; the patient moans and mutters incoherently, or he is restless, irritable, and easily disturbed. At first, he replies coherently to questions when aroused, or his answers are rambling and inconsequent; after a time he falls into a state of more or less complete **unconsciousness.**

It is less frequently, but in no small proportion of cases, much like **that of** chronic alcoholism—a form of *delirium tremens.* In spite of his extreme prostration, the patient **is restless** and fidgety; he sleeps **little** or not at all; he glances furtively from side to side, and makes eager but purposeless attempts to leave his bed; his tongue is protruded tremblingly, and there are muscular tremors of his limbs. The pulse is weak and frequent, the impulse and first heart-sound impaired, and the skin leaky.

Again, the delirium may be active and noisy. The patient shouts and screams. He tosses about and seeks with violent efforts to get out of bed. His strength is surprising. He has to be controlled by force, for which stout attendants are sometimes necessary. In this state the patients often show a suicidal disposition, and require careful watching, especially at night. Indeed, the mental state in typhus fever is so peculiar, that it is in no case safe to leave the patient alone. Patients who are quite rational to all appearances during the day, may wander about in delirium at night, and the semblance of reason at one period of the day may be followed by suicidal mania in the course of a few hours. The last form of delirium is called *delirium ferox;* it is much less common in typhus than the low, wandering form which has been called "typhomania," and is apt to be more or less transient and to terminate in profound exhaustion or even fatal collapse; in other cases it subsides into typhomania.

The mental state is peculiar, but that it differs essentially from that of the delirium in other fevers or acute diseases is not apparent. Dr. Murchison has collected, in "The Continued Fevers," an interesting account of the delusions and fancies of patients in the delirium of typhus, to which the reader interested in this branch of the subject is referred. The instances cited bear out the statement of Griesinger,[1] that the mind in the delirium of typhus is very often concerned with a limited number of constantly recurring alarming fancies.

Da Costa observed a fatal case in the wards of the Pennsylvania Hospital, in which delirium was absent altogether. The patient's mind remained clear to the last hours of life.

Wakefulness, drowsiness.—During the first two or three days the patient may be dull and inclined to sleep, but wakefulness is a prominent and distressing symptom in most cases until the beginning or middle of the first week, particularly at night. Inability to sleep and restlessness are also common in children. It is a curious fact, noted by many observers, that patients who sleep sometimes for several hours together, will often insist that they have not closed their eyes, although in all other respects rational.

To this sleeping, without being aware of it afterward, has been applied the awkward and useless term *coma vigil.* This term has also been applied, perhaps more correctly, to a condition of great gravity, "in which the patient lies with his eyes wide open, gazing into vacuity, his mouth partially open, his face pale and devoid of expression; the pulse rapid and feeble, or imperceptible; the breathing scarcely perceptible; and the skin

---

[1] Virchow's Handbuch. Band II., Abtheil. 2. Erlangen, 1864.

cold and bathed in perspiration. He is evidently awake, but he is indifferent and absolutely insensible to all that is going on about him." The condition so graphically described by Jenner is not infrequently observed shortly before the fatal issue in typhus. The term coma vigil is not sufficiently significant to describe it, nor is it sufficiently explicit, being indiscriminately applied to two widely different states, to designate either the one or the other. It belongs to a large class of meaningless words which, being neither descriptive nor explicit, nor generally understood or understandable, might well be spared out of medical literature, which they cumber.

Drowsiness, more or less marked, not infrequently alternates with wakefulness and delirium; but about the middle of the second week that indeterminate condition between sleep and waking, to which the term somnolence has been applied, commonly supervenes. It may follow prolonged wakefulness and mental excitement, or may occur without them. It is more or less profound according to the gravity of the case, and deepens by imperceptible gradations; first into stupor, then to coma.

Debility is one of the most characteristic features of typhus. It comes on early, and is in all cases marked. The patients are obliged to take to their beds from sheer weakness within the first day or two of their illness. So great is the loss of muscular power that the patients are unable to walk or rise without assistance, or, in many cases, even to turn in bed. As a general rule, the prostration increases till the ninth or twelfth day, when it is often complete. The excessive effort of the maniacal paroxysms in the acute form of delirium is apt to be followed by a corresponding prostration, so great in some instances as to prove rapidly fatal. The loss of strength sometimes is not very great until the beginning of the second week of the disease, when it develops suddenly with dangerous symptoms of extreme debility. This class of cases is rare.

Except when changed by restlessness or delirium, the attitude of the patient is, by reason of the loss of muscular power, that which is described as the dorsal decubitus. With increasing weakness he tends to sink toward the foot of the bed.

Paralyses of certain groups of muscles occur; hence the involuntary discharges and the retention of urine that so often attend the height of the disease. Dribbling of urine may result not only from paralysis of the neck of the bladder, but it may also arise as a consequence of overdistention of the bladder from paralysis of its muscular coat. In typhus and in all low fevers, the routine exploration of the suprapubic region by palpation and percussion is imperative, as in this way retention of urine, that by reason of the dribbling might be otherwise overlooked, is often detected, and may be relieved by the use of the catheter. Neglect of this precaution is apt to result in uræmia with coma and convulsions, or, more remotely, in catarrh of the bladder or ulceration of its mucous membrane.

Murchison states that there is occasional paralysis of the orbiculares muscles, and that, in consequence of the inability to close the eyes, ulceration and sloughing of the cornea may take place.

Among the disturbances of the functions of the nervous system is tremulousness. In few severe cases is trembling of the hands and tongue wholly absent during the period preceding the crisis or the fatal issue. In some cases—especially in those who have habitually indulged too freely in the abuse of alcohol, the aged and the very infirm, the whole body is observed to be in a state of tremor. Oscillatory motions of the eyeballs (nystagmus) and choreic movements of the limbs have been recorded. Subsultus tendinum, spasmodic twitchings of the muscles of the face, carphology or grasping in the air, and picking at the bed-clothes, also belong to the motor disturbances of the gravest cases of typhus. Hiccough occasionally occurs.

Much more rarely, and in grave cases only, tense contractions of groups of muscles are met with. The flexors of the forearm, and of the thighs and legs, are among the groups apt to be thus affected. Trismus, strabismus, and, in very rare instances, opisthotonus, have been seen.

General convulsions are met with, according to Murchison, about once in one hundred cases. They are due to uræmia in the vast majority of cases, and rarely appear earlier than the middle or end of the second week. An unusual tendency to stupor, and a marked diminution in the quantity of urine secreted, as a rule, precede for three or four days the fit, which is apt to be followed by death, or by coma that continues till death occurs. Life is rarely prolonged beyond three or four days after the occurrence of general convulsions. Fatal uræmic convulsions may, however, occur in cases that have apparently pursued a mild course, or even after the patient has entered the stage of convalescence.

The organs of special sense are to some extent implicated in the processes of typhus. During the first week the eyes are watery; later in the course of the attack they are dry. The conjunctivæ are commonly deeply injected from the earliest days of the attack. Most observers insist upon the fact that the discoloration of the conjunctiva is of a darker hue in typhus than in ordinary inflammations of the eye. The pupils are commonly contracted, sometimes in grave cases to a mere point, and are not infrequently insensible to the stimulus of light. Dilatation of the pupils, with failure to respond to light, is very rare, and occurs only in the most profound stupor or coma. Intolerance of bright light is common.

During the first four or five days, patients very often complain of noises in the head, and associate them with dizziness. After the fourth or fifth day, decided or even complete deafness is common; it often extends into the convalescence, but is not persistent. It was regarded as a favorable symptom by the earlier observers. This opinion is not borne out by recent observations. It is not a nervous symptom, but is probably due to catarrhal processes in the middle (or external) ear. Intolerance of sound is far less common.

Coryza forms part of the general catarrhal disturbance which attends the development of typhus.

Epistaxis is very rare in most epidemics of typhus. Most writers do not allude to it. Murchison met with it in about a dozen instances out of seven thousand cases; Barrallier 97 times in 1,302 cases, and Jacquot in about one-fourth the cases among the troops in the Crimea, where, however, typhus was very often complicated with scurvy. On the other hand, it was present in one-fourth the cases observed in Philadelphia by Da Costa, who states that hemorrhages from other sources, as the gums or bowels, did not occur in his cases, and that sordes were not unusually common. When epistaxis is present, it comes on early, and the cases in which it occurs are, for the most part, severe.

The taste, as in all fevers, is perverted. Sweet things are disliked; acids are often grateful.

The general sensibility of the surface is not infrequently increased. It is important to distinguish between hyperæsthesia of the surface of the abdomen and tenderness of the internal organs upon pressure. Toward the close of grave cases, general anæsthesia is said to be sometimes present. The mental state of such patients renders this observation in most cases doubtful.

THE PHENOMENA OF THE FEVER.

The temperature range of typhus, as Lebert has pointed out, lies midway between that of enteric and that of relapsing fever. With the former it shows a rapid, progressive increase of body-heat during the first days of the attack, and a continuous high range marked by moderate morning remissions. The range of typhus, however, differs from that of enteric fever in the much more rapid rise to its maximum, the shorter duration of high temperature, the marked fluctuations early in the second week, and its critical termination.

The curve of typhus resembles that of relapsing fever in that both rise rapidly to the maximum at the outset, remain continuously high, and terminate abruptly. The rise in relapsing fever is, however, much more abrupt, the range higher, the course shorter, and the crisis on the fifth or seventh day more precipitous. In no case does typhus show, after an intermission of several days, the sudden, intense, febrile relapse which is characteristic of relapsing fever.

Considerable discrepancies are apparent upon an examination of the records of the studies made by many competent observers on the temperature of typhus. Doubtless some of the statements made are based upon too limited a number of observations, possibly others are the result of the investigation of cases in which the typhus temperature has been modified by complications, or other sources of error may have been overlooked. It is, however, probable that the discrepancies in the records of different observers are due in great part to differences in the temperature range of typhus in different epidemics.

Without going into a detailed comparison of the recorded observations, we may regard the following statements as representing the main facts.

The temperature rises rapidly from the onset of the disease, and in average cases reaches its maximum at from the fourth to the seventh day, or about the period of the appearance of the eruption. Occasionally, the maximum is attained as early as the third day, or, and this is especially so in very severe cases, not until the ninth or tenth day. The maximum is commonly between 40°—41° C. (104°—105.8° F.); it rarely reaches 41.5° C. (106.7° F.), except in children, and it may not exceed 39.5° C. (103.1° F.)

Even on the first evening it may reach 40°—40.5° C. (104°—104.9° F.). On the evening of the fourth day it is seldom below 40.5° C. (104.9° F.), much more commonly 41° C. (105.8° F.), and not rarely higher, while the average morning temperature at this period is 39.5°—40° C. (103.1°—104° F.). Exceptionally, an evening temperature of 39° C. (102° F.) has been encountered on the third day; but this is not of itself, by any means, of favorable prognostic import.

After reaching its maximum, the temperature falls off to a very slight extent in about two-thirds the cases, but remains about the same in the rest, and there is otherwise little change for several days, until about the seventh or eighth day—more rarely, as late as the tenth, when there is commonly, except in the severe cases, a slight remission, which, in the very mild cases, may be followed by complete defervescence, but after which the temperature commonly rises again, and then gradually but slowly falls until the twelfth or fourteenth day, when it rapidly subsides to the normal, or in many instances even slightly below it.

The morning remissions are less marked than in enteric fever; they vary from one day to the next, but the usual difference is 0.5°—1° C. (0.9°—1.8° F.) for the second week, though the same curve may show greater or less variations, especially as the period of the crisis draws near. Except during defervescence, a higher morning than evening temperature is very rare. A curve in which the morning fall is absent may be looked upon as an unfavorable indication, and the same may be said of a sudden fall of temperature, with a rise in the pulse or without improvement in the other symptoms. A high range of temperature in the first week is apt to be followed by severe cerebral symptoms in the second. The absence

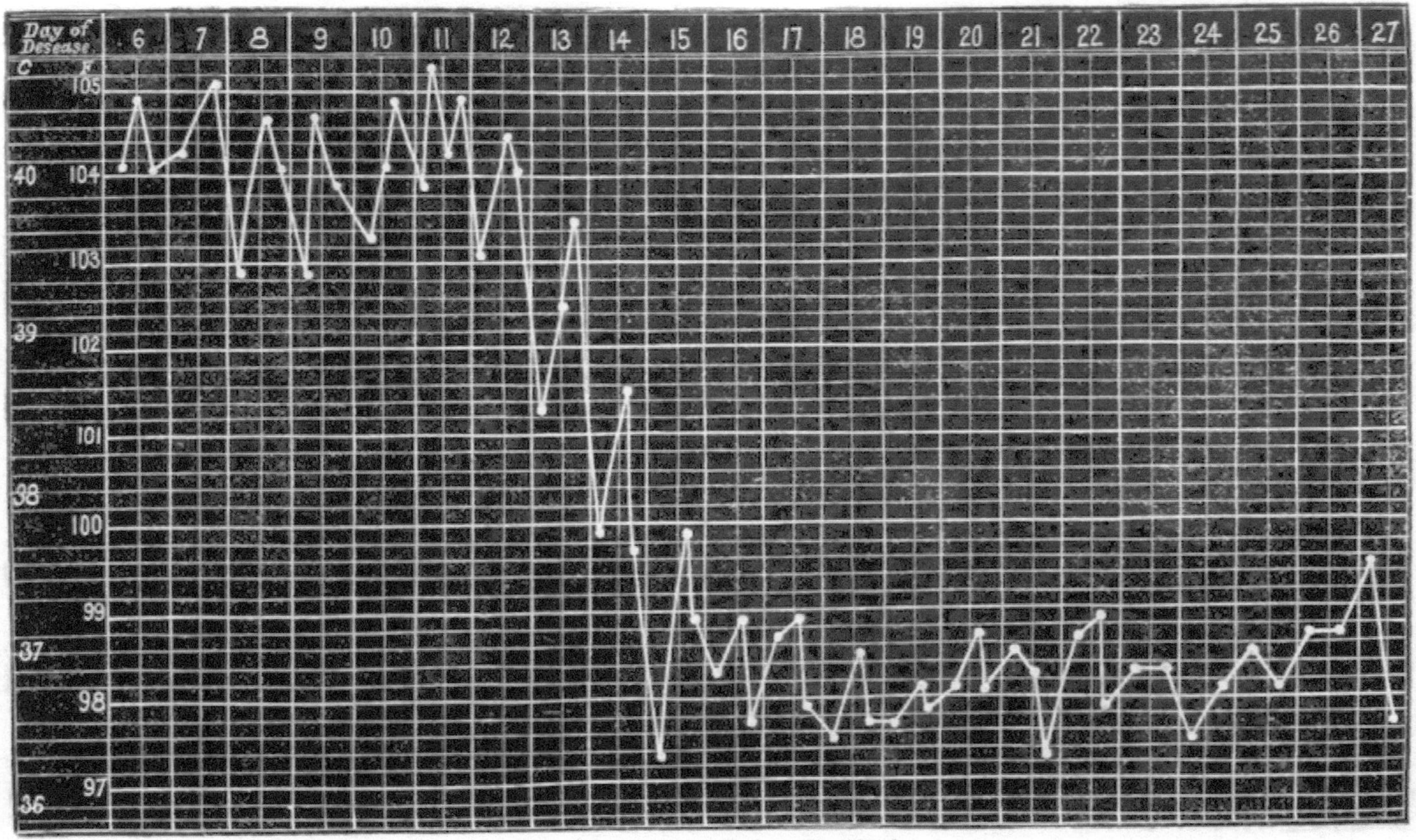

FIG. 19.—Temperature in **Typhus.** (Murchison.)

of a distinct, though slight remission, about the seventh or eighth day, is of unfavorable prognosis.

A fall of temperature before the crisis is common. This fall usually occurs in the morning of the day preceding the crisis. It may amount to $1.5°$—$2°$ C. ($2.7°$—$3.6°$ F.), or even to $2.5°$ C. ($4.5°$ F.), but the temperature rises again in the evening, to fall permanently on the following, the critical day. In rarer cases a decided rise, amounting to $2°$—$2.5°$ C. ($3.6°$—$4.5°$ F.) precedes the crisis, and there are cases in which a gradual abatement, with a progressively lower morning and evening temperature from day to day, occurs. Finally, there are cases in which no change in temperature precedes the crisis, which sets in suddenly and progresses with rapidity.

The critical defervescence may be completed within twelve hours. Much more commonly it occupies one, two, or even three days. It usually begins in the evening, only exceptionally during the course of the day. The fall in temperature amounts to $2°$—$4°$ C. ($3.6°$—$7.2°$ F.).

Recovery commonly takes place after the crisis; but in some instances the patients have fallen into fatal collapse after it, or death may occur in consequence of some complication.

The fall is usually to the normal, but not infrequently to a point a little below it, $36.5°$—$36°$ C. ($97.7°$—$96.8°$ F.). The evening after the lowest point is reached there is usually a slight rise, to be followed by a fall to below the normal the following day, and a subnormal temperature is often present for several days in the convalescence, liable, however, to occasional transient subfebrile exacerbations in consequence of complications, or without assignable cause.

In very rare instances an attack of typhus has been protracted into the third or fourth week, coming slowly to an end by true lysis.

A rise in temperature of from $1.5°$—$3.6°$ C. ($2.7°$—$6.4°$ F.) takes place just previous to, or at the time of death, in uncomplicated cases.

The pulse from the beginning of the attack is frequent, varying between 110 and 130, and keeping pace with the severity of the general symptoms, and in the main with the temperature range. The evening rate is usually slightly in excess of the morning, and the daily variations are inconsiderable. Sometimes it increases from day to day until death or recovery. Although a rapid pulse is commonly present in severe cases, a slow pulse is by no means invariably the indication of a mild case. The pulse is sometimes slow in cases characterized by extreme prostration, and death has taken place in cases in which the pulse at no time exceeded 100 (Murchison).

A gradual at first, but toward the end of the defervescence a rapid fall in the pulse-rate, is commonly the attendant symptom of improvement. With a temperature below the normal, in the first days of convalescence, there is usually a pulse lower than normal in frequency (50—70). If it remain frequent, particularly while the patient is in bed and quiet, this

TYPHUS FEVER. 273

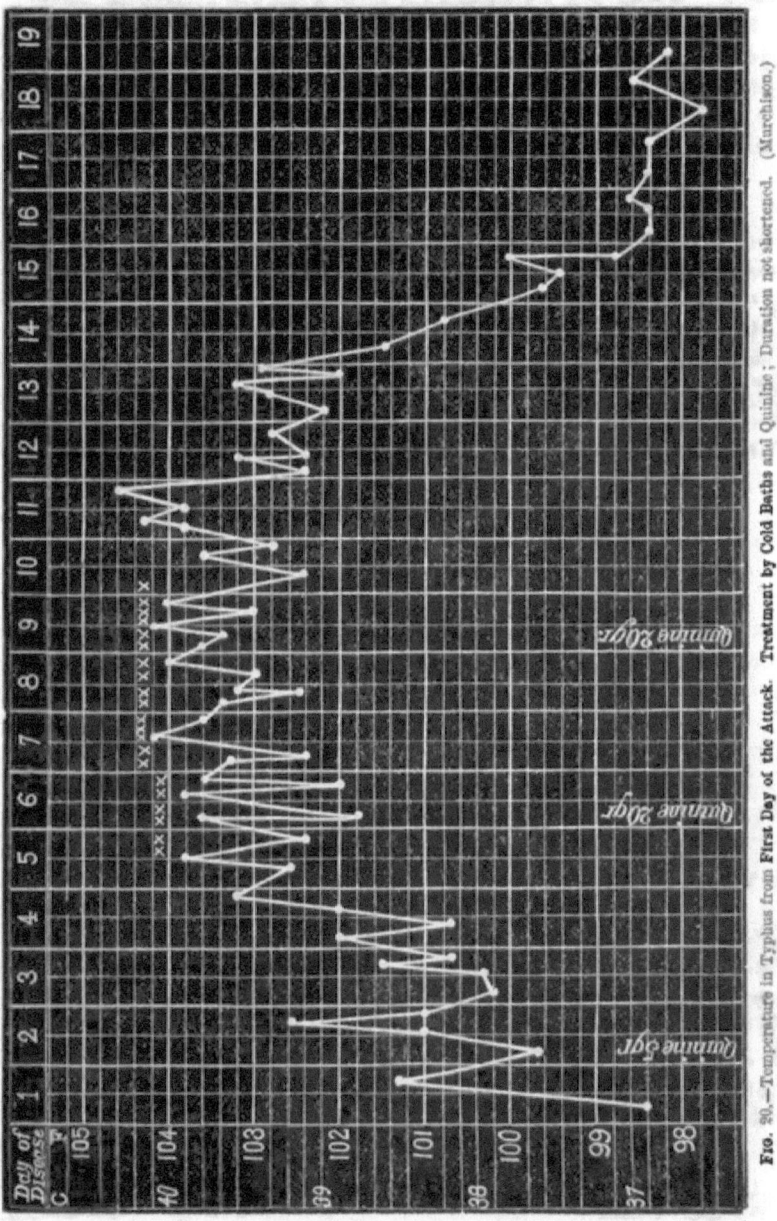

FIG. 90.—Temperature in Typhus from First Day of the Attack. Treatment by Cold Baths and Quinine; Duration not shortened. (Murchison.)

18

may be in consequence of some complication; and a decided rise, after it has fallen with the temperature, is almost always indicative of a complication, which is often pulmonary.

During the second week, if the depression be very great, the frequency of the pulse may be accelerated, while the temperature is slightly lowered; but, upon the whole, the pulse increases in frequency as the temperature rises, and falls with the defervescence.

In the beginning of the attack the pulse is full, soft, and compressible; in young and vigorous persons it may for a time be somewhat tense, as well as full, but this is highly uncommon. As the disease progresses and the strength of the patient becomes from day to day more impaired, the pulse diminishes in force as it rises in frequency, becoming smaller and weaker, until at length it is perceived with difficulty, or not at all. This feeble pulse of typhus is not only greatly modified in frequency, but also in force by changes from the horizontal to the semi-erect or erect position, both during the fever and in the convalescence. It is not rarely irregular or intermitting, often undulatory, but less commonly dicrotous in typhus than in enteric fever.

The impulse of the heart is almost invariably, except in the mildest cases, enfeebled from the fifth or sixth day of the disease. This enfeeblement is progressive, and for several days during the height of the disease the impulse is not infrequently absent altogether. Coincidently with the diminution of the impulse, the first sound becomes progressively less distinct, and may at last be inaudible. It is occasionally replaced by a soft, systolic murmur of hæmic origin. The second sound remains distinct. With convalescence, the impulse and the cardiac first sound only gradually regain their normal character.

The radial pulse is not always in correspondence with the impulse of the heart as regards force. The former may be small, weak, or even imperceptible, while the cardiac impulse is excited and so strong as to distress the patient, and the systolic sound but little enfeebled.

These phenomena relating to the heart are due in part to impaired innervation, and in part to the degenerative changes that take place in the muscular tissue of the organ. They are of the utmost importance clinically, as affording the most reliable guide to the administration of stimulants, both as regards the time and the amount.

To the enfeeblement of the circulation so characteristic of typhus are to be referred the duskiness and lividity of the face and extremities, often seen in the fully developed disease, the tendency to venous and arterial thrombosis described in rare instances, and the occurrence of embolism of the larger arteries, with resulting gangrene of an extremity.

The urine varies in quantity with the amount of fluids taken. During the first week, it is often reduced from twenty-five to fifty per cent. In the later periods of the disease it is not infrequently increased. In severe cases partial or total suppression may occur.

Notwithstanding the large amount of water drunk, and the dryness of the skin and absence of diarrhœa, the amount of urine in the whole

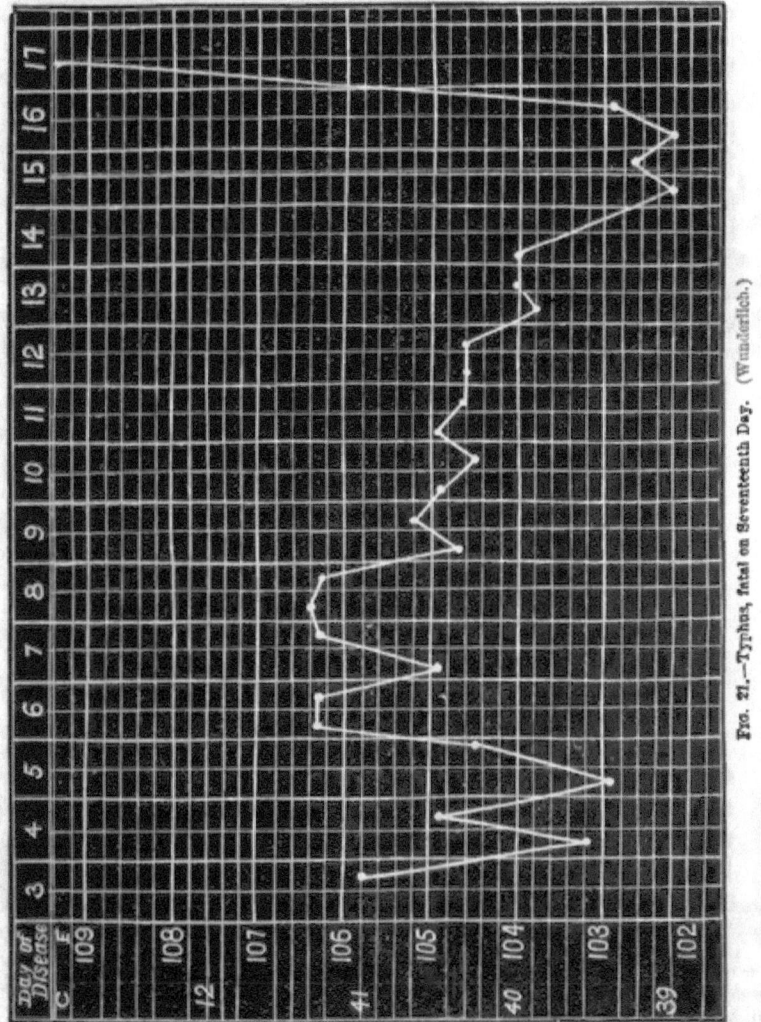

FIG. 21.—Typhus, fatal on Seventeenth Day. (Wunderlich.)

course of the disease is in most cases decreased. There appears to be a retention of water in the system. At the commencement of convalescence the secretion is often greatly augmented.

It is commonly high-colored at first; it may remain so until the crisis.

As the quantity increases, either at the height of the attack or upon the beginning of convalescence, the urine may become very pale. When partial suppression occurs, it is often of a dirty brown color like porter, with a copious deposit of renal epithelium and blood-elements.

The specific gravity is usually high in the beginning of the attack, and lower later in its course and in convalescence, varying with the amount of urine passed.

The reaction is at first decidedly acid; later it is often neutral, or even alkaline, when freshly passed, and deposits an abundance of urates and phosphates.

The chlorides gradually lessen; at the beginning of the second week they are reduced to a trace, or, in severe cases, are altogether absent. This is not wholly due to the non-ingestion of salt, since in health all chlorides may be withheld, yet for a considerable time, the urine will continue to contain them. With the approach of convalescence they reappear in some quantity in the urine without change in the diet, and gradually increase from day to day. Their disappearance is due to the processes of the fever, and takes place in cases uncomplicated by diarrhœa or by pneumonia.

The daily excretion of urea is at first considerably above the normal amount, notwithstanding the decrease in the quantity of food, and this increase is proportionate to the intensity of the fever and the consequent tissue-waste. The increase has been present in the earliest days upon which the urine has been examined. The daily excretion is very variable in the second week. In some cases it gradually diminishes, in others it remains high until the crisis. During the early days of convalescence the quantity of urea falls below the physiological standard, notwithstanding a generous diet, and gradually rises again as health and strength are regained.

Urea has been repeatedly found in the blood of persons who have died of typhus, with marked cerebral symptoms, even although there have been no disease of the kidneys and no diminution in the amount of urine (Murchison).

Uric acid is usually increased.

Albumen has been found to be present in the urine in a considerable proportion of cases, by all observers who have made it the subject of investigation. This proportion has greatly varied in different epidemics; sometimes constituting the greater number or even nearly all of the cases examined, at other times amounting to a very small percentage of them. The albumen varies in amount from a mere trace to an abundant deposit; it may appear early in the attack, or not until a day or two before the crisis; and finally, it is often transient, lasting at the most three or four days, sometimes persistent from the day of its first appearance, commonly the sixth or seventh day of the disease, until death or recovery. The cases in which albuminuria is early, or copious and persistent, are almost always severe. At the same time it does not arise as a symptom in very many severe or even fatal cases. The albuminuria of typhus is due to the altered condition of the blood, **which induces** hyperæmia of the kidneys, that may proceed to actual inflammation of **their structure.** With the albumen, and in many cases when its presence cannot be detected, **renal** epithelium, and hyaline, granular and epithelial casts are found. Casts are more frequently present **than** absent in the urine of severe cases (Da Costa). In cases of great severity, blood and blood-casts **are** met with. After death **the** kidneys have frequently presented the appearances of acute nephritis, where no history of previous renal disease existed. Previously existing nephritis, in a limited proportion of the cases, will account for the albuminuria.

Sugar **was** found in minute amounts in the urine of nine out of fourteen typhus cases, in which it was sought for by Dr. George Buchanan, who states that it appeared at any period between the sixth and twenty-seventh days, and only lasted a day or two. It was of no clinical significance.

SYMPTOMS MANIFESTED BY THE SKIN.

The eruption of **typhus is prominent among the clinical phenomena of** the disease. **It is very rarely absent. The statistics of the London Fever Hospital show that, of** 18,268 cases admitted to that institution in **twenty-three years, the eruption was noted in** 17,025, or in 93.2 per cent.; **and Dr.** Murchison informs us that these figures exaggerate the proportion of the cases in which it was absent, it being, in certain cases where it was **present, noted as** absent by resident medical officers, who were not sufficiently vigilant, or were new to their work. He further states that in the year **1864, when** the records were kept with unusual care, the eruption was **noted as present in all but 55** out of 2,493 cases, or in 97.77 per cent., and that some **of** the cases in which it was not found were admitted after the patient had passed through the attack, so that the probability that the eruption had been present and had disappeared is to be regarded.

All observers agree that the eruption is absent in a very small proportion of the cases, and that it is of diagnostic importance. Both the mottling, or subcutaneous rash, and the distinct measly eruption, are darker and more distinct upon the dependent parts of the body. **The** back should, therefore, always be carefully examined **in** case of doubt.

The character of the eruption **has already been** described **in the** consideration of **the** clinical history of **this fever. The** mottling and **the** distinct rash usually exist together; **but in** the lighter cases, and particularly in children, the former mainly constitutes the eruption, while in older persons the distinct rash is very prominent. The proportion of cases in which the eruption is altogether absent is much greater under the age of fifteen than over it. In children true petechiæ rarely appear; but they have been observed at all periods of life, from earliest infancy to extreme old age.

A copious eruption, deep in color and early becoming livid or petechial, generally accompanies the severe cases of the disease. The extent and distinctness of the eruption, and in particular its lividity and the abundance of the petechiæ, have been regarded in all times as proportionate to the general severity of the case.

The absence of the eruption, in the rare cases in which it is wholly absent, is not always of itself of favorable import. Lebert states that in his experience such cases have proved very serious, and in several instances have terminated fatally. On the other hand, Murchison has found the cases without any eruption mostly mild.

The eruption shows itself on the fourth or fifth day, as a general rule; it may, however, appear upon the third day, or not until the end of the first week after the beginning of the fever. The cases in which it appears later than the sixth day are extremely rare. It is first seen upon the sides of the chest or abdomen, or in rarer cases upon the backs of the hands or the wrists, and spreads rapidly over the trunk and extremities, respecting only the neck, face, the hairy scalp, and the palms of the hands and soles of the feet; but even these exceptions are not constant, and cases are not very rarely observed, especially in childhood, in which the distribution of the typhus-eruption is not less extensive than that of measles. A diffuse, erythematous blush is not infrequent during the first day or two after the appearance of the eruption. If the spots are faintly raised above the surface of the skin, their grouping presents a close resemblance to measles—a resemblance heightened by the conjunctival injection, and the nasal and pulmonary catarrh, which are also at this time well developed.

The eruption of typhus never appears, as does that of enteric fever, in successive crops. It requires in most instances, apparently from its very abundance, a variable time, often forty-eight or even sixty hours, for its full development; but the spots that appear upon the second or third day are superadded to those first seen, and are due to the same cause. After the rash is completely established, it is permanent, no new spots of the same kind appearing. Its average duration is from eight to eleven days; it disappears, as a rule, with the defervescence. In some cases, and particularly when it has consisted only of a faint mottling, it lasts but a brief time—from a few hours to a day or two—and wholly vanishes several days before the termination of the fever. Where the eruption is very dark, and where true petechiæ are abundant, the discoloration of the skin persists for some days into the convalescence, and only gradually fades.

The course of the typhus-eruption is as follows: at first the lesion consists of a hyperæmia of the cutaneous capillaries, which, being in part diffuse, is manifested by that mottling or marbling of the surface, described as the subcuticular **rash**; and being in part localized, shows itself in the pale, dirty pink or florid spots, faintly raised above the surface, disappearing upon pressure, and variously grouped, which have already been described as the rubeoloid or measly eruption. In the course of two or three days these spots are found to be no longer elevated; they have lost the brightness of their color, and, in consequence of the transudation of blood-coloring matter from the vessels, appear as reddish brown or rust-colored **stains in** the skin, not distinctly marginate, but fading obscurely into the tint of the surrounding surface. The hyperæmia has now given place to pigmentation, which cannot, however, be called petechial. The spots no longer disappear, although they grow less distinct under pressure. This change does not usually affect all the spots. Many of them, under ordinary circumstances, gradually disappear about the middle of the second week of the fever, or at the approach of the crisis. In the second week a minute extravasation of blood appears at the centre of some of the pigmented spots, and in certain cases petechiæ **appear**, not as a step in the evolution of the typhus-eruption, but as such from the beginning. The blood-stainings in the site of the spots, **and the** true petechiæ, are alike uninfluenced by pressure. The duration of the various stages of the eruption **is by no means constant.** In truth, it may come to an end without passing into the rusty color; still less before the appearance of petechiæ. In some cases the eruption is rusty, or livid and petechial, from a very early stage, or, in fact, from its beginning.

If death occur while the eruption is pinkish or florid and disappears on pressure, no trace of it is seen in the dead body; but if it has become rusty, or if petechiæ have developed, the eruption is persistent, and upon examination **of sections of the** skin it **is found to be** stained by the coloring-matter of **the blood.**

The subcuticular mottling usually disappears after a few days, while the spots grow darker and more distinct.

Hence, as Murchison has pointed out, the eruption of typhus is pale and blended in the early stages, darker and more spotted in the later periods.

True petechiæ are neither essential nor peculiar to typhus. In many cases they do not occur at all—in few before the last stages.

The erythematous blush which attends the outbreak of the eruption usually quickly subsides. **In the later periods of** grave **cases the skin** grows dusky or livid, especially in the dependent portions of the body.

Vibices and larger transudations of blood beneath the skin (*ecchymoses*) also occur in grave cases, and especially where scurvy exists as a complication, as it did in the war of the Crimea.

Murchison mentions "*taches bleuâtres*" as occasionally occurring.

Sudamina are occasionally met with. They are more common before than after the fortieth year of age, and may occur either with or without marked sweating.

Desquamation follows the disappearance of the eruption in a limited number of cases. It is fine and bran-like, and proceeds from above downward, but may become coarse or membranous on the hands and feet. The nails show a white band and a furrow as the result of the disturbance of nutrition which attends the fever. There is in most instances more or less falling of the hair during the convalescence.

Urticaria occasionally makes its appearance in young persons about the time of the crisis, or early in the convalescence.

Herpes occurs exceptionally. It may precede the eruption and give rise to brief confusion of diagnosis. It also occasionally appears toward the termination of the disease.

Erysipelas has occurred with frequency in the wards of certain hospitals, and is due to hospital influence rather than to the processes of typhus. It is a serious complication, but not necessarily fatal. It appears usually upon the approach of or during convalescence, and is usually confined to the face and head.

These eruptions are accidental.

The general appearance of the patient ill of typhus fever is peculiar. It is often of itself so striking as to indicate, even to a person of limited experience, the nature of the disease, and in cases of doubt it is of diagnostic value. The expression of the countenance, the appearance of the skin, the attitude, considered together, constitute what may be called the physiognomy of the disease. From the first the face is the index of the grave derangement of the functions of the nervous system that are so prominent. The look is dull and heavy; as Da Costa has well said, it is "coarser" than in health. It is also most weary. As the case progresses the expression becomes blank and stupid, the eyes are half-closed, or widely open and staring at nothing, and the lips relaxed, parted, the lines indicating thought and emotion blurred or altogether blotted out. The facial expression varies with the form of the delirium. In typhomania it is feeble, fatuous, or silly; in the trembling, which so closely resembles the delirium tremens of alcoholism, it is eager, restless, suspicious, and when the delirium is active or acute, the expression is often bold and defiant. If it is at all anxious, it betrays the terror or anxiety due to some fixed, distressing idea with which the delirious brain busies itself, not the anxiety of suffering or suspense met with in many acute diseases, for the pains of typhus are not commonly acute, and it is rare for the patient to manifest concern as to the issue of his sickness.

A uniform general flushing of the face or cheeks is generally present.

It may be deepest over the cheek-bones, but it is never circumscribed. It is not pink or florid, but of a dull, dusky-red color, or it may be of an earthy or leaden hue. In the advanced stage of grave cases **the face is often** livid, particularly about the mouth and nostrils. The eyes are suffused and the conjunctivæ injected; sordes collect upon the lips and teeth.

The patient lies most wearily upon his back. In the **graver cases** he is unable to move in **bed, or to help** himself. He lies with his hands crossed upon his abdomen or extended at his sides, unconscious of or indifferent to what goes on about him.

SYMPTOMS REFERABLE TO THE RESPIRATORY SYSTEM.

The respiratory movements are moderately accelerated during the first week, ranging from twenty to twenty-four per minute in uncomplicated cases. With the advent of delirium and the increased frequency of the pulse which attends the developing prostration, the respiration becomes more hurried and shallower. In grave cases, in which the disturbance of the nervous system is profound, the respiration is sometimes **abnormally** slow. It is more commonly hurried, and may without pulmonary complication be irregular, sighing, or jerky.

Affections of the lungs, as in enteric fever, are very common. Of these, the most are to be regarded as complications, and will be considered under that heading. But the *bronchitis*, which is usually **present in the** first week, appearing coincidently with the nasal and conjunctival catarrh, seems worthy to be regarded as a symptom. It is at first attended with but slight cough; little expectoration, or none at all; no difficulty and but little quickening of the movements of respiration, and there are detected upon auscultation a few scattered, subcrepitant, sonorous or sibilant râles. During the second week the bronchitis may become diffuse, and extend to the capillary tubes. This it may do in an insidious manner, without much cough or other danger-symptoms except quickening of the breathing. In other cases the pulmonary symptoms become grave, or they may even predominate so that the case may assume the form that has been described as bronchial typhus. Physical exploration reveals subcrepitant râles in all parts of the chest, localized bronchial breathing with dulness, areas of very faint respiratory sounds, or extensive hypostatic congestions, or the signs of lobar pneumonia.

SYMPTOMS REFERABLE TO THE DIGESTIVE SYSTEM.

The affections of the digestive tract consist chiefly of perverted functions and of catarrhal conditions of the mucous membrane of the mouth, pharynx, stomach, and intestines.

The tongue is at first covered with a thick whitish or yellowish-white fur. It may remain thus furred and moist throughout the attack in mild cases; but, as a rule, it becomes, at the end of the first or the beginning of the second week, dry and brown or brownish along its middle. In severe cases it is very dry, often retracted into a globular mass, and is coated with a dry, dark crust, which sometimes cracks at several places. In severe cases attended with profound asthenia the tongue is sometimes moist throughout. The tip and edges are usually pale. The deep fissures so often met with in enteric fever, and occasionally in relapsing fever, are rare in typhus. The tongue is protruded tremulously, especially in the second week of the disease. It is sometimes protruded with difficulty, or not at all; apparently in some cases from sheer weakness, in others from dulness of intellect, and yet in others on account of its dry and firmly contracted state.

Sordes begin to collect upon the gums and teeth, and even upon the lips, about the beginning of the second week. The gums also bleed at times, but this phenomenon is usually associated with a scorbutic tendency.

Loss of appetite is complete throughout the attack. A desire for food is very often expressed immediately upon awaking from the sleep that marks the crisis, and it is not a very unusual circumstance for patients to ask for food shortly before the crisis, before any indication of improvement as regards the general symptoms is apparent.

Thirst is a constant symptom. It varies greatly in urgency, and is excessive in about one-fourth the cases. It is most prominent during the first week, and diminishes or altogether ceases upon the advent of the graver nervous symptoms that set in with the second week.

Difficulty in swallowing occurs in a small proportion of the severer cases. It first appears in the later periods of the disease, and may be due to the extreme dryness of the mouth and throat, or to spasm or paralysis of the muscles concerned in deglutition.

Nausea and vomiting are not common in typhus. Vomiting has been noted in from six to ten per cent. of the cases in some epidemics in which it was made the subject of special investigation. It is in most instances an early symptom, but may recur for a day or two preceding the crisis, and in some cases is confined to the period of convalescence. In dyspeptic persons it may continue throughout the attack. The matters ejected usually consist of thin, gastric mucus tinged with bile.

Vomiting may be the forerunner of the symptoms of uræmic toxication, such as general convulsions and coma. Occurring toward the end of the first or the beginning of the second week, it thus becomes a symptom of the gravest significance, and should direct the attention of the physician to the closest scrutiny of the amount and character of the urine, and particularly to the absence or presence of albumen, if these matters have not already been made, as they invariably should be, the **subject** of routine examination.

Tympany is rare in typhus. In those cases in which it occurs **it is a later** symptom, and is associated with grave depression of the nervous **system.** It may so distend the belly and interfere with **the descent of** the diaphragm as to seriously embarrass respiration.

Abdominal pain and tenderness are not common. Slight colicky pains may occur during the first week, and there is usually a little obscure tenderness in the hepatic region. Pain or deep tenderness localized in the area corresponding to the ilio-cæcal region of the gut does not occur.

The liver is slightly enlarged in rather less than one-third of the cases.

Enlargement of the spleen is discovered in the greater number of cases of typhus by physical examination during life, and this organ is found to be enlarged and softened in about three-fourths the cases examined after **death.** The enlargement is acute, and may be made out by the fifth day. It is less common than in enteric fever, and does not attain the limits common in relapsing and malarious fevers.

Constipation is very common **in typhus. In many** cases, however, the bowels are moved regularly, and exceptionally **there is diarrhœa.** The state of the bowels varies **in different epidemics.** Lebert mentions that *diarrhœa* was frequent in the typhus epidemics **observed by him at** Breslau, and Da Costa and other observers of typhus in the United States **have** noted it as a frequent symptom. Spontaneous diarrhœa is not common in the lighter cases. Of 31 cases noted by Da Costa at Philadelphia, this symptom was present to a marked degree in 13, or 41.9 per cent. **Murchison,** on the other hand, informs us that, of 1,782 cases collected **from** various sources, diarrhœa occurred in only 184, or in 10.32 per cent., while **in 959 of 1,**739 cases, or 55.14 per cent., there was obstinate constipation; and of 14,589 cases admitted **to the** London Fever Hospital during nine years (1863-70), only 734, or 5 per cent., suffered from diarrhœa. Of the 734 patients in whom diarrhœa occurred, 178, or 24.25 per cent., died, while the death-rate of patients in whom diarrhœa did not occur was only 18.14 per cent. Diarrhœa may be present in the early days of the attack, or may occur at any time during its course, either spontaneously or as a result of purgative medicines. It is also observed at the crisis. It is usually mild, but may in rare cases be very trouble**some, or** even so excessive as to endanger the life of the patient by the **increased** prostration to which it gives rise. Involuntary discharges occur in cases **of the** most serious kind, **and** commonly upon the approach of death.

The stools, when there is diarrhœa, are usually of a dark, greenish brown color, and of a less fluid consistency than in enteric fever; in some cases they are of a light color and watery. In Da Costa's cases the stools were small, thin, and feculent, often offensive, and of a yellowish color. They are not commonly attended by pain or tenderness, although in rare instances the movements are preceded by colicky pains.

Complications and Sequels.

The complications of typhus **are numerous.** They vary in different epidemics, but sometimes appear to be determined by individual peculiarities, different members of the same family presenting, when attacked by the disease, the same complications, such as convulsions, palsies, gangrene, and the like. The fatal termination is not infrequently due to some complication; and the occurrence of a complication may postpone the critical defervescence or arrest it altogether, and thus prolong the attack to an unusual length, and cause it to end, in cases ultimately favorable, by a gradual defervescence (lysis).

The convalescence may be interrupted and greatly prolonged by the development of sequels.

Affections of the respiratory tract are common and serious in typhus.

Acute laryngitis, leading speedily to œdema of the glottis, is of occasional occurrence. This may occur of itself as a secondary affection, or it may be led in by ulceration of the larynx, by a post-pharyngeal abscess, by enlargement and suppuration of the parotid or submaxillary glands, or it may occur in consequence of erysipelas. Its advent, either with or without preceding inflammatory processes in contiguous **structures,** is insidious. Slight huskiness may be quickly followed, **after a brief** period, by laryngeal breathing and the signs of impending asphyxia, rendering prompt laryngotomy or tracheotomy necessary to save life.

The laryngitis **is sometimes** croupous, and diphtheria **of the larynx and pharynx occur.**

Laryngeal ulceration **is less common than in enteric fever.**

The obscure onset of the pulmonary complications of typhus has already been alluded to. It is of the utmost importance that systematic physical exploration of the chest be made from day **to day.** The gravest chest-complications may be developed with but little cough, little **or no** expectoration, and no complaint of pain whatever. The debility of **the** patient and his blunted perception serve **to mask the special symptoms of lung-trouble.** Hurried respiration and increased duskiness of the face are the danger-signals.

Bronchitis has already been spoken of. It is, in fact, a symptom rather than a complication. The great danger lies in its tendency to become diffuse and to extend into the finer tubes, and thus, leading to atelectasis **and secondary lobular pneumonia,** to destroy the patient by cutting off **extensive areas of** breathing surface.

True lobar pneumonia is infrequent in typhus. It is manifested by the usual signs, dulness—crepitus, bronchial respiration, and rusty sputa.

Gangrene of the lung also occasionally occurs; it is manifested by the peculiar and horrible fœtor, supervening upon the signs of acute inflammatory processes affecting the lungs, the altered look of the patient, and a simultaneous aggravation of all the symptoms. It is usually fatal.

Pleural effusions, both serous and purulent, occur in rare instances. Unless the chest be systematically explored they are apt to be overlooked, as they come on insidiously, without pain, and do not greatly embarrass the respiration until they have attained considerable volume.

Phthisis is sometimes lighted up during the attack of typhus or in the convalescence. The catarrhal pneumonia persists, and rapid emaciation, night-sweats, and muco-purulent sputa occur.

Blood-spitting is a very rare occurrence in typhus fever. Murchison points out the fact that it may occur in consequence of the pulmonary hyperæmia in a previously diseased lung, or by reason of the existence of the hemorrhagic diathesis.

Other hemorrhages are not uncommon in grave cases—not alone when, as so often has happened in camps and armies and in times of protracted scarcity of food, scurvy complicates the disease, but also when no such predisposing cause is present. Bleeding from the nose, the gums, the bowel, the urinary passages, and the vagina, as well as the spitting and vomiting of blood, have been observed. Slight wounds and superficial excoriations may give rise to serious or even fatal hemorrhage. These events are unusual. But, in many epidemics, large subcutaneous extravasations of blood are common, and after death similar effusions are found beneath the mucous and serous membranes, in the intermuscular planes, and within the substance of the muscles.

Boils occasionally break out in numbers during the convalescence. They constitute a troublesome sequel.

Among the rarer complications of typhus is *pyæmia* with *purulent deposit in the joints.* It begins with severe chills, followed by great prostration, rapid and feeble pulse, and acute swelling of the joints with tenderness and redness. There is commonly jaundice and sweating. The smaller joints also, are often implicated. After death the joints are found to contain pus, but abscesses in the internal organs are rare. This complication usually appears at the time of the crisis, or early in the convalescence.

Phlegmasia dolens was noted as of common occurrence in the convalescence in those epidemics in which bleeding entered largely into the treatment. This complication has been very rare in recent epidemics.

When *erysipelas* occurs, it usually comes on late in the fever or early in the convalescence. Much more rarely it occurs shortly after the beginning of the disease. It commonly commences about the root of the nose or the lobe of one ear, and spreads over the face and scalp, sometimes leading to the formation of abscesses of the eyelids and of the integuments elsewhere. The pharynx and larynx are very often implicated in the erysipelatous process, and œdema of the glottis sometimes results. Erysipelas occurs less frequently in other parts of the body. Many cases of erysipelas arising at the same time, where typhus fever patients are crowded together in a hospital ward, are to be attributed rather to bad hygienic influences than to the typhus processes.

Diffuse inflammation of the subcutaneous tissues, resulting in purulent infiltration, occasionally occurs, most commonly in the lower extremities. It is attended by the symptoms of serious constitutional disturbance and pain in the affected part.

Enlargement and suppuration of the parotid gland occur early in many epidemics. Sometimes, however, they are met with about the time of the crisis, and again they may not be developed until convalescence. They occur at all periods of life, but are proportionately more common after the thirtieth year. The tumefaction forms rapidly and suppuration speedily follows. Resolution, however, may occur without the formation of matter. The connective tissue overlying the glands is largely involved in the suppurative process. The parotitis is often associated with facial erysipelas or with extensive inflammatory œdema of the neck. This is a very dangerous complication. These inflammatory swellings occur also in the submaxillary glands, in the mammæ, the glands of the axilla and groin, and less frequently in the extremities. Their number is sometimes limited to one or two; sometimes they are numerous. They occasionally result in extensive gangrenous ulcers. In many epidemics they are absent. Parotid buboes and other inflammatory swellings occurring in typhus, suggest a relationship between this disease and the true plague. Murchison suggests that typhus is probably the plague of modern times.

Bed-sores are not very frequent in uncomplicated typhus. They are apt, however, to occur in cases protracted by other complications. They appear in those parts of the body subjected to pressure, the most common situation being over the sacrum. They also occur in the heels, upon the back of the head, at the trochanters, and over the vertebra prominens. They protract the duration of the illness, and may bring about a fatal termination of the case by exhaustion or by septicæmia.

Parts not subjected to pressure may become gangrenous in consequence of arterial thrombosis. The death of the tissues is usually preceded by darting pains and the signs of arrested circulation in the part, namely, numbness, coldness, and livid discoloration. The feet and ankles are apt to be involved, less frequently the nose, the penis and scrotum, and the external genitalia in the female.

In severe epidemics many observers have noted the tendency of wounds and ulcerated surfaces to become gangrenous in typhus-patients and even in those not suffering from the disease. Under such circumstances gangrene has resulted from the application of blisters and sinapisms.

Perforating ulceration of both corneæ occasionally occurs, in consequence of the exposure of the globe from the eyelids being kept constantly open.

Noma or *cancrum oris* occurs in some epidemics. It is fortunately a very rare complication, but is fatal in most instances. It is more frequent in children than in adults.

Necrosis is a rare result of typhus, as of other severe fevers. Murchison saw in one instance extensive necrosis of the fibula follow an attack of typhus. It is probably secondary to arterial thrombosis.

Pericarditis and endocarditis are extremely rare.

In rare instances *mental feebleness* follows the attack; but, as a rule, the intellectual faculties are restored shortly after the crisis. Maniacal attacks sometimes occur during the convalescence. They are usually transient.

Palsy may occur as a sequel of typhus fever. It may involve both lower extremities or one-half the body. Numerous examples of hemiplegia are recorded, and there are not a few cases of right hemiplegia with aphasia to be found in the literature of the subject. In other cases the palsy is restricted to individual muscles or groups of muscles. In very rare instances paralysis of one side of the face has been observed.

The paralysis following typhus usually terminates, in the course of some days or weeks, in recovery. It may, however, be persistent.

The deafness which so frequently attends the fever usually passes away in the early days of convalescence. Inflammation of the external auditory meatus, or of the middle ear, may give rise to permanent impairment of hearing upon one or both sides, and suppurative inflammation of the ear may be remotely followed by secondary inflammation of the meninges, as in scarlet fever.

Transient dimness of vision is occasionally noticed after severe attacks (Murchison).

It remains to notice some of the complications due to derangements of the digestive tract. Murchison saw one case of *acute glossitis* in his great experience. The patient recovered after free incision of the tongue. The occurrence of *diarrhœa* has already been spoken of. (See p. 283.)

Dysentery has prevailed in some epidemics, side by side with typhus, and especially in many outbreaks in camps and besieged cities, and has, under such circumstances, become a frequent complication.

Jaundice, a frequent symptom in relapsing fever, is very rare in typhus. When it arises, it is due either to congestion of the liver, or to gastro-duodenal catarrh occurring as complications; or it may be one of the group of symptoms belonging to pyæmia; or, finally, it may appear about the time of the manifestation of the typhus-rash, as one of the expressions of the overwhelming action of the poison upon the blood. It is then to be regarded as an ominous indication.

If *menstruation* occur during the course of typhus, it may be profuse and even endanger life. Murchison states that he knew of one case in which death was due to flooding.

Pregnancy affords no exemption from the attack of typhus. Pregnant women, even in the more advanced stages, may suffer from typhus without miscarriage; and when this accident does occur, it is not necessarily fatal to either the mother or child. During nine years (1862–70) 107 typhus-patients in the London Fever Hospital were known to be pregnant. Of this number 49 aborted from the tenth to the fourteenth day of the attack; of those who aborted 9 died. The remaining 98 recovered (Murchison).

In respect of the tendency to miscarriage and the danger to the life of the child, typhus is in strong contrast with relapsing fever.

VARIETIES.

The general characters of typhus present but little variation. The picture seen at the bedside has in the main the same general outlines and coloring in all epidemics. Variations in the groupings of the symptoms, and lightening of the tints or deepening of the shadows in particular cases, are to be attributed to differences in the constitution and habits of the patients, to differences in the circumstances under which epidemics arise, and, above all, to the complications which are so numerous, so common, and frequently so important in determining an unfavorable ending. The varieties of typhus that have been described by various authors, who have depended upon the prominence of certain symptoms or groups of symptoms as a principle of division, is very great. Among them may be mentioned inflammatory typhus, nervous or ataxic typhus, adynamic typhus, ataxo-adynamic typhus, catarrhal, scorbutic, purpuric, and dysenteric typhus, and a further exercise of ingenuity could almost indefinitely increase the list. A more useful distribution of the cases into varieties is that based upon the course and development of the attack considered in its completed clinical history. The cases, thus regarded, readily arrange themselves into the following five groups:

1. Common Typhus.
2. Fulminant Typhus.
3. Walking Typhus.
4. Mild Typhus.
5. Abortive Typhus.

1. *The common form* has already been sufficiently indicated in the foregoing sketch of the clinical history of the disease, of which it forms the basis.

2. *The fulminant form* (*typhus siderans*) is characterized by the furious onset of the attack, the intensity of all the symptoms, the extreme to which the temperature abruptly rises, the early appearance of grave cerebral symptoms, and finally by the rapidity with which death occurs, this event taking place within the first three or four days of the attack, or even in a few hours from its beginning. This form is rare.

3. *Walking typhus* (*typhus ambulatorius*) begins insidiously or in an intermittent manner, so that the patient is not compelled to keep his bed at first, or rises and goes about after an initial paroxysm. The case, apparently benign at first, after a time assumes the gravity characteristic of typhus, the patient suddenly falling into a state of extreme prostration or even dangerous collapse, or becoming delirious and manifesting a suicidal tendency

4. *Mild typhus* (*typhus levissimus*) is a form unattended by severe symptoms of any sort. The nervous phenomena seldom exceed headache and mild delirium limited to the night. The temperature scarcely rises above 39° C. (102.2° F.). The eruption is little marked, or absent altogether. The defervescence occurs commonly on the seventh day, and is usually completed in the course of the eighth day; it is often attended by herpes. This form, which is to be distinguished from abortive typhus, occurs more or less frequently in all epidemics. It also occurs at times and places where typhus is endemic rather than epidemic, and, in the absence of the eruption, is apt to be confounded with simple continued fever or febricula. It is apt to occur chiefly among young persons, and among individuals, who, although exposed to the contagion of typhus, live under favorable hygienic conditions. Closely allied to this form is the condition described by Jacquot as *typhisation à petite dose*. This condition is attended by malaise, slight fever, loss of appetite, a sense of bodily and mental fatigue, broken rest and some mental confusion, but does not pass into actual typhus.

5. *Abortive typhus* (*typhus abortivus*). This form is only to be recognized by its occurrence during the epidemic prevalence of typhus. It begins abruptly with headache, epigastric distress, a chill or chilliness, followed by decided fever. Pains in the back and limbs also occur. There is mental dulness, a foul tongue, constipation. The attack presents, in short, all the symptoms characterizing the onset of the prevailing epidemic disease. At the end of the second, third, or fourth day, however, there is a critical defervescence accompanied by sweating, diarrhœa, or, in some instances by vomiting. Convalescence then occurs.

Prognosis and Mortality.

A knowledge of the duration of the disease is important with reference to the prognosis. The mean duration of typhus fever is about fourteen days; mild cases may end in permanent improvement at the close of the first or beginning of the second week. The duration of average cases is from thirteen to fifteen days. Uncomplicated cases rarely exceed twenty days. If the defervescence be postponed to the end of the third week, it is in consequence of some local complication. The mean duration of 500 cases that terminated in recovery was, according to Murchison, 13.43 days, while the mean duration of 100 fatal cases was 14.6 days, but in all of the fatal cases protracted beyond the twentieth day the result was due to some complication. The attack may, however, be prolonged to four, five, or six weeks, but this is always in consequence of complications. The convalescence may be stated to be about as long, as a rule, as the attack, so that an interval of from four to six weeks from the beginning of the attack usually elapses before the patient is able to return to his customary avocation. The length of the attack varies somewhat at different periods of life, being shorter, as a rule, in childhood and youth than in middle or advanced age. The mean duration of the attack has been found to be longer at the beginning of an epidemic than toward its close.

True relapses in typhus are extremely rare.

The death-rate varies greatly in different epidemics. Lebert estimated it to be six or seven per cent. in the Valais epidemic in 1839, while Jaccoud states that in the Crimea and in Algiers the mortality has reached from fifty to fifty-five per cent. Outbreaks in camps and armies, and those following severe famine, have invariably been attended with a higher death-rate than those affecting the civil community under more fortunate circumstances. Griesinger computes the mean mortality at from fifteen to twenty per cent. Not only does the mortality vary in different epidemics, but it also varies in different years of the endemic prevalence of typhus. Thus, we find, upon consulting the statistics of the London Fever Hospital, that the average mortality for 23 years was 18.92 per cent., or, deducting 686 cases fatal within forty-eight hours, 15.76 per cent.; but that, in the year 1857, the mortality was 25.18 per cent., in 1858 it was 60 per cent., in the year 1859 it was 33.33 per cent., while in 1860 it was 40 per cent., the admissions in these years being respectively 274, 15, 48, and 25.

The mortality in any community is lower than the mortuary reports of its hospitals would indicate. Cases occurring in the higher walks of life, among children, and mild cases, are less likely to become the subjects of hospital statistics than those occurring among the destitute, those enfeebled by privation, and the aged and infirm inmates of poor-houses, and the like. Murchison estimates that the mortality in London, allowance being made for these sources of fallacy, does not probably exceed ten per cent.

Sex influences the mortality. In childhood the number of deaths is comparatively greater among females than among males; but, after the fifteenth year, typhus appears to be somewhat more fatal in males than in females. The difference in adult life is probably due to the greater prevalence of the alcoholic habit in the male sex, and the greater consequent liability to morbid conditions of the liver and kidneys.

Age is of great importance as influencing the death-rate. In childhood and youth, typhus is by no means a fatal malady. In old age, on the contrary, it is most mortal, and the death-rate progressively rises from the earliest to the latest periods of life.

I have rearranged one of the tables of the London Fever Hospital statistics given by Dr. Murchison, so as to show the number of admissions in each of the decades of the years of life in 18,268 typhus-patients, and the corresponding deaths and percentage of mortality:

There were under 10 years 1,430 cases,   58 deaths, or   4.05 per cent.
Between 10 and 20   "   5,121   "   171   "   3.33   "
   "   20 " 30   "   4,127   "   510   "   12.35   "
   "   30 " 40   "   2,976   "   690   "   23.18   "
   "   40 " 50   "   2,546   "   906   "   35.58   "
   "   50 " 60   "   1,231   "   630   "   51.17   "
   "   60 " 70   "   588   "   383   "   65.13   "
   "   70 " 80   "   116   "   88   "   75.86   "
   "   80 years and upwards,   3 "   3   "   100.00   "
Age doubtful,   130 "   8   "   6.15   "

In the Breslau epidemic of 1868–69, Lebert found that the disease became more fatal with advancing years, as is shown by the following table:

Age.	Percentage of all cases.	Percentage of total mortality.
Under 15 years..................	15.2	2.7
From 15 to 20 years..............	16.1	3.16
From 20 to 30 years..............	22.8	15
From 30 to 40 years..............	23	26
From 40 to 50 years..............	13.4	24.1
From 50 to 60 years..............	7.4	20

As in other epidemic diseases, the mortality is greatest at the beginning and height of epidemics, and gradually declines as the number of cases diminishes.

Race and nationality exert but little influence upon mortality. In Philadelphia, in the epidemic of 1836, the mortality was greater among the blacks than among the white population (Gerhard).

Among individual peculiarities unfavorably influencing the prognosis, are intemperate habits, diseases of the kidney, gout, obesity, and mental depression. Fatigue and privation before and at the beginning of the attack, increase the danger. Nursing women, according to Dr. Murchison, are prone to a high degree of anæmia, and in them the chances of death by asthenia are increased.

During the attack a presentiment of death is of ominous prognostic import. It is not, however, a necessarily fatal sign. It is apt to be present among persons of cultivated intelligence, and in particular among medical men. The danger may be said to be in general terms proportionate to the severity of the cerebral symptoms and to the early date of their appearance. Thus, severe headache, constant and high delirium, profound stupor, indicate great danger, and the earlier their appearance the greater their significance. Extreme prostration, especially early in the course of the disease, is a bad sign. So also is tremulousness, twitching of the muscles and tendons, and grasping in the air. That condition of complete unconsciousness with wide open eyes, to which the name *coma vigil* has been applied, is almost always the forerunner of death. Sleeplessness, alternating with delirium, and protracted some days without relief, is most unfavorable. An extremely rapid pulse is unfavorable, especially if it be undulatory, small, or irregular. Death sometimes occurs, however, in cases in which the pulse has not risen above 100. A moderate fall in the pulse is usually of favorable import. A very faint, or inaudible systolic heart-sound, and a feeble or imperceptible impulse, are indicative of danger. As a general rule, the copiousness of the eruption is in proportion to the danger of the case, particularly if it be at the same time dark or livid. On the other hand, cases attended by a scant eruption of light color or by none at all, are commonly favorable. To this statement, however, we must make the exception that Lebert states that he has seen severe and even fatal cases without eruption. If the temperature rise very high, above 41° C. (105.8° F.), or if it fails to fall during the second week, this is to be regarded as of unfavorable import. The presence of complications influences the prognosis unfavorably. Among the more serious of the many complications of typhus may be mentioned previously existing or recent renal disease, pyæmia, parotid and other bubonic swellings, gangrene, bed-sores, erysipelas, and diseases of the respiratory tract.

Death may take place at any period. It commonly occurs toward the end of the second week, that is, about the period of the crisis, in uncomplicated cases, by asthenia in consequence of degeneration of the cardiac muscular tissue; or by coma resulting from the retention in the blood of the waste products due to the fever-process; or, where pulmonary complications exist, in consequence of asphyxia.

In the majority of cases the fatal result is due to some complication.

Anatomical Lesions.

Emaciation is usually not marked in the cadaver, unless death has taken place after the termination of the second week.

Post-mortem rigidity is usually of short duration, and decomposition of the body takes place more rapidly after death from typhus than in most other diseases occurring at the same time of the year. The integuments of the dependent parts of the body show more or less extensive and deep discoloration. The whole surface, including the face, is not unfrequently livid. If death occur during the fever, traces of the eruption often persist.

The muscles have usually lost their normal red color, and show a dirty, brownish red, or grayish red discoloration. Upon microscopical examination, the muscular tissue of the heart and the voluntary muscles are found to have undergone granular and waxy degeneration, particularly if death has occurred after the end of the second week. Extravasations of blood are met with in the substance of the rectus abdominis and other muscles.

The mucous membrane of the digestive tract shows no characteristic lesion. The stomach, in a large proportion of the cases, presents no changes excepting those incident to slight catarrh, namely, patches of vascular injection and softening. Minute ecchymoses are also occasionally encountered. The enlargement and ulceration of the solitary and agminate glands of the intestine, which are constant in enteric fever, are not encountered in typhus. Exceptionally the glands are slightly more prominent than in health, as is the case after death in various other diseases, and Lebert states that the solitary glands, as well as Peyer's patches, are occasionally the seat of minute, isolated, superficial ulcers, especially in the vicinity of the ileo-cæcal valve. The mesenteric glands are exceptionally slightly enlarged.

Degenerative changes in the kidneys are not uncommonly found, as the result of pre-existing renal disease, at the examination of the bodies of those dead of typhus. The traces of recent disease are also of very common occurrence. The kidneys are usually hyperæmic, the cortex is swollen, opaque, and fatty, and the tubules, upon microscopical examination, are found blocked with granular epithelium sometimes commingled with blood-corpuscles. Occasionally the kidneys are decidedly enlarged and gorged with blood, presenting the appearances met with in acute, scarlatinal nephritis.

Enlargement of the spleen is the rule, being present in about three-fourths of the cases. This organ is often softened; in many instances it is pulpy, and not rarely diffluent, when the capsule is divided. Extravasations of blood into the tissue of the spleen are not uncommon. When the softening is less marked, the corpuscles of Malpighi are enlarged and unusually distinct.

The liver is not unfrequently slightly enlarged. It is hyperæmic if death occur before or at the time of the critical defervescence. If the sickness has been prolonged into the second week, the liver is often pale, fatty, and friable. Its cells are found to contain an excessive amount of fat-globules and fine granules. In rare instances the liver, even when examined shortly after death, has been found emphysematous, crepitating, and containing a frothy liquid with bubbles of gas, portions of it floating when thrown upon the surface of water. This condition is due to rapid local decomposition. The pancreas has frequently been found hyperæmic and slightly enlarged.

Peritonitis is among the rarest of the complications of typhus fever. In one instance it was due to the bursting of a softened embolic deposit in the spleen. Occasionally subperitoneal ecchymoses are encountered. Pericarditis and endocarditis are exceedingly rare. There is usually a considerable amount of serum in the pericardium. The changes in the heart are similar to those of enteric fever. It is soft, flaccid, easily torn, and the muscular tissue is of a brownish yellow color. These changes are most marked when death has taken place late in the course of the disease. In some instances they are restricted to the left side of the heart.

The blood found in the heart and larger vessels is sometimes fluid, sometimes coagulated into a black, pultaceous clot. Pale coagula are very rarely found.

The bronchial tubes exhibit the signs of recent catarrhal inflammation; they are almost constantly injected, and contain a secretion varying from thin mucus to thick pus. Patches of atelectasis are common as a result of capillary bronchitis. The dependent portions of the lungs are usually deeply hyperæmic. Pulmonary œdema is frequent. True pneumonia is occasionally met with, and gangrene of the lung is far from uncommon.

Subpleural ecchymoses occur. Pleurisy is rare. When present, it is commonly fibrinous, and probably secondary to inflammatory processes in the lungs. When serous, it is apt to become purulent. The bronchial glands are sometimes swollen.

The brain and nervous system, notwithstanding the prominence of the nervous symptoms of the disease, present no characteristic changes. If death take place early, there is vascular injection of the membranes and some hyperæmia of the substance of the brain. If it take place late, the membranes and brain-substance are anæmic, and there is an accumulation of serum in the subarachnoid space and in the ventricles. Hemorrhage into the arachnoid space belongs to the rarer lesions of typhus. It gives rise to the formation of a delicate, filmy clot, usually extended over some part or the whole of the convexity of the brain. Its source has not yet been discovered. The quantity of fluid present within the cranium in **typhus** is not greater than that usually found in persons of an advanced age, or who **have died** from wasting diseases (Murchison). The fluid occupying the ventricles and subarachnoid space is not to be looked upon, any more than is the vascular injection previously spoken of, as the result **of inflammation.** The former accumulates mechanically to fill up the space within the cranium occasioned by wasting of the brain-substance: the latter **is** likewise passive or mechanical, and is not greater or more common than in death from other acute, febrile diseases. Moreover, there is no direct relation between the amount of **the** vascular injection, or the quantity of the fluid, and the **gravity** of the symptoms during life.

Anatomically, typhus fever presents no characteristic lesion. At the **autopsy we encounter** the changes due to prolonged, intense pyrexia, namely, nutritive disturbances, a tendency to fatty degeneration of the muscles and the glandular viscera, and disintegration of the blood, with diminished volume of the brain-substance and increase of intracranial **fluid.** To these we must add frequent and grave lesions of **the** organs of respiration, and, in the greatest number of cases, enlargement and softening of the spleen.

Diagnosis.

The direct diagnosis of typhus fever **must remain** doubtful in **most cases** until the appearance of **the eruption. Of diagnostic importance are the** abrupt onset, headache, extreme lassitude, the tendency of the fever to rapidly augment. If the disease be **epidemic, or** if the patient presenting these symptoms be known to have been exposed to the con**tagion of typhus,** the diagnosis is probable. It becomes certain upon the appearance **of the** eruption. **The critical defervescence** about the fourteenth day is characteristic.

The differential diagnosis between typhus and the following diseases, namely: enteric **and** relapsing fevers, tropical remittent **fever,** cerebrospinal fever, measles, the plague, and alcoholism, requires some words of consideration.

The chief points of contrast between *enteric fever* and *typhus* and *relapsing* fevers are arranged in a tabular form upon page 338. **They are** so manifest that it would appear impossible to confound these affections, yet it is to be borne in mind that these three **fevers have been regarded** as essentially the same until within a few decades, and that they are now very generally regarded, by continental physicians, as varieties of a common fever. Moreover, all the symptoms are not always present; and exceptionally a symptom characteristic **of one** may appear in another of these diseases, as, for example, we may meet with constipation in typhoid fever and with diarrhœa in typhus. Even the eruption may be mixed, as is seen in the following case, reported by Da Costa:

A boy, sixteen years of age, **was received into** the Philadelphia Hospital with evident signs of the beginning of a fever of **low type. A** day or two after his admission, and corresponding, as nearly as could **be** ascertained, to the fifth day of the disease, an eruption showed itself all over the **body.** It was dark-colored, petechial in its aspect, and did not disappear on pressure. Associated with it were drowsiness and constipation. In a few days more, however, the symptoms changed: the dark eruption faded and rose-colored spots were perceptible on the chest and abdomen; diarrhœa **set in, and the fever ran** its course to a favorable termination—with the character of **typhoid, just as at the outset it** had assumed the character of typhus.

Similar cases have been recorded by other competent observers, among whom may be named Murchison, Peacock, and T. J. Maclagen.

Remittent fever, in its ordinary form, as met with in this country, bears but little resemblance to typhus; but the malignant remittents of tropical and subtropical climes occasionally present strong resemblances to it. They are attended with great prostration, low, muttering delirium, dry, brown tongue, a feeble pulse, contracted pupils, and in some instances by petechial eruptions. Typhus is, however, rare in the countries where remittents of this form make their home. Remittent fever is not contagious. It is apt to be associated with pure intermittents and other forms of disease due to malaria. Moreover, true remissions do not occur in typhus. And the peculiar eruption of typhus is never met with in remittent fever. The enlargement of the spleen in malarious diseases is not only much greater, but it is also more dense than that of typhus. Finally, the course of typhus fever is uninfluenced by antiperiodic remedies.

The differential diagnosis between cerebro-spinal fever and typhus is to be found elsewhere. (See page 96.)

Measles and typhus in children present some points of resemblance, the most important of which relate to the appearance of a somewhat similar eruption about the fourth day of each disease. The eruption of measles is, as a rule, brighter in its tints, and the pre-eruptive stage of measles lacks the intensely febrile character which belongs to that of typhus. Moreover, in measles, coryza and cough are constant, whereas the more serious bronchial affection of typhus is insidious and often attended with but little cough. Furthermore, the eruption of typhus passes through a typical course, subsiding speedily into maculæ or stains which do not fade upon pressure. The diagnosis, if doubtful, may be simplified by an examination of other individuals in the household of the affected person. In this country measles is peculiarly a disease of childhood, whereas typhus is apt to attack the adult members of the household before the children.

The resemblance between typhus and the *true plague,* as it is known to us by description, is most close. Both are highly contagious diseases, of abrupt onset, attended by grave cerebral symptoms and petechial eruptions. In the plague, however, nausea and vomiting, a pale face, an anxious expression, blood-spitting, and the early appearance of glandular swellings with suppuration and the simultaneous improvement in the cerebral symptoms, are met with. Moreover, the duration of the plague is much shorter than that of typhus, death taking place commonly between the third and fifth day, and convalescence beginning about the close of the first or the beginning of the second week.

Certain forms of alcoholism are attended by a trembling delirium which, in itself, cannot be distinguished from the *delirium tremens* met with at times in typhus. The history of the case is of diagnostic value. Moreover, in alcoholism the tongue is moist, the skin is leaky, there is no eruption, and the temperature is but slightly elevated, if it be elevated at all.

Treatment.

Prophylactic treatment is of importance, both as regards communities and individuals. It embraces the measures belonging to personal and public hygiene. Thus, typhus is no longer popularly known as ship or jail fever, for public opinion and legal enactments have enforced the observance of decent sanitary regulations where large numbers of ignorant or destitute individuals are crowded together in confinement. Improvement in the sanitary condition of these localities, formerly infested with typhus fever, has led to its disappearance as an endemic disease, while the great epidemics of typhus have become, even in the wars of recent times, almost unknown.

The armies of the two greatest military movements since the Napoleonic campaigns, namely, the American and the Franco-Prussian war, escaped its ravages.

It is impossible to prevent the importation of the germs of typhus fever into any locality. The object of prophylaxis is to reduce to a minimum the conditions which favor the outbreak and extension of the disease. These conditions are overcrowding, destitution, and their attendant evils—filth, both of person and dwellings, foul air, and the like. Typhus never makes its first appearance in the cleanly and well-ventilated homes of the opulent classes. If it extend to them at all, it is by spreading from less fortunate localities. The deduction from this observation is obvious. Both in season and out of season, but especially wherever typhus has shown itself, the strictest hygienic regulations ought to be enforced. Upon the outbreak of an epidemic, the isolation of the first cases in hospitals, and the thorough cleansing and ventilation of the houses and rooms from which they are removed, with general sanitary measures to obviate the predisposing causes of the fever in the affected neighborhood, are important.

If it is possible, the infected buildings should be thoroughly fumigated with sulphur, ventilated, whitewashed, and allowed to stand unoccupied for a considerable time. The clothes and belongings of the patients should likewise be disinfected by prolonged exposure to heat, or to the fumes of burning brimstone, or by boiling in water containing carbolic acid. The infected bedding should be subjected to the same treatment, and the materials used for filling mattresses and pillows should be burned. Absolute cleanliness in the sick-room is to be insisted upon. The excretions should be promptly disinfected. Persons in attendance upon the sick must be allowed opportunities for proper rest and exercise in the open air, and they should be made aware of the importance of thorough ventilation in diminishing the danger of contagion.

The general management of typhus fever is the same as that of enteric, regard being had to the early and grave adynamia which so often characterizes the affection under consideration.

Hygienic measures relate to ventilation, to cleanliness, and to diet. Typhus cases, when treated in hospital, should be placed in large rooms by themselves, and never more than four or six together; the windows, even in winter, should be kept open so as to secure careful and thorough ventilation. All observers insist that bad air is more to be dreaded than cold. When cases are treated at their homes, as is necessary in the well-to-do classes, similar regulations are to be observed; and in particular, all unnecessary furniture, and all curtains and hangings which are liable to interfere with ventilation, on the one hand, and to absorb and retain the contagion, on the other, are to be taken away. Quietude is to be observed, and all visits, except such as are absolutely necessary, are to be prohibited.

Moderate quantities of milk and arrow-root may be given, as a meal, morning and evening; and during the twenty-four hours, at intervals of two hours during the day and three during the night, milk, or, in many cases, milk-punch, broths, or light soups, may be given alternately. If the patient be asleep at the time the food is due, it is important that he be not disturbed. After an hour, or an hour and a half, if he do not awake, he may be aroused and an increased amount of nourishment at once administered. The thirst which distresses him may be satisfied with water in abundance; or, if the patient prefer it, he may occasionally have weak lemonade or carbonic acid water. Cold sponging is usually agreeable to the patient, and is useful for purposes of cleanliness. It is important to control the temperature by means of antipyretics. *Systematic cold bathing* is stated by Lebert to be not only well borne, but to meet with no opposition from the patient as soon as a few baths have been taken. The rules to be observed are the same as those laid down for the hydrotherapeutic treatment of hyperpyrexia in enteric fever. (See page 229.) In view of the marked depression so characteristic of typhus fever, it is probable that the abstraction of heat by means of cold baths will be found less applicable in this disease than in other febrile diseases in which adynamia is less prominent. Wet packs may **sometimes take** the place of bathing, or bags of ice, applied to the abdomen and **to the head**, will be useful in reducing, at all events to some extent, the temperature. Cold-water injections are admissible in typhus fever, and are satisfactory in relieving constipation. Large doses of quinine, 1.0—2.0 grammes (15—30 grains) may be employed to **reduce temperature.** For this purpose digitalis may also be given; and **quinine, digitalis,** and hydrotherapeutic measures, may **be used together.** Thus, the infusion of digitalis may be given **at intervals of four hours, with** an antipyretic dose of quinine once in twelve or twenty-four hours, and an occasional resort to the wet pack, or the application of ice-bags. *Stimulants* are very generally required, but they are to be used with judgment. Most children, **and a** large number of the adult cases, may be satisfactorily treated without them. Alcohol is seldom required before the appearance of the eruption; it is most useful in the second **week, and often necessary** upon the approach of the crisis, even in cases where **it has not before been required.** Old people, and those previously greatly debilitated, almost invariably require alcoholic stimulants **in the beginning of the disease.** Persons of intemperate habits also commonly **require alcohol from the onset of** the disease, and in greater quantities **than those unaccustomed to** its use in health. Stimulants must **be promptly given in cases of great** prostration, with low delirium and a tendency to coma ; also where the systolic heart-sound is faint, or when the pulse is frequent, feeble, or undulatory. Here, as in all low fevers, the first sound of **the heart and the character of the pulse are the best indications for the administration of** stimulants, both as regards the time and the quantity. Delirium in itself is not necessarily an indication for the administration of **stimulants. If, however, the patient become calmer and more rational** under their use, they are to be **continued or increased.** No definite rules **can be laid** down as to **the quantity to be administered in the twenty-four hours.** It varies from an ounce or two, given in tablespoonful doses when the symptoms seem to call for it, **to twelve** or fourteen ounces distributed throughout the twenty-four hours, **at regular** intervals. Sound whiskey or brandy are the most satisfactory forms in which alcohol **is given.** The **English** writers recommend **the malt liquors, and state that they are** agreeable to many patients. Where circumstances **permit its use, champagne may be given.** Among the diffusible stimulants **useful in typhus are** ammonium carbonate, chloroform, camphor, and turpentine. When **the prostration is extreme and the** patient is unable to swallow, brandy, or whiskey, or ether may be hypodermically **administered. As will commonly happen in the** severest cases, the condition of the patient may render it impossible to give the necessary food by the mouth. Under such circumstances, an endeavor to support the patient's strength and to prolong life must be made by means of rectal alimentation and medication.

It is of the utmost importance that the patient's strength be husbanded from the beginning of his sickness. All mental and bodily effort is to be avoided. It is a common observation that those who struggle against the disease in its early days usually suffer from great prostration later. The patient should betake himself to his bed as soon as the fever appears. If there be marked prostration during the first week, and under most circumstances during the remainder of his illness, the patient should not get out of bed for any purpose. Many persons, particularly men, object to using the bed-pan. In typhus it is, in many cases, imperative that the patient do not assume the upright position. Fatal syncope may result. The management of the patient in delirium will often tax the patience and tact of the nurse to their utmost. In most cases restraint by forcible measures is unnecessary; it is always a last resort, and to be deplored.

When we come to speak of the treatment of typhus fever by *medicine*, we find that no drug or course of medication is adequate either to arrest or to shorten the course of the primary disease. The sufferings of the patient may be mitigated, and the attendant complications to some degree warded off or controlled, when they arise, by a watchful attendance and judicious therapy; but this is all. No cure for typhus is known.

The mineral acids occupy the highest rank among the medicines employed in the treatment. *Nitro-muriatic acid* is usually preferred in this country. Da Costa gave it in three to five drop doses, either alone, or alternately with *turpentine*, when pulmonary complications were present. If the latter became prominent, the acid was discontinued and turpentine given in connection with *quinine*, and applied externally in the form of stupes to the chest.

Most of the complications, the pneumonias, and parotid swellings, were treated by turpentine and quinine, and stimulants.

Lebert, who regards drugs, as such, unnecessary, gives the following as a placebo:

℞. Acid. phosphor. dil............ 4–8 c.c. fl. ʒ j.–ij.
Syrupi.................... 32 c.c. fl. ℥ j.
Aquæ..................... 160 c.c. fl. ℥ v.
M. Signa.—A tablespoonful q. s. h.

Headache is to be treated by cold applications, or by external warmth, or, if it be distressing and the subject young and robust, by two or three leeches applied to the edges of the hair on the temples.

Sleeplessness, nervous excitement, and delirium require, in many cases, special treatment. Much may be done toward their alleviation by the skilful management of the sick-room and the personal ministrations of the nurse. Chloral, opium and its derivatives, especially in combination with minute doses of belladonna or atropine, and the diffusible stimulants, are to be employed. Camphor, alone or in combination with morphia, or with tartar-emetic, opium, and musk (Graves), is highly lauded against the subsultus, tremors, and delirium. It is necessary to call attention to the danger of chloral in large doses where there is any tendency to failure of the heart.

Stupor, if marked, may be treated by small cupfuls of strong coffee, repeated at intervals of three or four hours. Cold affusions may become necessary. Stimulating rectal injections containing turpentine may be administered. They are additionally useful as provoking the action of the bowels.

The complications of typhus are to be treated in accordance with general therapeutic principles. It is probable that suppurative inflammation of the parotid and its overlying tissues, and similar processes elsewhere in the body, will prove less generally fatal than formerly, if promptly treated by the evacuation of the pus, and antiseptic dressings in accordance with the rules laid down by Prof. Lister.

Convalescence is commonly steadily progressive, and rapid, unless complications retard it. For a few days the diet is to be restricted, both as regards quantity and kind, as the appetite is often ravenous and there is danger of excess. Barks, iron, quinine, and perhaps best of all, cod-oil, should be given if the patient remain feeble and pale. Sleeplessness is to be obviated by judiciously chosen hypnotics. The patient must be warned against the risks incurred by too great haste to resume his occupation.

VI.

RELAPSING FEVER.

DEFINITION.—An acute, contagious fever, rarely occurring except as an epidemic, and in seasons of scarcity of food; it consists of: (*a*) a febrile paroxysm, characterized by abrupt onset, active fever, a moist, white tongue, epigastric tenderness, vomiting, and often jaundice, enlargement of the liver and of the spleen, and the absence of eruption, and terminating suddenly with free perspiration about the fifth or seventh day; (*b*) an interval of complete apyrexia; and (*c*) an abrupt relapse, on or about the fourteenth day from the beginning of the disease; this relapse runs a course similar to that of the initial paroxysm, and comes to an end by crisis on or about the third day. Convalescence usually ensues upon the termination of the relapse, but a second, third, or even fourth relapse, has been observed. Fatal termination infrequent; enlargement of liver and spleen, but no specific lesion, found upon examination after death.

SYNONYMS.—Febris recidiva; Typhus recurrens; Five days' fever, with relapses; Short fever; Five days' fever; Seven days' fever; Seventeen days' fever; Fièvre à rechute; Typhus à rechute; Das recurrirende Fieber; Wiederkehrendes Fieber; Rückfalls Fieber; Armentyphus; Die Hungerpest; Tifo recidivo.

Epidemic fever of Edinburgh; Scotch epidemic of 1843; Epidemic fever of Ireland; Epidemic remittent fever; the Silesian fever of 1847; Dynamic or inflammatory fever; Synocha or relapsing synocha.

Mild yellow fever; Bilious relapsing fever; Gastro-hepatic fever; Remitting icteric fever; Bilious typhoid; Famine fever.

Typhinia.

The term Relapsing Fever is derived from one of the most constant and certainly the most striking peculiarity of the disease. It has passed into general use. It is sufficiently distinctive, and is not open to the objection of embodying any theory regarding the origin or nature of the fever.

The synonyms are very numerous; they have been suggested by the consideration of various particulars relative to the special epidemic prevalence of the disease, to the duration of the attack, to symptoms that are prominent, or to a supposed relation between it and typhus. The circumstances forming the basis of most of them will appear in the course of the following account of the affection.

Historical Sketch.

Dr Robert Spittal[1] in 1844 called the attention of the medical profession to the fact that relapsing fever is not a new disease. He showed that the epidemic at that time prevailing in Scotland was exactly the same, in all its important features, as an epidemic described by Hippocrates as having occurred more than twenty centuries ago in the island of Thasos, off the coast of Thrace. The points of resemblance between the ancient and the modern epidemics are: the occurrence of relapses after an intermission of five or seven days, the crisis, the copious perspiration, epistaxis, jaundice, a tendency to miscarry, and the like.

Apart from the occasional mention, by several writers, of the occurrence of one or more relapses in the course of continued fever, in the epidemics which they have described, and which may or may not refer to cases of relapsing fever prevailing in connection with typhus, there is no definite account of the epidemic prevalence of the fever under consideration prior to the year 1739. Dr. Rutty[2] chronicled a disease prevalent in Dublin in that year, in the following words: "The latter part of July and the months of August, September, and October, were infected with a fever, which was very frequent during this period, not unlike that of the autumn of the preceding year, with which compare also the years, 1741, 1745, and 1748. It was attended with an intense pain in the head. It terminated sometimes in four, for the most part in five or six days, sometimes in nine, and commonly in a critical sweat; it was far from being mortal. I was assured of seventy of the former sort at the same time in this fever, abandoned to the use of whey and God's good providence, who all recovered. The crisis, however, was very imperfect, for they were subject to relapses, even sometimes to the third time; nor did their urine come to a complete separation. Divers of them, as their fever declined, had a paroxysm in the evening, and in some these succeeded pains in the limbs." A little farther on, after speaking of the fever of the summer of 1741, the same author says: "It seems also not unworthy of notice, that through the three summer months there was frequently here and there a fever, altogether without the malignity attending the former, of six or seven days'

---

[1] The Antiquity of the Fever prevalent in 1843: Edinburgh Monthly Journal of Medical Science. Vol. iv., 1844.

[2] A Chronological History of the Weather, Seasons, and Diseases in Dublin, from 1725 to 1765. By John Rutty, M.D. London, 1770.

duration, terminating in a critical sweat, as did the other also frequently; but in this the patients were subject to a relapse, even to a third or fourth time, and yet recovered."

During the epidemic of fever which prevailed in Ireland from 1797 to 1803—a period of great destitution among the lower classes—many cases were observed, the account of which fully corresponds to relapsing fever. We are informed by Drs. Barker and Cheyne[1] that "the fever of 1800 and 1801 very generally terminated on the fifth or seventh day by perspiration; that the disease was then very liable to recur; that the poor were the chief sufferers by it; and that it was much more fatal among the middling and upper classes in proportion to the number attacked."

Relapsing fever prevailed from time to time during the first sixteen years of this century, in Ireland and elsewhere, while in all probability the widespread epidemic of 1817–19 was largely composed of it. This great outbreak began in Ireland during a period of great scarcity of food, and was carried into England and Scotland by the migration of the Irish poor, who flocked into the large towns, condensing their population, and introducing habits of uncleanliness and improvidence with the seeds of disease.

The distinction between the two forms of fever, namely, relapsing and typhus, which composed this epidemic, had not then been made; they were regarded as modifications of one disease, and it was a general impression that the relapsing fever could produce common typhus, and *vice versa*. Dr. Murchison[2] has, however, shown by a critical study of the symptoms, the death-rate, and the results of treatment by bloodletting, that relapsing fever mainly constituted the epidemic, and has probably correctly inferred from the circumstance that the rate of mortality increased in many places with the advance of the epidemic, that the proportion of typhus to relapsing cases was greater toward the close of the epidemic than at its commencement.

From the subsidence of the epidemic of 1817–19 till 1826, there is no record of the occurrence of relapsing fever. In that and the two following years a great outbreak of fever raged first in Ireland, and later in Scotland and England. It followed commercial distress, and was chiefly confined to the largest towns. Dr. O'Brien,[3] who observed it in Dublin in

---

[1] An Account of the Rise, Progress, and Decline of the Fever lately Epidemic in Ireland, etc. By F. Barker, M.D., and I. Cheyne, M.D. 2 vols. London and Dublin, 1827.

[2] A Treatise on the Continued Fevers of Great Britain. By Charles Murchison, M.D., LL.D., F.R.S. Second edition. London, 1873.

[3] Medical Report of the House of Recovery and Fever Hospital, Cook Street, Dublin, etc.: Transactions of Kings and Queens' College of Physicians in Ireland. Vol. v. Dublin, 1828.

1826-27, states that "at the conclusion of the spring and commencement of the summer (1826) a vast body of artisans residing in the Liberties of Dublin were thrown out of employment, and actually labored under all the miseries of artificial, yet positive famine, being destitute of the means of purchasing food." This epidemic, like that of 1817-19, was composed of relapsing and typhus fevers, the former being more prominent in Ireland, and at the beginning, and the latter being much more common toward the end of the epidemic, which at the last was almost exclusively made up of it. It was in this epidemic that a correct distinction between the two fevers was first drawn. Dr. O'Brien, in the report above cited, wrote as follows: "At the commencement of the epidemic two species of fever were distinguishable in the wards of this hospital, which, to use the words of Sydenham, we shall call the fever of the old and the fever of the new constitution. The first was the ordinary typhus of this country, marked by its usual protracted periods, running on to the eleventh, fourteenth, seventeenth, or twenty-first days. This species of fever was far inferior in numerical amount to the other, but far more fatal. The other species of fever, or that of the new constitution, which constituted the bulk of this epidemic, was one of short periods, terminating in three, five, seven or nine days, but the second of these periods was the most frequent. The patient was destined, perhaps, to be harassed by one, two or three relapses, which prolonged the whole duration of his illness beyond that of the most protracted typhus—in fact, the liability to frequent relapses was one of the most striking characteristics by which this fever was distinguished from all previous epidemics, at least, which happened in our time."

From 1828 till toward the end of the year 1842, relapsing fever seems to have disappeared from the British Islands. So little was it known or thought of during this interval, that when it next appeared it was looked upon as a new disease by many of the physicians who first encountered it.

In 1842, there arose an extensive outbreak of fever, which differed from those that had preceded it in neither beginning in Ireland, nor in spreading to that country. The first cases were observed on the east coast of Fife, and not in the large cities. The earliest cases in Glasgow occurred early in the autumn, and the disease became generally prevalent in December. The cases steadily increased till October, 1843, when the epidemic began to abate. The number of cases is estimated at 33,000, or between ten and eleven per cent. of the entire population. In Edinburgh and in Aberdeen the disease made its appearance in February, 1843, and increased till October, after which it abated by degrees, and in the following April had nearly ceased. It prevailed generally over Scotland, and was not restricted to the large cities. In England its distribution was chiefly restricted to the large cities. In this epidemic cases of relapsing fever largely preponderated; typhus was, with the exception of a few localities at Dundee, comparatively rare, and everywhere the latter fever increased with the progress of the epidemic. This fact was clearly established by the statistics of the hospitals and infirmaries of Glasgow and Edinburgh, where the distinction between the two fevers was more clearly recognized.

Cases of relapsing fever were occasionally encountered from the time of the subsidence of this epidemic until 1846. In the last months of that year there appeared in the British Isles an epidemic of fever of great magnitude and severity. It arose in Ireland after the failure of the potato crop, and at a time of great consequent famine and destitution. At the end of the year it reached Glasgow; Edinburgh in March; it fell upon Liverpool in January, 1847, upon London in March, upon Manchester in April. It prevailed very generally over Ireland, and in the large towns of Scotland and England. It reached its height in the autumn of 1847, but did not wholly disappear till the end of 1848. In this epidemic the cases of typhus constituted by far the greatest number of the sick. Enteric fever was also observed, and relapsing fever was common. The greater preponderance of the last-named fever in the early part of the epidemic was noted by nearly all observers. In the greater number of typhus cases which this epidemic presented, it had very much the same relation to the epidemic of 1843 that the epidemic of 1826 had borne to that of 1817–19 (Murchison).

From the time of the epidemic of 1846, relapsing fever has gradually subsided. In 1851, there was a local increase of it in London, where the disease was almost exclusively confined to Irish people, all in a state of destitution and mostly recently arrived from their own country. At the same time the fever prevailed to some extent in Edinburgh. In 1853, it was common in Ireland. In 1855, it disappeared, and, as Murchison informs us, for fourteen years not a case of relapsing fever was observed in any hospital of Great Britain, while in Ireland it seems also to have been unknown.

In 1868, relapsing fever reappeared—this time in London, where it attained its maximum in December, 1869, and declined gradually till June, 1871, when it came to an end. During the time of its prevalence in London the fever showed itself also in other large cities of England and Scotland. In this, as in former epidemics, the cases occurred chiefly among the poorest classes of the population; most of the patients were in an extreme state of destitution.

In 1846–47 there prevailed in Upper Silesia—a province of Prussia—and elsewhere in Germany, an epidemic of fever which resembled that then prevalent in the British Islands. It first occurred among the Silesians, a people whose condition closely resembles that of the Irish peasantry, and appeared in a time of severe famine. It consisted partly of relapsing fever and partly of typhus. Griesinger[1] states that relapsing fever also probably **formed** part of the epidemic which prevailed in Bohemia during the same year, and that he encountered it in 1851, in Egypt, associated with "other forms of typhus." During the summer months of 1855, it prevailed among the British troops in the Crimea.

In 1863, relapsing fever broke out at Odessa; in 1864, it appeared in St. Petersburg, where it was mostly restricted to the poorest class of the people. Overcrowding of the wretched habitations of the poor had resulted from the influx of recently liberated serfs to the capital in search of work; food of all kinds was at the same time high and bad. The pestilence followed. It was, **as** in the epidemics which have been described as occurring **in Great Britain** and elsewhere, composed of mixed cases of relapsing fever and typhus. The number of cases of the former relative to those **of** the latter fever **was** much greater at the beginning of the epidemic.

Toward the close of the year 1867, relapsing fever and typhus again appeared together as an epidemic in Silesia. They prevailed in East Prussia, and spread in 1868 to Breslau, Berlin, and other large cities of Germany. This epidemic was generally ascribed to destitution and want of food, but whether it was an extension of the epidemic in **Russia or not,** is not clear (Murchison).

About the close of 1872, relapsing fever again broke out in Berlin and in Breslau. This epidemic prevailed until the close of the following summer.

The disease reappeared in Berlin in 1879, and is now (spring of 1880) to **a** slight extent prevalent in that capital.

The geographical range of relapsing fever is much wider than it was formerly thought to be. It has been observed in greatest frequency, and **has** assumed its greatest importance, in the British **Isles.**

Of the numerous epidemics that have swept over these islands, all but two have originated in Ireland. Of these, the earlier, which occurred in 1843, and has **been** described as the Scotch epidemic of that year, arose in Scotland and implicated Ireland **but little,** if at all; the second was that of 1868, which first appeared in London at a period when there was no relapsing fever in Ireland. The fever was then prevalent upon the Continent, and it is probable that this outbreak was an extension of the epidemic in Germany.

Beyond the limits of the epidemics already mentioned as having prevailed in Upper Silesia, North Germany, in Poland, **the** Crimea, and **widely** in Russia, relapsing fever **is not** known to have occurred in **Europe, except** in a few isolated **and** restricted outbreaks. Such minor epidemics have likewise been observed in Siberia, Algiers, and on the island of Réunion. This fever has **also occurred in** India and other tropi-

---

[1] Virchow's Handbuch der Pathologie und Therapie. XI. Band, Abtheilung 11. Erlangen, 1864.

cal countries, and chiefly in connection with typhus. Relapsing fever has never appeared as an indigenous disease in America. It has been imported on several occasions during its epidemic prevalence beyond the Atlantic, but its outbreak has been, for the most part limited, and restricted to the cities of the seaboard. In June, 1844, fifteen cases of relapsing fever were transferred from a Liverpool packet arriving in Philadelphia with Irish immigrants, to the Philadelphia Hospital. The disease did not spread to the attendants or the other patients in the hospital, although the cases were not isolated. Two women, sisters of one of the immigrants, who had been for a long time resident in the city, contracted the fever and were afterward admitted to the hospital. No other cases occurred.[1] The disease appeared in a like restricted manner in New York in 1848,[2] and in 1850-51, Professor Austin Flint[3] observed fifteen cases of fever among recently arrived Irish immigrants in the hospital at Buffalo, which, upon subsequent examination of the notes, proved to be undoubted instances of relapsing fever. At the same period more numerous cases were observed in Canada. In 1869, it again appeared in America, the first cases being observed in Philadelphia in September,[4] and in New York in November.[5] The greater number of the patients was among the poor of the Irish and German population. The disease was thought to have been imported, but from what source remains unknown. Dr. Parry, who first encountered it in Philadelphia, and who carefully investigated this outbreak, states that every attempt to trace the origin of the earliest cases failed. The disease spread slowly among the most destitute of the population, reaching its height about the middle of the following year (1870) and then rapidly subsided. It had entirely disappeared by the end of the second quarter of 1871. Isolated outbreaks, traceable in almost every instance to the epidemics in New York and Philadelphia, occurred during the same period in Boston and in Washington, D. C., and at several intermediate points.

Relapsing fever in all the great epidemics has been associated with typhus. As a general rule, the former disease has supplied the greater proportionate number of cases at the beginning or in the early part of the epidemics, and has gradually disappeared till, toward the close of the epidemic, typhus alone prevailed.

---

[1] Meredith Clymer, M.D.: Notes on the History of Relapsing Fever. New York Med. Jour., March, 1870.

[2] Relapsing Fever and Ophthalmitis Post-febrilis in New York. By A. Dubois, M.D., Trans. Amer. Med. Assoc., 1848.

[3] Clinical Reports on Continued Fever, Based on an Analysis of One Hundred and Fifty-two Cases. Buffalo, 1852.

[4] Observations on Relapsing Fever, as it occurred in Philadelphia in the Winter of 1869 and 1870. By John S. Parry, M.D. Amer. Jour. Med. Sciences, Oct., 1870.

[5] On Relapsing Fever: A Lecture by Prof. Austin Flint, M.D. New York Med. Journ., March, 1870.

Etiology.

I. Predisposing Causes.

Climate has no direct influence upon the development or propagation of relapsing fever. It has been observed in India and in Egypt, as in Moscow, St. Petersburg, and in Siberia, upon the continent of Europe, as in the insular climate of Great Britain. Nevertheless, the disease has occurred most frequently and has prevailed most fiercely, since the time of its recognition as a distinct affection, within certain restricted geographical limits. These boundaries include the British Islands, among which Ireland stands forth prominent as the seat of origin of by far the greater number of the epidemics which have invaded the sister isles.

The season of the year has little or no influence upon the epidemic prevalence of relapsing fever. Epidemics arise, advance and subside, uninfluenced by the season. In widespread visitations of the disease it has broken out in one place at one time of the year, at another a little later, and still later at a third, while it has reached its maximum and has declined at the different places at different periods of the year, and to all appearances wholly uninfluenced by them.

Relapsing fever has also prevailed in seasons remarkable for the amount of rain, and in seasons of prolonged drought.

Age acts as a predisposing cause, seeing that the disease is very frequent in childhood—one-third of all the cases—and that, after the early periods of life, the greatest number of cases—in fact, more than one-fourth of all cases—occur between the twentieth and thirtieth years. From thirty-five to fifty, the liability rapidly declines, and after fifty relapsing fever is comparatively rare, although it cannot be said that the liability ceases. Of 2,111 cases received into the London Fever Hospital in twenty-three years (1848–70), according to Dr. Murchison, 114 males and 81 females—in all 195, were beyond fifty years of age.

Sex cannot be said to exert any direct influence. Statistics of 10,333 cases derived from various sources, and tabulated with reference to sex by the author last named, show 6,175 males and 4,158 females. This difference is to be attributed to the fact that males constitute by far the largest proportion of the vagabond classes, from which are chiefly derived the cases of relapsing fever which form the basis of hospital statistics.

Occupation and mode of life.—No occupation predisposes to relapsing fever, nor does any in itself confer immunity from it. As in the case of other directly contagious diseases, nurses and other attendants upon the sick, including medical men, are exposed to the constant danger of contracting the disease. It has been a common observation that in all epidemics a large proportion of the cases admitted to the hospitals have been wandering musicians, peddlers, beggars, and tramps.

The mode of life and the social condition of the individual exert, beyond all question, a powerful influence. Destitution, filth and overcrowding are strong predisposing causes of relapsing as well as of typhus fever. These conditions not only favor their outbreak, but they also conduce in the highest degree to their spreading. All accounts of epidemics of both these diseases state that they have arisen among, and have been for the most part confined to, the poorest of the population and to the most crowded districts of great cities. In the instances where persons living in affluence have been attacked, by reason of the contagious character of the fevers, the spread of the diseases among them has been limited. Epidemics have in no case arisen among the better classes of society. With very few exceptions, Irish writers insist upon the connection which exists between fever and famine. Failure of the crops, or want of food depending upon lack of money to buy it—a state of artificial famine—has preceded almost every epidemic of relapsing fever. As has been already stated, many cases were observed during the great fever-epidemics which attended the closing years of the last and the first years of the present century, in which the symptoms closely corresponded with relapsing fever. This was a period of great want. Before the outbreak of the epidemic of 1817, the inhabitants of Ireland had been in a state of extreme starvation, due to a succession of bad harvests and other causes. The Scotch epidemic of 1843 was not preceded by failure of the crops; yet the condition of the poor was deplorable, and had been for some years the subject of appeals to the authorities. Murchison states that between 1840 and 1843, four public subscriptions, amounting to twenty thousand pounds sterling, had been raised in Edinburgh alone to relieve their pressing necessities. Upon the appearance of the fever the poor alone suffered. It is stated that of the poor scarcely a single person escaped, while some of the medical men practising among the better classes did not meet with a case. The epidemic of 1847 made its appearance at a time of extreme destitution and misery among the lower classes of Great Britain and Ireland; and the appearance of relapsing fever in Silesia in the same year followed upon a succession of three bad harvests, which had brought the inhabitants to such want that numbers died of starvation alone. The epidemic of 1864–65 in Russia was restricted to the poorest and most destitute of the people, and occurred at a time when provisions of all kinds were high in price and of poor quality.

The foregoing facts form the basis of the opinion long entertained that scarcity of food is the exciting cause, or one of the exciting causes of relapsing fever. This opinion was generally held by the earlier writers upon the subject, and has been, in recent years, most ably and learnedly advocated by Murchison. It is certainly in accord with the general statement that no great epidemic has ever arisen among a well-fed populace, nor spread to any great extent among the prosperous classes of an infected locality, and it appears to explain the apparently independent origin of the fever after the lapse of years and at distant points.

Against it are, however, arrayed the following facts:

Persons are constantly, and communities occasionally, exposed to great want without fever resulting.

When the disease attacks the well-nourished, it runs a similar course and presents the same symptoms that characterize it among the destitute.

There is no direct evidence to show that starvation either occasions the symptoms of relapsing fever, or that it gives rise to any infecting principle capable of producing this or any other specific fever in the starving individual, or in those brought in contact with him.

Epidemics have prevailed among the poorer classes in communities where no general scarcity of food existed.

Lebert[1] states that the theory that looks upon relapsing fever as a famine fever is not borne out by the various epidemics that have occurred at Breslau, nor by the aspect of the relapsing fever patients he examined, who presented on an average a well-nourished appearance. Parry writes that the patients whom he saw in the outbreak in Philadelphia in 1869–70, appeared to be, with a single exception, well-fed, and were even fat, and that all his patients were able to obtain a plentiful supply of milk, meat, eggs, or any other article of diet that was ordered. The cases did not occur among the unemployed and vagrant, but in the families of those who held positions in the neighboring stores and factories, and many of whom had been so employed for years. At the time that the disease made its appearance wages were high, the crops had been abundant, breadstuffs were cheap, and potatoes were plenty. He attributed the disease to overcrowding and the "notoriously small breathing-space allotted to each individual in the houses of the poor," and refers to the observations of Muirhead,[2] who advances the same view and denies the potency of starvation as the cause of the fever. Dr. Bennett[3] states "that he had charge of the Fever Hospital in 1846, where relapsing fever largely prevailed, and he could say that in not one case had it been traceable to starvation."

Finally, the discovery by Obermeier,[4] in 1873, of minute organisms in the blood of relapsing fever patients, points to the nature of the morbific principle which is the exciting cause of the disease.

---

[1] Ziemssen's Cyclopædia. Vol. i. Article on Relapsing Fever.

[2] Relapsing Fever in Edinburgh. By C. Muirhead, M.D.: Edin. Med. Journ., July, 1870.

[3] Edin. Med. Journ., Aug., 1870.

[4] Dr. Otto Obermeier: Vorkommen feinster, eine Eigenbewegung Fäden im Blute von Recurrenskranken. Centralblatt für die Med. Wissensch., No. 10, März, 1, 1873.

Before taking up the special consideration of the exciting cause, we must state our view that, whatever may be the form which that cause assumes outside of the human body, whatever its mode of preservation and of conveyance from person to person and from place to place, it finds in the bodies of the destitute and famished the most favorable circumstances for its lodgement and development, and in the **neglect of personal hygiene and the foul** overcrowding begotten of destitution, the most favorable conditions for its rapid dissemination, and that destitution is, therefore, the most powerful of the predisposing causes of the disease.

II. THE EXCITING CAUSE.

Relapsing fever is due to an infecting principle communicable from the sick to the well, either directly or indirectly, by means of the atmosphere, various fluids, and even solid substances. The nature of this poison is no longer unknown. Obermeier's discovery of minute spiral filaments in the blood of relapsing fever patients threw a flood of light upon this subject. Since the date of that observation, Lebert and his assistants, Weigert and Buchwald, have found, by systematic examination of the blood of relapsing fever cases, that these protomycetes are never absent during the initial febrile paroxysm nor in the relapse, although they diminish very rapidly in numbers after defervescence. They have been repeatedly observed and described by other microscopists, and quite recently Paul Guttman,[1] upon examination of the blood of 280 cases of relapsing fever, found them in every case during the period of the fever. He states that the numbers of the characteristic "spirilli" of Obermeier were not always proportionate to the intensity of the attack nor the elevation of the temperature, that they were sometimes abundant in cases attended by moderate rise in temperature, and rare where the rise was great. They were not seen, upon repeated examinations of the blood, during the period of apyrexia; and it is of interest to note that, when in the intermission the temperature rose in consequence of a complication—as, for example, pneumonia, spirilli were absent. He regards their parasitic nature as established beyond question. What becomes of the spirilli in the blood of the apyretic period is not known. No trace of their remains is discoverable. Guttman regards their rapid disappearance as the more remarkable in view of the facts that they can be preserved as microscopic preparations for a long time (nine months or longer), and that they may be recognized in the blood thirty-six hours after death. Up to the present time all attempts to cultivate the spirilli outside the human body have been unsuccessful.

Dr. Guttman also describes very minute moving corpuscles, which are

---

[1] Zur Histologie des Blutes bei Febris Recurrens. Virchow's Archiv, LXXX., 1880.

found in the blood of relapsing fever patients both during and between the paroxysms, but in rather greater numbers during the febrile periods. They are from one-thirtieth to one-twentieth the size of a red corpuscle, and of a round or oval shape. They are not peculiar to the blood of relapsing fever, but occur also in that of patients suffering from other acute febrile diseases, as pneumonia, scarlet fever, measles, enteric fever, typhus, diphtheria, and erysipelas, and in smaller numbers even in the blood of persons in health. They were successfully cultivated in Pasteur's fluid, and appear to be micro-organisms (mikroparasiten) derived from the atmosphere.

Lebert describes the spirilli as follows: "They are exceedingly slender, never exceeding in diameter 0.001 mm., and in length, 0.15 to 0.2 mm. Their form is spiral. In their interior I have been unable to make out either fat-particles, sheaths, or structure of any kind. Their motion is very lively, rotary, twisting and rapidly progressive, but soon ceases under the ordinary conditions of microscopic examination. . . . Thus far we have sought in vain for this organism in the secretions and excretions, as well as in the internal organs; it is probable, however, that in the future it may also be found in these localities."

Up to the present time this variety of protomycetes has never been found in any other disease.

The conclusion that this parasite has to do with the causation and development of relapsing fever is inevitable. It constitutes the contagium. These spiral filaments, communicated from individual to individual, spread the disease. Finding in the human body the conditions favorable to their development, they multiply indefinitely. The functional perturbations to which their presence gives rise constitute the phenomena of the fever. It is probable that, under favorable circumstances outside the body, their existence may be prolonged through a considerable length of time. This existence may be latent, yet capable of assuming the most energetic activity when introduced into the human body. Such being the case, the possibility of transmission to remote points follows, and the rise of epidemics at considerable intervals of time and at points far distant from each other is comprehensible without the assumption of the independent origin of the germs of disease, or the new development of an old poison. It is much more in accordance with the general laws of organic development to accept a continuous concealed existence of the germs, than to have recourse to spontaneous generation to account for their development (Lebert).

The origin of an epidemic is due to the importation of the *materies morbi* in the person and belongings of a patient, or in other materials from an infected locality, or else to circumstances calling into activity germs that have maintained a latent and harmless existence during a more or less extended lapse of time. The history of relapsing fever in all great epidemics points to scarcity of food and its attendant evils as the conditions favoring the activity of relapsing fever germs.

When the disease has appeared in any locality it spreads with great rapidity by contagion, but in every community it forms centres of greatest prevalence. These foci are determined by the dense crowding together of the poor in the most wretched quarters of cities, and by impure drinking-water and stagnant water in the neighborhood of dwelling-houses.

The rapidity of the spread of relapsing fever in single houses, or within limited districts, is in proportion to the number of the inhabitants and the amount of intercourse between the sick and those surrounding them. It has been a matter of common observation that when the disease has made its appearance in a house inhabited by several families, the occupants of one apartment have been seized one after another, or nearly at the same time, that those dwelling upon the same floor have been next attacked, and afterward the neighbors upon the other floors in the order of the intimacy of the intercourse between families. Reid,[1] of Glasgow, has placed upon record two observations which illustrate the above statement. The first is the account of the introduction and spread of the disease at the Dalmarnock colliery in 1843.

> The colliers, comprising forty different families, occupied a large tenement standing alone in the midst of open fields. It consisted of three stories, entered by three separate stairways. In May an Irish family took possession of a single apartment on the uppermost story, the youngest child being at the time sick of the fever. On the second of June the father sickened, and afterward successively every member of the family. The fever then spread from room to room, and in the space of two months attacked twenty-two persons on this story, the other inhabitants remaining all this time exempt.
>
> In the second instance the disease was introduced from a neighboring village into a house of two apartments occupied by eleven persons. All of these were attacked, and every one suffered the relapse; but in the adjoining house, with a similar entry and separated only by a brick partition, where the occupants were nearly equally numerous, and from their circumstances and habits equally susceptible, all escaped.

In this connection it is proper to call the attention of the reader to the statement already made, that the attendants upon the sick are very liable to contract the disease. This liability increases in proportion to the closeness of the association between the attendant and the patients, required by the duties of the former. Thus, male and female nurses, and the resident physicians in hospitals, are much more frequently attacked than the visiting physicians. In fact, in general hospitals it is only those who are brought into close relation with relapsing fever cases, or who wash

---

[1] The New Form of Fever at present Prevalent in Scotland. By W. Reid, M.D., London Medical Gazette, vol. xxxiii., 1843.

their clothes or bedding, that contract the disease. Nurses in the medical wards into which fever cases are not admitted, and those in the surgical wards, as a rule escape.

My colleague, Dr. Morris Longstreth, at the time when, as residents in the Pennsylvania Hospital, we had the opportunity of observing relapsing fever in the wards during the epidemic of 1869-71, contracted the disease. The records of every epidemic abound in instances of the communication of the fever from patients to their attendants. Dr. Welsh[1] wrote in 1819, as follows: "When acting as clerk to Dr. Hamilton, in the Royal Infirmary, in the course of four months my three colleagues, two of the young men in the apothecary's shop, two housemaids and thirteen or fourteen nurses, caught the disease, and the matron and one of the dressers died of it. Since I left the infirmary, three more of the gentlemen acting as clerks, one of the young men in the shop, and many more of the nurses, have caught the infection, but the number I do not know. Since Queensbury House was opened, on February 23, 1818, my friends, Messrs. Stephenson and Christison, the matron, two apothecaries in succession, the shop-boy, washer-woman and thirty-eight nurses, have been infected; four of the nurses have died. With the exception of two or three nurses, who have been but a short time in the hospital, I am now the only person in this house who has not caught the disease within the last eight or ten months. Several students, whom curiosity led too near the persons of the patients, might be adduced as additional evidence. When it begins in a family, we always expect more than one of them to be affected. I could mention instances of four, five, six and seven being sent to the hospital out of one family; eight, nine and ten out of one room; twenty and thirty out of one stair; and thirty and forty out of one close; and this all in the course of a few months." Writing of the fever of 1843-44, Dr. Wardell[2] states that "most of the medical officers connected with the Edinburgh Royal Infirmary and additional fever hospitals were seized with it; eight of the resident and clinical clerks in quick succession became affected, and out of that number no less than six were yellow cases, and thus obviously in danger of their lives. The majority of the nurses and domestics took the disease, and of the former, at one time no less than nineteen were laboring under it. Some of the dispensing physicians and other practitioners took the disorder, as also several of the clergy and visitors of the sick, whose duties brought them to the bedsides of the patients. The few cases occurring among the higher classes resident in the new town were generally to be traced to the in-

---

[1] A Practical Treatise on the Efficacy of Bloodletting in the Epidemic Fever of Edinburgh, illustrated by Numerous Cases and Tables, extracted from the Journals of the Queensbury House Fever Hospital. By Benjamin Welsh, M.D. Edinburgh, 1819.

[2] The Scotch Epidemic Fever of 1843-44. London Medical Gazette, xxxvi.-xl. 1846-47.

fluence of contagion, the parties affected having had either immediate or indirect communication with those suffering under the disease." Cormack,[1] in his account of the same epidemic observed: "Almost all the clerks and others exposed to the contagion have been seized. Dr. Heude, and his successor Mr. Reid, in the New Fever Hospital, Dr. Bennett, my successor there, Mr. Cameron and his successor, Mr. Balfour, in the adjoining fever-house, as well as most of the resident and clinical clerks in the Royal Infirmary, have gone through severe attacks during the past summer and autumn. Hardly any of the nurses, laundry-women, or others coming in contact either with the patients or their clothes, have escaped; at one time there were eighteen nurses off duty from the fever; and of those who have recently been engaged for the first time, or of those who have hitherto escaped, one and another is from time to time being laid up." Murchison informs us that, "in the London Fever Hospital, during the years 1869–70, twenty-seven of the nurses and officers, and five patients contracted relapsing fever. One nurse, who had been in the hospital for nearly twenty years, and had passed through typhus, had a severe attack of relapsing fever shortly after the first cases of the disease were admitted."

Persons in health, from localities where the disease is unknown, are attacked upon coming in contact with the sick in an infected community at a distance. The pestilential centres of relapsing fever are in all instances limited to the quarters of cities and like districts inhabited by the poor, while persons living in easy circumstances and in opulent neighborhoods, under favorable conditions of public and personal hygiene, as a rule wholly escape. This immunity ceases, however, upon their visiting the sick. On the other hand, relapsing fever is, in every epidemic, liable to be imported by infected persons into localities before exempt.

The history of the march of the disease in the epidemics of Great Britain sufficiently illustrate this statement. It is also stated that it was carried in this way from St. Petersburg to other cities in Russia, and most writers are agreed that the American outbreaks were due to importation from the other shores of the Atlantic, although it was not possible to trace its route. It is certain, however, that in several of the local epidemics outside of Philadelphia and New York the disease was brought from those cities by persons who had been in contact with the sick.

The following striking example of the contagion is narrated by Parry:

"A man left Philadelphia about February 1st, remaining for two months in Western Pennsylvania. During his absence his health was good, and he had no known opportunity to take any disease. On returning home he spent several days with a

---

[1] Natural History, Pathology and Treatment of the Epidemic Fever at present prevailing in Edinburgh and other Towns. By J. Rose Cormack, M.D. London, 1843.

friend in the second paroxysm of relapsing **fever**. He then went to his brother's, and ten days after reaching the city was seized with mild relapsing fever, and was **sick** five days. In about two weeks his brother's wife was taken and had it severely. Subsequently another case occurred in the same family. During the remission this same man went to his brother-in-law's in a distant portion of the city. Here he had the relapse, which lasted four days. This family consisted of six persons, four of whom were children. Only one of the six, the mother, escaped. It is worthy of note that **the** youngest **children, who were most exposed by being with** their uncle, and **who were aged respectively four and six years**, were taken **first on** the eleventh and twelfth days after their relative **reached the house**. The older ones, who were nearly **grown up and engaged at work during the day**, did not **take sick** until the younger ones were in the relapse."

In hospitals the nurses and attendants have never contracted the disease until after the admission of relapsing fever patients.

Without doubt it is in many instances communicated **from the sick to the** well by direct contagion, that **is,** by **actual contact**. Hence, it spreads rapidly in chambers and houses occupied by large numbers of destitute persons, and in the lodging-houses frequented by the vagabond poor.

But it is also largely communicated by **fomites**. In this way only can be explained **the great** liability of the laundry-women in hospitals to contract the **disease** without direct contact with the sick, and under circumstances that **render it in the highest degree improbable that the poison** reaches them **either by means of the atmosphere or of drinking-water.** Parry relates the following instances in which relapsing fever was transported to a distance by infected clothing:

"A family lived in **a healthy neighborhood and were in comfortable circumstances.** One of the sons was **employed in a factory, where they procured a new hand, who, it** was afterward learned, **had just left a hospital where he had been ill with relapsing** fever. From him the son purchased a pair of overalls and carried them home. On April 19th one of the sisters washed them. She was taken ill with the fever on May 1st. At the same time this garment was handled by two other sisters, who fell sick on the 2d and 3d of **May, respectively."**

"A woman learned **through** the newspapers that her husband had been picked up ill in the street and taken to the Philadelphia Hospital. He had not been home for some time before. On March 21st or 22d she sent a friend to the hospital to learn his condition. She found him dead, it was stated, from relapsing fever. She went to the dead-room and identified his body, which was not brought away for burial. She carried his clothing to her own home and placed it in a room next to her children's bed-room, with an open door between them. Four cases of the disease afterward occurred **in** the family. On April **7th**, a boy was taken, April 25th a girl, April 29th **another** girl, and on May 3d **her husband.** There had certainly been no cases in the immediate neighborhood before that time."

It is in the highest degree probable that the disease can **be communi**cated by means of the atmosphere. But the distance to which the poison can be transported in this way in sufficient concentration to produce the disease cannot be very great. Those only who are in close communication with the sick, or who visit them in their ill-ventilated quarters, or who reside near at hand, suffer. With free ventilation the disease almost ceases to be communicable (Murchison). **Lebert deems** it worthy of remark that in all epidemics occurring in his wards, in which thorough ventilation is maintained summer and winter, cases of contagion have been exceedingly rare.

The danger of contracting the disease through **the atmosphere appears** to increase with the length of the exposure. In a few instances the disease **has** seemed to follow promptly upon exposure. The poison in these **cases must** have been very concentrated. As a rule, the resident physicians in hospitals are more apt to contract relapsing fever than dispensary physicians who visit their patients in their badly ventilated houses, and, remaining but a short time, have constant opportunities to breathe an uncontaminated air in passing from house to house. The length of time necessary to contract relapsing fever by exposure to the atmosphere of the sick-room without actual contact is longer than in the case of typhus. Finally, Lebert ascribes great importance to drinking-water as a carrier of the infecting principle. The pathogenetic protomycetes thriving in it may infect many persons in the same house at the same time, or in rapid succession, as is seen in cholera. The researches of this observer, in 1868 and 1869, show that in 27 per cent. the interval between new cases in the same house was only 1 day; in 16 per cent., 2 days; in 11 per cent., 3; in 5 per cent., 4; in something over 6 per cent., 5; in 6 per cent., 6; and in 4 per cent., 7 days. In other words, 75 per cent. occurred within the first week, and 54 per cent. within the first three days. It follows from these figures, he adds, that too much stress must not be laid upon the transmission of the disease from individual to individual by direct contagion, and he regards the simultaneous or nearly simultaneous infection of several persons by means of drinking-water as the most probable explanation of the facts. He informs us that the nidus of typhus and relapsing fever in Breslau was in a quarter of the city supplied by such impure drinking-water that a whole fauna and flora might be found in it.

The period of incubation of relapsing fever is variable. In some rare instances it has been absent, the symptoms following immediately upon the first exposure to the contagion. According to Murchison, it varies from five to sixteen days. Parry estimates it to be from seven to fifteen days. Lebert states that it is from five to seven days. The number of accurate observations bearing upon this point is limited.

No immunity from subsequent attacks is experienced by those who have suffered from relapsing fever. Observers have recorded the occurrence of second and even third attacks in the same individual, within the course of several months, in almost all epidemics. In this respect relapsing fever presents a striking contrast to typhus, and, in fact, to most of the other infectious fevers. Dr. Christison[1] remarks that, during the epidemic of 1817–19, he experienced no fewer than three separate attacks in his own person, within fifteen months.

As has been already pointed out, there exists a remarkable association of relapsing fever with typhus in epidemics. Prior to 1843, the former fever was looked upon as a mild form of the latter. To Dr. Henderson,[2] of Edinburgh, is due the credit of having first pointed out their essential difference. He showed that they were characterized by different symptoms, and stated his belief that they arose from different poisons. His views, which were confirmed by many other observers at that period and since, were based upon the two-fold proposition that, first, the one fever under no circumstances gave rise by communication to the other; and secondly, that an attack of typhus never conferred immunity from relapsing fever, any more than the latter afforded protection from typhus.

Henderson and others found that only in the rarest instances, cases of relapsing fever and of typhus fever occurred at the same time in the same house. On the other hand, numerous excellent observers have recorded instances of the association of the two fevers in the same house and even in the same room. This discrepancy is to be readily explained by the manner in which the two diseases are associated in most epidemics. In circumscribed localities there was the same sequence of the two fevers as was found in studying the history of wide-spread epidemics: at first, relapsing fever only; then relapsing fever and typhus together; and, last of all, typhus alone (Murchison).

Cases in which relapsing fever follows upon typhus in the same individual are rare; but the instances in which the order of events has been reversed, and the latter has followed relapsing fever in the course of a few weeks or months, have been so numerous as to attract the attention of most observers. Lebert collected accurate statistics of fifty-three cases of relapsing fever in which an attack of typhus occurred at an interval of from several weeks to a few months later. The subjects were mostly between fifteen and sixty years of age. The mortality of typhus in those cases was 7.55 per cent., half the death-rate of the other typhus cases. Whether this lowered death-rate was the result of the chance association of favorable cases, or of an influence on the part of the forerunning relapsing fever poison, which rendered that of typhus less dangerous, remains unsettled.

The peculiar relationship of relapsing and typhus fevers, both as regards the individual and as regards the community, point to an affinity

---

[1] On the Changes which have Taken Place in the Constitution of Fevers and Inflammations in Edinburgh during the Last Forty Years. Edin. Med. Journ., Jan., 1858.

[2] On Some of the Characters which Distinguish the Present Epidemic Fever from Typhus. Edin. Med. and Surg. Journ., vol. xli., 1844.

between them that, in spite of their essential difference, cannot be accidental.

CLINICAL HISTORY.

Relapsing fever is divisible into **four distinct stages**. These are, in ordinary cases: the **primary paroxysm, the intermission, the relapse, and convalescence.**

The attack begins abruptly—a prodromic stage being, as a **rule, absent.** If prodromes occur at all, they are of short duration, and consist of general malaise, dull pains in the head, wakefulness, loss of appetite, and the like.

The speedy onset of the disease is characteristic. On waking in the morning, or in the middle of the day while engaged in their ordinary pursuits, more rarely later in the course of the day, or at night, the patients are seized with high fever, ushered in with a sense of chilliness in about half the cases, and with a decided chill in a much smaller proportion of them. When the disease begins with a rigor, it recurs in some instances irregularly during the first two or three days; and, as sweating is often, though by no means in all the cases, present during this period, a superficial resemblance to the paroxysms of intermittent fever may arise. The sweating usually breaks out upon the face and upper parts of the body, while the rigor continues without the intervention of a distinct hot stage.

In other instances sweating does not occur till the second or third day, when it may be profuse and continue for several hours, without relief to the headache or other symptoms. The skin is, during the paroxysm, frequently bathed in sweat, while the temperature remains high. There is debility from the onset, and this, with the giddiness, headache, and pains in the joints and muscles, compels the patient to betake himself to bed at once. In the lightest cases he is able for a time, or even throughout the attack, to continue his avocation.

In a little time after the initial symptom the skin becomes dry and very hot; there is intense thirst and great aggravation of the pains; appetite is lost, and nausea and vomiting are common, sometimes persistent. The vomited matters consist of a greenish fluid. The temperature rises rapidly. The morning following the onset it may exceed 39°—40° C. (102.2°—104° F.), and, assuming an irregular and faintly marked intermittent type, it mounts, in the course of a few days, some degrees higher—41°—42° C. (105.8°—107.6° F.). The pulse is frequent, usually exceeding 110, often 120, and occasionally beating as often as 140—160 per minute. The difference in the frequency of the pulse in the morning and in the evening is but slight. It is of moderate fulness and tension, often quick, sometimes dicrotic.

The tongue is usually moist, and covered with a white or yellowish—

white fur of varying thickness ; it is apt to continue thus coated throughout the paroxysm ; in a small proportion of the cases it becomes dry, or shows a dry, brownish streak in the middle. The bowels are constipated, or rarely there is slight and somewhat persistent intestinal catarrh.

In a varying proportion of cases, but without great frequency, jaundice appears during the course of the first paroxysm. There is no characteristic eruption ; sudamina appear late ; herpes facialis occasionally occurs. The skin is, as a rule, moist after the first few days. In many cases it remains dry until **the crisis.**

As early as the second day a feeling of distress in the upper part of the abdomen is complained of. This approaches more nearly to actual pain in the left hypochrondium than in the right (Lebert). Physical examination reveals enlargement of the liver and a rapidly progressive in**crease** in the size of the spleen, which not infrequently reaches below the **ribs.** There is marked tenderness in the epigastrium and in the splenic and hepatic areas.

At the commencement of the attack, pains in the back and joint-pains are marked. To these are speedily added distressing muscle-pains in all parts of the body, **as well as in the upper and lower extremities—but most severe, as a rule, in the calves of the legs. These pains are described by** the patients as stabbing, burning, grinding. **They** are present when the **body is in repose,** but are aggravated both by movement and by pressure. After the first days the headache lessens, but the muscular pains persist. The patients lie motionless to avoid the increase of pain which change of position induces.

Sleeplessness is a distressing symptom. **Pain prevents sleep. The mind** is clear. The expression lacks **the dulness of typhus and enteric** fevers ; delirium is rare.

Epistaxis occurs, but with no great frequency. It is more common in childhood than in adult life (Parry).

The urine presents the characters of febrile urine in general. It moreover not infrequently contains albumen. When jaundice is present, it contains bile-pigment.

Upon the fifth, sixth, or seventh day, as a rule, but sometimes as early as the third, very rarely as late as the tenth, the sickness apparently **comes** to an abrupt end. The symptoms, in some instances having even augmented in severity, suddenly cease. The change is mostly attended **by a** critical discharge, usually by a profuse sweat, sometimes by diarrhœa, more rarely by bleeding from the nose, rectum, or vagina. In rare instances the crisis is preceded by a brief, violent delirium. The temperature, usually **during** the course of the night, falls to a point below the normal standard, the pulse becomes much less frequent, the skin cool. The breathing, **which has been hurried,** becomes normal, the pains in the muscles, **and** the headache **lessen greatly,** or cease altogether. Thirst no longer torments the patient, the tongue cleans, appetite returns, the liver decreases in size, and the spleen contracts almost as rapidly as it augmented in volume; epigastric tenderness disappears, and jaundice, if present, begins to fade. To all the evidences of a severe, even alarming illness, have rapidly supervened a condition of comfort and apparently almost complete convalescence. But for a feeling of weakness, the patient regards himself as well. His strength augments from day to day, and he arises and moves about—often, if in hospital, insisting upon going to his home, in disregard of the warnings that he will suffer a relapse.

During the intermission, in most cases, the convalescence is rapid and the patient in truth resumes the appearance of health. The appetite is usually excellent. In many cases, however, there is a notable slowness of the pulse—40 to 68; in not a few the first sound of the heart is faint, sometimes almost inaudible, while the second is relatively intensified. Great muscular weakness, and even paresis of the lower extremities, have been observed at this period.

The spirilli of Obermeier, constant during the periods of pyrexia, are not now found upon microscopic examination.

The period just described usually lasts about a week. In some instances it does not exceed four or five days; in others, it may extend to two weeks, and in very rare cases the first paroxysm has comprised the whole of the attack, not being followed at all by a second pyretic period.

Between the twelfth and twentieth days from the beginning of the attack, but in by far the greatest number of cases, on or about the fourteenth day, the patient, unexpectedly to himself and with the same suddenness as before, again falls ill. Commonly in the night, but sometimes during the day, the relapse sets in. Its advent is attended by chilliness or a decided rigor, or it may be marked by fever without either. The symptoms are a repetition of those of the primary paroxysm. There are the headache, the pains in the back and limbs, the hot skin, the abrupt high fever, the rapid action of the heart, the furred tongue, vomiting, constipation, tenderness in the epigastric zone, that characterize the earlier sickness. The liver and spleen again undergo rapid augmentation in volume, and upon microscopic examination the spirilli are found in numbers not less than before. With the approach of convalescence their number again diminishes and they finally disappear.

It may be stated that, as a general rule, the symptoms of the relapse are less severe than those of the first febrile period; exceptionally they are more so. The type of the fever of the second paroxysm is more distinctly remittent than that of the first, marked remissions occurring in the morning, decided exacerbations toward night.

The length of the relapse is usually about three days; it is occasionally almost abortive, not exceeding a day; at other times it may be extended to five days or more.

The second crisis, like the first, commonly sets in during the night, and is attended by abundant sweating and a fall of the temperature below the normal, with a corresponding decrease in the frequency of the pulse.

The second defervescence is in some cases also preceded by a brief but marked intensification of all the symptoms.

Occasionally a second relapse, attended with symptoms similar to those of the first and lasting two or three days, occurs on **or about the twenty-first day**. Less frequently a third, and still **less frequently a fourth relapse**, has been observed.

At the termination of the disease the condition of the patient is comparatively comfortable. The fever **ceases, the** pains disappear, appetite is, in most cases, speedily regained; but the loss of strength and the emaciation are such **that** a number of weeks must elapse before the sufferer is sufficiently restored to health to resume his ordinary avocations. The whole period, from the beginning of the sickness till complete convalescence, is, upon an average, six weeks. Anæmic murmurs often persist for a still longer period.

The death-rate varies between two and four per cent., **differing in different epidemics.** Death may occur from the intensity of the fever and **the** consequent exhaustion, usually at **the close of the relapse, or by progressive exhaustion, after several relapses.** Occasionally a sudden fatal termination takes **place at the crisis, by failure of the heart.** Death is **due, in** some instances, to suppression of urine, with **coma and convulsions.** It may also result from pyæmia following softening and **abscesses of the** spleen, and the last-named lesions have by rupture caused fatal peritonitis.

Pregnant women almost invariably abort or miscarry during the course of relapsing fever. This accident exceptionally occurs in the first paroxysm, **commonly in the second.** The fœtus, even at the approach of term, perishes, and the **life of the mother is often,** though not invariably, lost.

Death is frequently the result of this and other complications, particularly pneumonia, or of the aggravation of previously existing severe disease. In relapsing fever, as in other epidemic diseases, abortive cases are not infrequently encountered. Cases of this kind may terminate with a single febrile paroxysm, attended with symptoms of moderate intensity, sometimes indeed so light as not to compel the patient to take to his bed, or a second paroxysm of short duration and little severity occurs.

In view of the protomycetic basis of the disease, it is difficult to com**prehend** the varying intensity of the attack. The numbers of spirilli discoverable in the blood have not always been proportionate to the severity of the symptoms. There is, doubtless, a different degree of tolerance of the presence **of this** particular parasite in different individuals. In the words of **Lebert,** "it is possible that, according to the predisposition, a **grave difference may result as regards the pyrogenetic products.**"

Cormack,[1] in his description of the epidemic of relapsing fever in 1843, referred the cases to two general groups. Of these, the first he called the *ordinary* or *moderately congestive form*. This included the common, mild, and average cases, which were rarely fatal except in consequence of some complication. The second he termed the *highly congestive form*. In this form a deep, persistent, purple color of the face, intense jaundice, marked enlargement of the liver and spleen, hemorrhages from the mucous tracts, drowsiness, delirium, and subsultus were prominent symptoms. The paroxysms in the graver form were separated by a period of remission rather than by a distinct intermission. These cases were rare, but often fatal, the patient falling into a condition of collapse, which often lasted for some days before death occurred. This form corresponds with that which has been described by recent observers as "*bilious typhoid.*" It has occurred with varying frequency in many of the epidemics of relapsing fever, and has had much to do in determining the high death-rate in some of them, notably in the Russian epidemic of 1864–65. Bilious typhoid, which has not occurred in any of the outbreaks of relapsing fever in America, was first fully described by Griesinger,[2] who observed it at Cairo in 1851, and gave it this name.

Analysis of the Principal Symptoms.

Symptoms Referable to the Nervous System.

Headache is an early and persistent symptom. It subsides with the crisis only to recur with the relapse, in which it is, however, often somewhat less intense. It is commonly frontal, sometimes general, and is throbbing or darting in character. In rare instances it is mild, and ceases after a day or two.

Vertigo is very common. It occurs as an early symptom, and patients often declare that it is the giddiness rather than the fever that forces them to take to bed speedily after the onset of their illness. This symptom continues throughout the primary paroxysm, and returns in the relapse; it causes the patients to stagger like drunken persons when they attempt to stand or walk.

Delirium is rare. When it exists it is transitory, but is apt to be active and noisy. It occurs for the most part in hysterical or intemperate persons. In most cases the mind is unclouded throughout the attack.

Stupor and coma occasionally come on rather suddenly at or soon

---

[1] J. Rose Cormack, M.D.: The Natural History, Pathology, and Treatment of the Epidemic Fever at present prevailing in Edinburgh. Edinb., 1843.

[2] See Virchow's Handbuch der speciellen Path. und Therap. Band II., Zweite Abtheil. ; also Dr. Van Harlingen's translation in Lebert's article in Ziemssen's Cyclopædia, vol. i.

after the crisis, in consequence of suppression of urine. They may be attended with general convulsions. The patient may sink into persistent stupor, with dry, brown tongue, muttering delirium, and the attendant symptoms that make up that condition known as the "typhoid state," in consequence of the intensity of the fever. This is rare, but when it takes place the crisis does not occur, and the condition is one of the greatest danger.

Insomnia is often marked, and occasions great distress; it is in large measure due to the pains. In the cases observed by Dr. Parry it was a much more prominent symptom in the early than in the later months of the epidemic. Sleeplessness attended in many of his cases the primary paroxysm, the remission, the relapse, and the period of convalescence; it did not yield to the administration of hypnotics. Toward the end of the outbreak it was a less grievous symptom, and was easily controlled by remedies.

Debility is an early symptom. That it should become marked toward the end of the sickness is apparent from a consideration of the symptoms that attend the attack. In most instances, however, the patient is able to get out of and into bed again, and to help himself.

Pain.—Among the more characteristic and distressing symptoms of relapsing fever are the severe pains in the muscles and joints complained of by almost all the patients. Pain in the back is severe during the first few days. The other pains are also present from the beginning; they continue throughout the paroxysm and the relapse. In many cases they are also present during the intermission; at this period they are apt to be more distinctly articular, but are unattended by swelling or by any grave difficulty of treatment. The muscle-pains are seated in the neck, chest, and abdomen, as well as in the extremities; they are usually most severe in the lower limbs, and in particular in the calves of the legs. They arise spontaneously, but are also excited by pressure and by voluntary movement, and compel the patient to preserve as nearly as possible a motionless attitude in bed. They are described as resembling the neuralgic pains that follow unaccustomed or over-prolonged use of certain groups of muscles. Subsiding during convalescence, they leave behind them marked muscular weakness.

Muscular palsies will be considered among the sequels. Retention of urine and involuntary evacuations are very rare. When they occur, it is in consequence of sudden syncope or cerebral complications attendant upon uræmia. Involuntary fecal discharges are sometimes due to extreme diarrhœa in grave cases. Tremors are not observed except in the subjects of previous alcoholism.

THE PHENOMENA OF THE FEVER.

The temperature rises with great rapidity, and attains a height infrequent in the other fevers. Its course is characteristic of the disease. During the initial rigor it is often as high as 39° C. (102.2° F.), and within twenty-four hours it attains 40°—41° C. (104°—105.8° F.). The maximum of temperature may be attained upon the first day, during the mid-course of the paroxysm or shortly before the crisis; the last is the most frequent, and at this period the temperature occasionally runs up rapidly in the course of a few hours. The curve is irregularly remittent, the morning temperature being from 1° to 1.5° C. (1.8°—2.7° F.) lower than that of the midday or evening. Occasionally the remissions are much more marked, but the variations are neither constant for different days nor for the same hour of successive days; in some cases they do not amount to more than a few tenths of a degree, and it is not uncommon to note a higher temperature about noon, or early in the afternoon, than in the evening. In rare cases the diurnal curve shows no remission whatever, the evening and the morning temperature being alike. The remittent type is most constantly present, and is most distinctly marked in children.

In no other disease is so decided and so rapid a critical defervescence met with as in relapsing fever. It is always sudden, very frequently preceded by an increase in the severity of the symptoms, and sometimes ushered in with a chill. It commonly occurs in the evening or toward morning, and is complete as a rule in the course of a few hours. The temperature falls from 3°—6° C. (5.4°—10.8° F.) in cases that may be spoken of as average instances, and it is not uncommon to observe a fall of even 7° C. (12.6° F.) within a short time. Murchison informs us that falls of 13° F. in six, and 14.4° F. in twelve hours, have been noted. A comparison of the temperatures of the febrile paroxysm and the fall during the defervescence, indicates a subnormal temperature as almost constant at the termination of the crisis. This is found to be the case. It is not rare to find the temperature at this period as low as 36°—35° C. (96.8°—95° F.), or even much lower. According to the author last referred to, 94° F. and even 92° F. have been recorded, and, in one instance where collapse supervened, a rectal temperature of 90.6° F. was observed. After two or three days it rises to the normal in the morning, and becomes subfebrile in the evening, and then, becoming that of health for a time, it again rises slightly upon the approach of the relapse.

A temperature of 39° C. (102.2° F.) or more, attends the onset of the relapse, in which the same rapid rise to a great height, and an even more rapid fall to below the normal standard than in the primary paroxysm, are encountered. The maximum temperature of the whole attack is not in-

RELAPSING FEVER. 327

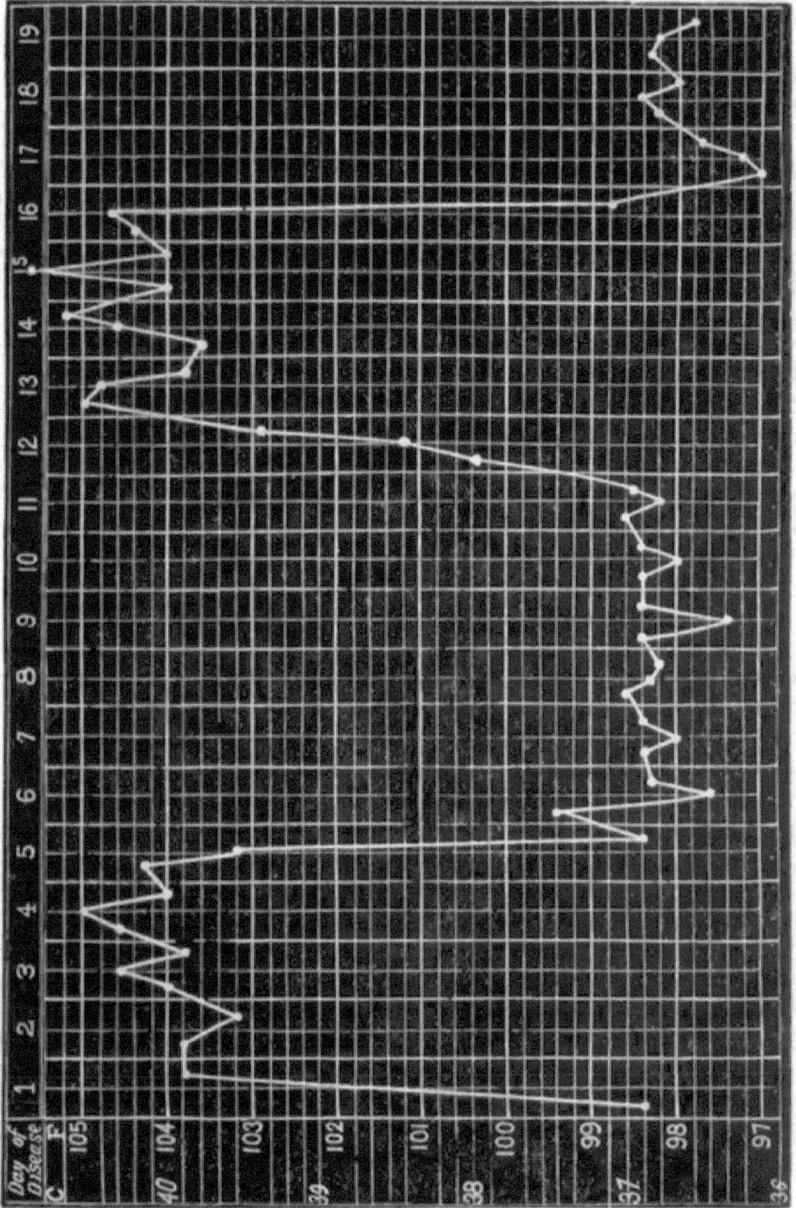

FIG. 22.—Temperature in Relapsing Fever from First Day of the Attack. (Murchison.)

frequently met with in the relapse. The crisis is accompanied by free perspiration. If other relapses follow, the temperature curve attending them is the same, but the fever is of shorter duration.

The maximum temperature of relapsing fever varies from 41° to 42° C. (105.8°—107.6° F.) in cases that cannot be looked upon as exceptional. A temperature of even 42.5° C. (108.5° F.) has been observed. These excessively high temperatures, if not long continued, are not attended with great danger to the patients, nor do they give rise to cerebral symptoms. In this respect relapsing fever differs from other diseases characterized by intense pyrexia.

The foregoing statements are based upon observations of temperatures taken in the axilla.

The pulse is always frequent. It is commonly above 112, but may vary from 90 to 120, or even beat as rapidly as 160 or 170 per minute. It is more frequent by 20 or 30 beats in childhood than in adult life. This frequency is attained very early in the course of the disease. It is not of unfavorable prognostic omen. The number of beats per minute increases toward evening and with a rising temperature. A gradual or progressive increase does not occur with the progress of the attack, although it is not uncommon to find a sudden increase in pulse-rapidity, as well as a decided sudden rise in temperature, immediately preceding the crisis.

With the defervescence there is a sudden fall in the pulse-rate to the normal, and often below it. In a few hours, declining a little before the temperature begins to fall, it may change from 140 to 48—54. During the intermission it is often abnormally slow, 40—60; but if the patient leave his bed it becomes more rapid, 100 or more upon his assuming the upright posture, and continues to beat rapidly. There is no constant ratio between the rate of the pulse and the temperature. Murchison states that there is less correspondence between them in the relapse than in the primary paroxysm, a pulse not exceeding 90 being sometimes met with where the thermometer marks a temperature of 106° F.

The pulse during the febrile paroxysms is often at first full and tense, but with the crisis it becomes small and feeble, and is often jerking and irregular; after the crisis it is compressible, and not seldom dicrotic. With convalescence, as the patient gains strength, the pulse resumes its normal character.

About the time of the crisis, and in particular immediately after it, the impulse and first sound of the heart are often very greatly impaired, and sudden death from syncope may take place. Within the course of a few days, and with the use of stimulants, the heart regains its power.

A soft systolic murmur is heard over the cardiac region during both paroxysms in frequent cases and sometimes in the intermission. Its area of greatest intensity is at the base; it is propagated in the course of

RELAPSING FEVER. 329

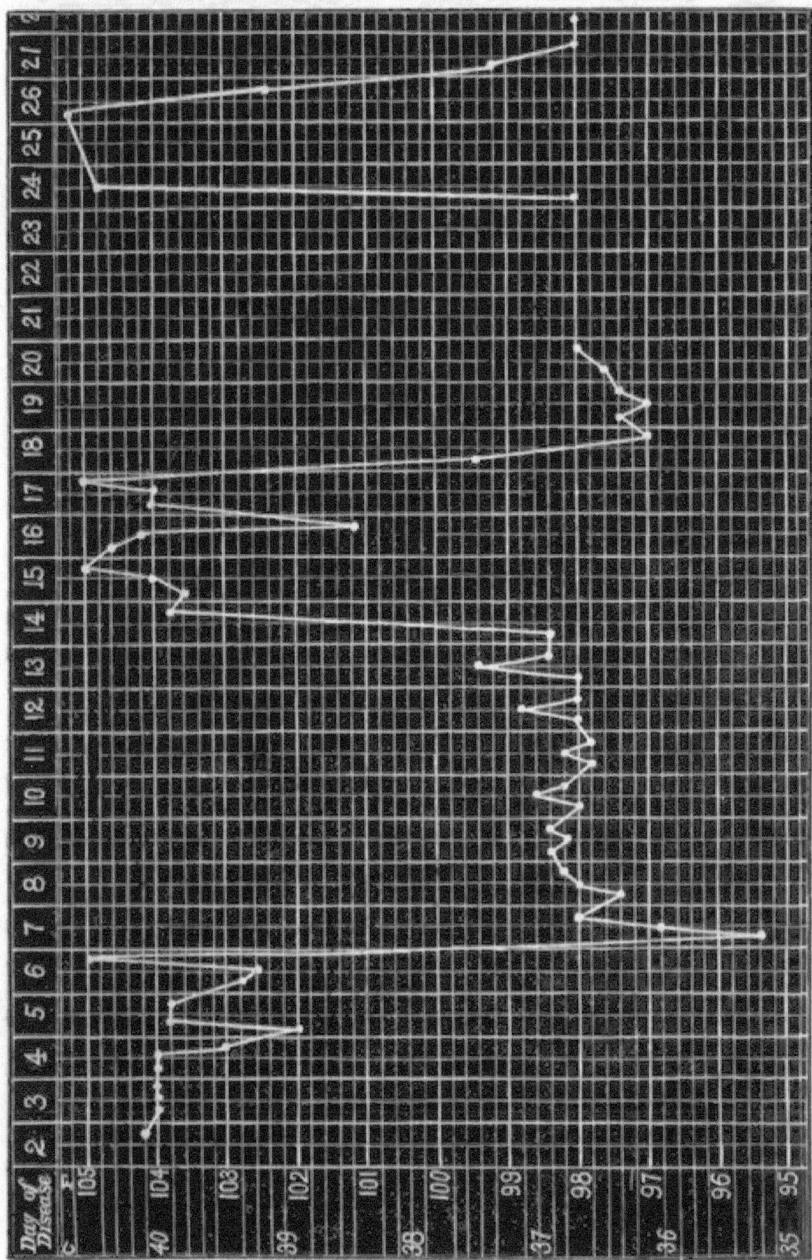

FIG. 23.—Temperature in Relapsing Fever; Enormous Fall at First Crisis; Two Relapses. (Murchison.)

the great vessels, and becomes faint or is lost entirely in the erect position.

The urine presents the characteristics of febrile urine in general. Its amount varies with the quantity of fluids ingested, and is influenced by the abundant sweating that occurs at the crisis, and, in many cases, during the progress of the febrile paroxysms. As a rule, it is diminished in quantity during the febrile stages, and of darker color and higher specific gravity than normal, and becomes normal or even increased in quantity shortly after the crisis. It is frequently cloudy, and deposits a sediment consisting of the urates; less often uric acid is present, and crystals of the oxalate of lime. It is commonly acid in reaction, but occasionally alkaline when passed. The triple phosphates are present in the latter case. In some cases the amount is greatly reduced immediately after the crisis, the patients being exceedingly weak and sweating profusely. Two cases under Parry's observation did not void more than an ounce in twenty-four hours for several days, yet there was no evidence of uræmia.

Albumen in small amounts is often present in the urine during the primary paroxysm, and Murchison reports a case in which copious hæmaturia occurred in both paroxysms, although the urine during the interval contained no trace of albumen. Recovery took place. Tube-casts are found along with the albumen. In the first paroxysm they are usually hyaline, in the relapse they contain granular matter and oil-particles. If pre-existing Bright's disease complicate the case, the character of the urine, especially as regards the quantity of albumen and the nature and abundance of the renal casts, will be modified. The opinion of Obermeier,[1] that acute desquamative nephritis is one of the ordinary phenomena of relapsing fever, calls for a closer examination into this point in future epidemics. It is not confirmed by other recent observers. In cases complicated with disease of the kidneys, and under other circumstances, marked diminution or suppression of urine has been followed by uræmic symptoms, such as delirium, stupor, coma, and convulsions. This condition is apt to supervene at or about the time of the crisis. Such patients have in many instances recovered after a copious discharge from the kidneys. Murchison states that he "has never known typhoid symptoms in relapsing fever without albuminuria or some other evidence of retarded elimination by the kidneys."

In those cases marked by jaundice, bile-pigments are found in the urine; the biliary acids have also been detected.

The skin in relapsing fever shows no characteristic eruption. The abundant perspiration gives rise to plentiful crops of sudamina. Herpes facialis occurs, but not frequently. Minute petechiæ are occasionally observed in delicate persons, and are apt to be most abundant upon the

---

[1] Quoted by Murchison.

lower extremities. Extensive desquamation sometimes occurs, and during the pyrexia the nutrition of the nails is impaired, as is shown by the development upon them of white transverse lines.

SYMPTOMS DUE TO DISTURBANCE OF THE DIGESTIVE ORGANS.

Thirst and *loss of appetite* are due to the fever; upon the defervescence the former, which is often excessive, ceases, and desire for food returns; with the relapse the thirst and anorexia reappear. In some instances an inordinate desire for food during the febrile paroxysms, and especially in the relapse, has constituted a remarkable feature of the disease. Patients with a temperature of 40° C. (104° F.) or higher, have, in some instances, begged for solid food and eaten it eagerly in considerable quantities, without apparent injury—a statement attested by numerous competent observers.

The tongue is usually indented at its edges by the teeth, and covered with a whitish or yellowish-white fur of varying thickness. In other instances the fur is of a brownish color from the beginning. The edges and a triangular space at the tip are sometimes clear and of a brighter red than normal. The papillæ are enlarged in some **instances, so that the** tongue may be likened to that of scarlet fever; less commonly the organ is red and glazed, especially in the relapse. As a rule, it is moist throughout the sickness; but it may show a dry, brownish streak down the middle about the third or fourth day. It sometimes becomes deeply fissured—a very painful symptom. In very severe and in fatal cases it becomes dry, brown, and crusted, and sordes collect upon the teeth and lips. The foregoing facts being considered, it may be stated that the tongue in relapsing fever presents no constant characteristic appearance.

Nausea and *vomiting* are common symptoms. They appear early and soon subside. In some cases, however, the vomiting persists to the end of the paroxysm, ceasing with the crisis, but returns with the relapse in some few instances.

The vomited matters consist of the substances taken into the stomach, of gastric mucus, and of bile. They are of a greenish or yellowish color, and are usually scanty. "Black vomit" was, in rare cases, observed in **some** of the earlier epidemics. Hæmatemesis has also been noted.

Pain and *tenderness* in the epigastrium are present in a large proportion of the cases. They are frequently associated with vomiting, but are **by no** means proportionate to its urgency. The pain is usually slight; it may, however, be so severe as to interfere with respiration (Murchison). It may be limited to the epigastrium, or extend across the epigastric zone. In the latter case it is most severe in the left hypochondrium, and is patho**logically referable** to the acute enlargement of the spleen.

Enlargement of the liver occurs in most of the cases. It appears later than the enlargement of the spleen, and is much less marked. It is attended with pain upon pressure in the hepatic region.

Jaundice appears in varying frequency in different epidemics. It has seldom been observed in **more than** twenty per cent. of **all cases, and is, as a rule, still less frequent.** It rarely appears earlier than the third or fourth day of the primary paroxysm, and in some instances not until the crisis. If it comes on in the first paroxysm, it usually fades rapidly during the intermission. With the relapse it may again deepen; it sometimes does not appear before this stage of the fever. It is usually slight and disappears in the course of a few days; in some cases, however, it is intense and persistent. It occurs in all ages, but is most frequent in the middle periods of life. Its presence imparts to the physiognomy an appearance not common in the fevers of temperate climates.

Jaundice is not in itself a dangerous symptom. In severe cases it is sometimes associated with albuminuria.

Enlargement of the spleen is a constant symptom. So rapid is the alteration in the volume of this viscus that the enlargement may often be detected within twenty-four hours of the beginning of the attack, and it not infrequently amounts to two or three times its normal bulk. It projects below the margin of the ribs, and may, even at its maximum, which is attained about the close of the primary paroxysm, give rise to visible bulging of the surface of the abdomen. It rapidly decreases during the apyretic period, but again enlarges in the relapse. During the convalescence it rapidly diminishes in size, but more or less enlargement may often be detected for a long time after the attack.

The stools may retain their normal color and consistence, but not infrequently they are darker than in health. Intestinal catarrh sometimes gives rise to more or less persistent diarrhœa.

Hemorrhages are met with. Epistaxis is not infrequent. It has been observed oftener in childhood than in adult life. It is sometimes so severe as to require plugging of the nares. Hæmaturia has already been alluded to. Intestinal bleeding may also occur, but is not a common accident of this disease. The catamenia occurring during the progress of relapsing fever are apt to be profuse. Severe uterine hemorrhage may occur in connection with abortion.

COMPLICATIONS AND SEQUELS.

Mild bronchitis is not uncommon. It usually requires no treatment, and interferes but little with the progress of the case or with recovery.

Pneumonia occurs as a complication in some epidemics. It usually appears in the course of the primary paroxysm or in the relapse. In the cases observed by Lebert in the epidemic of 1868–69, in Breslau, it showed a strong tendency to become double, and was in severe cases the cause of death. In rare cases pneumonia terminates in gangrene.

Pleurisy occasionally occurs. On the left side it may arise as a secondary lesion to splenic abscess.

Chronic pulmonary affections appear to be but little influenced by the disease.

When the subjects of *fatty degeneration of the heart* are attacked by relapsing fever, there is danger of sudden death from syncope. This untoward accident may take place in the first paroxysm, in the intermission, or in the relapse. It has been perhaps more frequently noted at or about the time of the first crisis. Sudden collapse and death from heart-failure may also occur in consequence of other forms of organic disease, and cases have been observed in which sudden death has occurred shortly after the patient has appeared to be doing well and the disease seemingly running a mild course, without the post-mortem discovery of any lesion adequate to account for it.

Acute laryngitis, with œdema, has in more than one instance necessitated the performance of tracheotomy in the course of relapsing fever. Dr. Begbie[1] mentions a case of this kind, in which the complication was ascribed to peculiar exposure to cold during the fever.

Gangrene of the feet, nose, ears, and lips have, in rare instances, occurred, in consequence of arterial thrombosis.

Splenic enlargement may persist for a considerable time after the attack, and is in such cases to be regarded as a sequel. It is of two kinds: first, painless and associated with profound anæmia; and second, tender upon pressure, and accompanied by fever of remittent type.

Abscesses of the spleen occur in rare instances. They give rise to pyæmic symptoms, and may be the cause of acute peritonitis or left pleurisy, or they may burst into the descending colon. The softened spleen may rupture during the paroxysm, and cause death by hemorrhage into the peritoneum. In view of the possibility of this accident, palpation of the splenic area is to be performed with great circumspection.

Anæmia is very commonly a sequel of relapsing fever. It is usually marked, sometimes attended with puffy eyelids and œdema of the lower extremities. It gradually amends, but in some cases is persistent. Anæmic murmurs are common.

Subcutaneous abscesses, parotid swelling, and *buboes*, occur, in very rare instances, during the convalescence.

Erysipelas also occurs, and is sometimes fatal.

Diarrhœa is not an uncommon complication and sequel. In some epidemics it has been the cause of a considerable proportion of the deaths. It is occasionally critical.

---

[1] **Reynolds' System**, vol. i., article Relapsing Fever. By J. Warburton Begbie, M.D.

Dysentery is also mentioned as a sequel. It is, in rare instances, the cause of peritonitis.

Pains in the muscles and *joints*, and *various neuralgias*, are very constantly annoying symptoms during the early days of the convalescence. With regaining strength and improved nutrition they pass away.

Local palsies are infrequent after relapsing fever. Paralysis of the deltoids, and the flexors of one or both forearms, has been observed. Paresis of the muscles of the upper and lower extremities has been noted with greater frequency. The loss of power comes on suddenly in the early days of convalescence, and is accompanied by numbness. It is transient, and disappears in the course of a week or ten days.

Lebert alludes to *hæmorrhagic pachymeningitis* as a sequel, and states that it was frequently encountered in the St. Petersburg epidemic.

Inflammatory affections of the internal structures of the eye, such as *iritis*, *choroiditis*, and *retinitis*, have occurred with considerable frequency during the late convalescence in some of the epidemics.

These affections never occur as sequels of typhus or enteric fever. They have been described by various authors under the name of "post-febrile ophthalmia," and, in particular, the accounts given by Mackenzie[1] and Dubois[2] are of interest among the earlier descriptions.

Quite recently, **Dr. Julius Trompetter**[3] reported that, in three hundred and twenty-five cases of relapsing fever in Breslau, twenty-one cases of choroiditis were observed; they were nearly all of the acute form. On admission to hospital, the patients mostly presented the characters of well-marked choroiditis in the form of cyclitis. Very frequently **hypopyon** appeared, without inflammatory phenomena on the part of the iris. Turbidity of the vitreous humor was ascertained to be present in all the cases, and the visual acuity was always considerably impaired at the commencement of the illness. The field of vision showed a limitation of the periphery in all directions. The course of the choroiditis was in general favorable; its average duration was from a month to six weeks. In two cases both eyes were affected. Dr. Trompetter believes that the affections of the eye in relapsing fever are due to embolism arising from partial necrosis and abscess of the spleen.

In a recent epidemic of relapsing fever, at Königsberg, **Dr. Luchbau** has also investigated the frequency of ear and eye complications. No less than three hundred cases were treated in the town hospital. Of this number only one hundred and eighty cases were, however, specially examined as to the existence of ear complications, and these were found in fifteen only, and in all the middle ear was the part affected. In most cases there was suppuration, and the pus was evacuated through the tympanic membrane. In most cases of disease of the middle ear in acute maladies the inflammation appears to arise by extension from the throat; but it was found that, in relapsing fever, pharyngeal catarrh is absent, as a rule, in the cases in which the middle

---

[1] W. Mackenzie, M.D.: Account of the Epidemic Remittent Fever at Glasgow in 1843, and of the Post-febrile Ophthalmitis. London Medical Gazette, vol. xxiii., 1843.

[2] Relapsing Fever and Ophthalmitis Post-febrilis in New York: Trans. American Med. Assn., 1848.

[3] Klinische Monatsblätter für Augenheilk., January, 1880.

ear suffers, and there was no evidence of disease of the Eustachian tubes. The prognosis is not unfavorable if prompt treatment is adopted.

Only six cases presented eye symptoms out of the hundred and eighty examined (three and a half per cent.). In three there was iritis, which was unilateral in every case. All these cases did well. In one case, however, some weeks later, the patient complained of failure of sight, and opacities were discovered in the vitreous. In two other cases optic neuritis occurred. In one the **affection was discovered in the first relapse.** The second relapse was severe, and some time afterward there was atrophy **of the optic nerves,** and vision was reduced to one-tenth. In the other case, the neuritis also occurred during the second febrile attack; a few days after it had ceased, the swelling of the optic papilla was discovered, dirty red in color, with arteries narrowed and **veins distended** and somewhat tortuous. Vision was reduced to one-third in one eye, **and one-fifth in the other.** Another patient came into the hospital during the first relapse with iritis and hypopion. The ocular trouble healed completely, but after the relapse the patient insisted on leaving the hospital and passed through the second relapse at home, **under** very unfavorable conditions. When it was over he returned to the hospital with double irido-cyclitis. Numerous thick flakes were seen in the vitreous, the fundus was very indistinct, but the papillæ were seen to be red and swollen, and there were numerous retinal hemorrhages. The account of these cases is published in the October number of Virchow's *Archiv*.[1]

Prognosis and Mortality.

In general terms **the prognosis in relapsing fever is favorable, the** death-rate being low.

Death occurs, not directly in consequence of **the fever, except** in rare instances, but by reason of some complication, as feebleness of the heart, uræmia, peritonitis, pneumonia, or abortion. It may take place during **the paroxysm, the intermission, the relapse, or after the** second critical **defervescence.**

Of 2,115 cases admitted to **the London** Fever Hospital from 1847 to 1870, according **to Murchison, 39, or** 1.84 per cent., or about 1 in **54,** proved **fatal.** Deducting from **this** number 10 cases fatal within forty-eight hours after admission, the death-rate was only 1.38 per cent., or less than 1 in 72. An analysis of the statistics of the Scotch epidemics made by the same author, give **for** one series of 6,300 **cases** a mortality of 260, or 4.12 per cent., or 1 **in 24.23 ; and for** a second series of 10,444 cases, 462 deaths, or 4.42 per cent., **or 1 in 22.6.** These two series of cases, taken in connection with the **statistics of** the London Fever Hospital for the period mentioned, give, in **a total of** 18,859 **cases, 761 deaths, a mortality** in England and Scotland of 4.03 per cent., or 1 in 24.78. Lebert informs us that in three epidemics in Breslau the **mortality did not rise above two** to three per cent.

In the Russian epidemic of 1864–65, of 12,382 cases, 1,574 terminated

---

[1] Lancet, Dec. 11, 1880.

fatally, being 12.7 per cent., or 1 in 7.86—the highest recorded death-rate in any epidemic.

The death-rate increases, in adult life, progressively with the age of the patient. During childhood and adolescence, relapsing fever is scarcely ever fatal. Reverting again to the statistics of the London Fever Hospital, we find that, of the 2,115 cases admitted, there were:

	Under 20 years,	804 cases,	3 deaths, or	0.37 per cent.
Between	20 and 30 "	562 "	4 "	0.71 "
"	30 " 40 "	322 "	8 "	2.48 "
"	40 " 50 "	232 "	6 "	2.58 "
"	50 " 60 "	119 "	9 "	7.56 "
"	60 " 70 "	66 "	7 "	10.60 "
"	70 " 80 "	6 "	2 "	33.33 "
Age doubtful,		4 "	0	"

The death-rate is, according to almost all published statistics, a little higher in the male than in the female sex. This is due to incidental circumstances. As in other epidemic diseases, the mortality is greatest at the outbreak and during the height of epidemics.

ANATOMICAL LESIONS.

No constant anatomical lesion is found after death. The spirilli are discoverable, in some instances, in the blood, if death takes place during the pyretic stages (Guttmann). But they have been sought for in vain in the spleen, lungs, and other organs, although the possibility of their existence can by no means be denied (Lebert).

The body is often emaciated; the skin, in addition to the cadaveric discolorations common after the infectious diseases, shows petechiæ, if they were present during life; the jaundice persists, and even deepens (Murchison). The color and texture of the muscles is unchanged; but, upon microscopic examination, there is not infrequently found, especially when death has taken place after a protracted illness, granular infiltration of the muscular fibres, amounting sometimes to fatty degeneration. Cadaveric rigidity appears early, and continues for a considerable time.

The stomach is usually normal, but small extravasations of blood are met with in the mucous membrane of this viscus, and in other mucous and serous membranes. This has been particularly observed in those cases in which urgent vomiting has preceded death, or in those characterized by black vomit.

The intestines are normal, except in cases in which diarrhœa or dysentery has occurred. After the former, injection of the mucous membrane, particularly toward the lower end of the ileum, is seen; after dysentery, the lesions peculiar to that affection are met with.

The solitary follicles are sometimes slightly enlarged. They are never ulcerated, nor are the agminate glands of the small intestine. Slight swelling of the mesenteric glands is sometimes found.

The liver is slightly or moderately enlarged, and deeply congested, especially when death occurs during the pyrexia. In rare instances, it is the seat of small deposits of a dull yellow color, softened in the centre. The gall-bladder is generally filled with a clear, viscid, yellow, or brownish bile.

The spleen is enlarged, sometimes to two- or three-fold its normal size. This change, except in cases that have resulted fatally at a late period, after the second defervescence, is met with in all cases. Its capsule is smooth, very tense, and clouded. Upon section the parenchyma is soft, in many cases almost diffluent. It may present a homogeneous appearance, or the Malpighian corpuscles may be seen with unusual distinctness. Minute roundish or irregular deposits of a dull yellow color, similar to those found in the liver, are frequently met with. They contain granular detritus, with cell-elements and free nuclei. These are also found in the lymph-follicles, and may be observed in different sections, in all stages, from simple follicular enlargement to the aggregations of detritus (Lebert). Wedge-shaped infarctions are occasionally met with, either firm or breaking down, but without demonstrable embolic origin. If the spleen be greatly softened, no decided structure can be recognized. If death take place some time after the termination of the relapse, in consequence of any complication, the spleen is found to be reduced in size, and its capsule shrivelled. In rare instances abscesses, due to the breaking down of the infarcts mentioned above, are found underlying the capsule and still more rarely the spleen is found to be ruptured.

The heart presents no change consequent upon the processes of relapsing fever, except, in some instances, after protracted illness, slight granular infiltration of the muscular fibres. Fibrinous coagula are found in the heart and great vessels, together with fluid blood.

The lungs show only those changes which attend the various pulmonary complications of relapsing fever. These are chiefly bronchitis and pneumonia. The latter is often double, and may in rare cases result in gangrene. The signs of recent pleural inflammation are rarely encountered. Hypostatic congestion is rare as compared with typhus or enteric fever.

Diagnosis.

If regard be had to the temperature, but little difficulty can attend the diagnosis of relapsing fever, even in the beginning of epidemics. The abrupt and unusual rise in temperature, the slight and inconstant morning remissions and frequent midday rather than evening exacerbations, the critical defervescence at the expiration of five or seven days, and the rapid decline of the temperature to a point below the normal, constitute a group of phenomena characteristic of this, and met with in no other disease. The **acute, progressive, and** extreme enlargement of the spleen, the **coincident, but less marked increase** in the size of the liver, the tenderness in the region of both these organs and in the epigastrium, and the muscular pains, are also diagnostic. Equally characteristic is the abrupt relapse, after an apyretic period of several days, with its **repetition of the symptoms of** the primary paroxysm and the extraordinary rise, high **range and sudden** fall of the temperature to a point below the normal.

Clinically, relapsing fever and typhus are widely unlike. Whether they are equally unlike etiologically, speaking in general terms, remains for future investigations to decide. The striking fact that the former, in all its great epidemics, has prevailed in connection with typhus, and commonly in a definite relation with it as regards the progress of the epidemic, being proportionately most common at the beginning of the outbreak, less so as the epidemic advances, and giving place wholly to typhus at its close—this fact, coupled with the well-established observation that relapsing fever patients are prone to typhus after convalescence, while typhus fever patients are little liable to suffer from relapsing fever within a short time, makes it appear most possible that in a broad sense these two fevers are due to closely associated causes.

Prevailing, as they so constantly did in the early epidemics, together as a pestilence—known by the simple designation of "fever," or "the fever," it was natural to regard relapsing as a mild form of typhus fever. The error, once established, was overthrown with difficulty—a difficulty to which the nosological method of the continental writers has contributed not a little. By this method the typhus, enteric, and relapsing fevers, and sometimes others—for the designation is an elastic one—are classed together as the common group of so-called "typhus" diseases, the first being regarded as "exanthematous typhus," the second as "abdominal typhus," and the last as "recurrent typhus." The following tabular arrangement of the principal phenomena of the three fevers just named, will serve to show how unlike they are in their clinical aspects, and, at the same time, to present in the most concise manner their more important points of differential diagnosis:

TYPHUS.	ENTERIC.	RELAPSING.
Essentially an epidemic disease, although endemic in certain **localities.**	An endemic disease, often sporadic, but occasionally appearing in circumscribed epidemics.	An epidemic disease, often **the congener** of typhus.
Highly **contagious.**	Not directly **contagious.**	Contagious.
Attack sudden, often without prodromes.	Attack generally insidious.	Attack sudden.
Course continuous.	**Continuous.**	Broken by a period of complete apyrexia.

TYPHUS.	ENTERIC.	RELAPSING.
Duration about fourteen days; rarely exceeds twenty days.	From three to four weeks.	Duration of primary paroxysm from five to seven days; of the relapse, about three.
Defervescence critical, or by very rapid lysis.	Terminates by prolonged lysis.	Ends abruptly by crisis.
True relapse so rare as to be almost unknown.	Relapses occasionally occur; they are irregular, inconstant, and accidental.	Relapse constant and an integral factor of the attack.
Face deeply **flushed**, dusky.	Face pale; if there is flushing, it is confined to the region of the cheek-bones, and is circumscribed.	Face often flushed; the color lacks the duskiness of typhus, and is not circumscribed, as in enteric fever.
Conjunctivæ deeply injected; pupils contracted.	Eyes clear; pupils often dilated.	Conjunctivæ slightly injected; pupils natural.
Delirium and **stupor** early and prominent.	Less constant, more gradual in development, and of longer duration.	Mind commonly clear throughout.
Abdominal symptoms absent; constipation the rule; meteorism rare.	Abdominal symptoms prominent. Diarrhœa and meteorism the rule.	**Pain and** tenderness in the epigastric zone. Constipation the rule, occasionally diarrhœa sets in at the crisis.
Intestinal hemorrhage extremely rare. Acute dysentery may occur during convalescence.	Intestinal hemorrhage not unusual.	
Epistaxis does not occur.	Epistaxis common.	Epistaxis occasionally occurs, especially at the time of the crisis.
Skin pungently **hot**, sometimes emitting a peculiar odor.	Skin hot; sometimes bathed in acid perspiration.	Skin hot; **profuse sweating** at crisis.
Eruption deep in **color**, copious, general **in its distribution**.	Eruption light red, sparse, discrete, commonly confined to particular regions of the trunk.	**No definite eruption.**
Emaciation slight.	Emaciation great.	Emaciation not marked, save when the patient has suffered from insufficient food prior to his illness.
Pneumonia and bronchitis of finer tubes.	**Bronchitis and pleurisy.**	Bronchitis common, but rarely severe. Pneumonia occurs.
Death not infrequent at end of first week, and often before the conclusion of the second.	Death usually takes place in or after the third week.	A fatal issue rare, except in consequence of complications.
No characteristic lesions found in the body after death.	Constant lesions of the ileum and the mesenteric glands.	Post-mortem appearances not characteristic.

Remittent fever is to be diagnosticated from relapsing fever by the marked differences in the range of temperature, the duration of the attack, the character of the crisis, the length of the intermission, the relapse, and the great contagiousness of the latter. Moreover, the circumstances under which the diseases appear and prevail in the community are of diagnostic value.

Treatment.

Prophylactic treatment must be based upon our knowledge of the predisposing as well as of the exciting causes of the disease. Upon the appearance of relapsing fever, renewed efforts must be made to relieve the sufferings of the poor, and chiefly to provide them with a sufficient quantity of wholesome food. As far as is possible, overcrowding must be diminished in the districts most liable to become pestilential centres of the disease. The drainage is to be looked to, and, if defective, temporary measures to drain away stagnant water must be immediately resorted to. All accumulations of filth and garbage must be at once removed. The system of visitation among the healthy, by laymen competent to instruct them as to the measures proper to be taken with the view of avoiding the disease, that was instituted in Paris during the cholera epidemic of 1849, is suggested by Lebert.

In view of the possibility of the introduction of the protomycetes by drinking-water, it should be subjected to boiling. Abundant ventilation is of the first importance. Contagious as relapsing fever is, it does not spread, even when cases occur, in the large and well-ventilated houses of the opulent, nor to any great extent in the roomy and properly aired wards of well-managed hospitals, except to those whose vocations bring them into close contact with the sick. As has already been stated, physicians visiting from house to house among the poor, remaining only a short time in the presence of the patients, and passing quickly again into the open air, are less liable to contract the fever than the resident physicians of hospitals, who pass from bedside to bedside, without the opportunity, for several hours at a time, of breathing an uncontaminated atmosphere.

Cleanliness of the abode and of the person is scarcely second in importance to abundant ventilation. The contagium is readily transmitted by means of the clothing and bedding of the sick. Soiled clothes should be thrown into boiling water as soon as taken off, and carbolic acid, or carbolic acid soap, used in the water with which they are washed.

If patients be removed to a hospital, or after convalescence has set in, the apartment should be fumigated by burning sulphur, thoroughly aired, cleansed, and whitewashed. It is obviously impossible to treat all the rooms in the densely crowded districts of cities in this manner, but, in proportion as these measures are promptly and generally carried into effect, will the spread of the disease be retarded. The bedding should also be subjected to the sun and air, and, if possible, fumigated; the cheaper materials used in filling mattresses, as straw, moss, fine shavings, and husks, should be burned.

As the result of the experience of all observers upon an extended scale, it may be stated that up to the present time no drug or method of treatment has been found to exercise any decided influence upon the course of the disease.

It is scarcely necessary to allude to *bloodletting*. From the day that relapsing fever was distinguished from typhus, it was clear that the critical defervescence on the fifth or seventh day of the short fever, ascribed to depletion, was, in fact, not the result of treatment at all, but an event of the natural course of the disease.

Repeated cold baths and *large doses of quinia* reduce the temperature, but neither affect the duration of the paroxysm nor prevent the relapse. Quinia has been tried in vain, in moderate and large doses, both during the pyretic period and in the intermission. *Arsenic* is likewise ineffectual.

The observations of Dr. Riess,[1] of Berlin, are of great interest. He found *sodium salicylate* very effective in reducing the temperature, and, given in large doses during the intermission, in lessening the severity of, and apparently even sometimes preventing, the relapse. These observations are to be tested by a more extended investigation of the value of this drug in future epidemics.

It has been suggested that the enormous development of the protomycetes in the blood during the paroxysm, and their disappearance during the intermission, are strong arguments in favor of the administration of parasiticides; that, with this view, a more systematic administration of the sulphites, and the disengagement of sulphurous acid gas in the air of the sick-apartment, should be attempted. Remedies of this kind had been tried without success before Obermeier's discovery. Parry, in 1870, administered, without in the least abating the violence of the course of the disease, the sulphites, the hyposulphites, and the preparations of chlorine. The destruction in the blood of enormous numbers of disease-producing parasites—so low in the scale of existence as to lie upon the most distant borders of independent life—so minute that they dwell in the ultima Thule of microscopic vision—by means of parasiticides administered in any amount short of compromising the integrity of the blood itself, is highly problematical. Meanwhile, relapsing fever must be treated on the *expectant plan*.

Rest in bed, quietude, abundance of fresh air, cleanliness, a carefully regulated diet consisting of milk, broths, meat-jellies, light farinaceous foods, or, if the patient craves them and can digest them, even the stronger soups, meat and vegetables, but always plenty of cooling drink, will in many cases suffice. The tendency of the disease is to recovery.

The patient must not be allowed to suffer from thirst. Let him drink freely. The best beverages are pure water, carbonated water, seltzer water, or milk diluted with any of these. If he prefer it, let him drink

---

[1] Berlin. klinische Wochenschrift, iii., 1879.

water acidulated with the juice of lemons or limes, or let him take ten or fifteen drops of dilute phosphoric or muriatic acid every three hours, in a wineglassful of water slightly sweetened, rinsing his mouth and teeth afterward.

Cold applications to the head, by means of ice and bran in bladders or caps of india-rubber, are useful in mitigating the headache. They should be applied only during the paroxysms of pain. The frequent resort to friction, with *anodyne liniments* will give relief to the muscle-pains. For this purpose—

  ℞. Chloral hydrate........ 16—32.00 gm.   ℥ ss.—j.
   Lin. saponis camph ..... 200.00 c.c.   fl. ℥ vj.
  M.

or a lotion consisting of equal parts of chloroform and olive oil, may be employed.

If the pains be very severe, the *hypodermic use of morphia,* alone or with *atropia,* will be required to relieve them.

Opium and its derivatives, by the stomach, appear to have in very many cases but little effect, either in relieving pain or producing sleep, in relapsing fever. Parry and other observers state that a remarkable tolerance for this drug was established during the attack.

Potassium bromide is useless.

Sleeplessness will yield to the administration of *chloral hydrate* in moderate doses. This drug is to be given with caution where the action of the heart is enfeebled.

An emetic, followed by *mild purgatives,* is of use in relieving the vomiting and pains in the epigastric zone. At the same time sinapisms, hot fomentations, or small blisters, should be applied. Carbolic acid may also be given for the nausea and vomiting. If the pain in the region of the spleen is very great, poultices should be applied, or frequently renewed cold-water applications may afford relief.

Alcoholic stimulants are to be given, not as a part of a general routine treatment, but as called for by the weakness of the patient and the character of the pulse, the impulse of and the first sound of the heart. At the time of the crisis they are of great benefit, and must be given during the first days of the remission, and again in the early convalescence.

If collapse threaten, it must be treated by prompt stimulation by alcohol, spirits of chloroform, ammonium carbonate, artificial heat, and so on.

Diarrhœa calls for the employment of astringents and opium.

Bronchitis, occurring as a complication of relapsing fever, is usually of a mild form, and does not require especial therapeutic intervention.

Pneumonia is to be treated upon general principles. It may be said that, almost without exception, intercurrent pneumonias call for increased stimulation.

Parry, after trying various drugs, found that quinia in combination with camphor was most useful, during **the intermission and in the early** convalescence, in relieving the **patient's sense of prostration and inducing** sleep.

The anæmia of convalescence from relapsing fever urgently demands an abundance of wholesome, nutritious food, the vegetable **tonics, such** as the best preparations of cinchona and **nux vomica, the best-borne** preparations of iron, and, if the pallor be protracted and the patient **take** it well, cod-liver oil.

Chronic enlargement of the spleen should be treated by quinia and iron, and externally by inunctions **of the red** iodide of mercury ointment.

VII.
DENGUE.

DEFINITION.—An acute, febrile affection, of short duration, due to an unknown external specific cause, and prevailing in extensive epidemics, which are chiefly confined to warm climates; it consists of two distinct, brief, febrile paroxysms, each attended by a different group of symptoms, and separated by an intermission lasting from a few hours to several days. The first is characterized by continuous high fever, distressing pains in the joints and muscles, interfering with motion, and occasionally by a cutaneous efflorescence; it usually terminates suddenly with some critical discharge; the second paroxysm is marked by a milder fever of remittent type, an eruption of different character, which is attended with intense itching and followed by desquamation, by some recurrence of the joint-pains, and by debility; it gradually subsides. The disease is extremely painful, but very rarely fatal; its morbid anatomy is therefore unknown.

SYNONYMS.—Febris exanthematica articularis; Exanthesis arthrosia; Exanthesis rosalia arthrodynia; Scarlatina rheumatica; Scarlatina mitis; Eruptive articular fever; Eruptive rheumatic fever; Rheumatic fever with gastric irritation and eruption; Eruptive epidemic fever of India; Epidemic inflammatory fever of Calcutta; Epidemic anomalous disease; Peculiar epidemic fever.

Dandy fever; Polka fever; La Piadosa; La Pantomina; Colorado; Bouquet; Bucket; Giraffe; Stiff-necked fever; Broken-wing fever; Break-bone fever; Toohutia; Three-day fever; Knockel Koorts; Aburuka-Bah (Father of the Knee); Date fever.

Dengue, pronounced *dangay*.

"This disease, when it first appeared in the British West India Islands, was called the *dandy* fever, from the stiffness and constraint which it gave to the limbs and body. The Spaniards of the neighboring islands mistook the term for their word *dengue*, denoting prudery, which might also well express stiffness, and hence the term *dengue* became at last the name of the disease."

This term, begotten of a misapprehension of a word applied to it in jest, has become the generally accepted designation of the disease.

To a similar origin are due many of the popular names by which it is known in the countries where it has prevailed. The **people are** often disposed to make a jest of epidemics not attended by danger to life, the more perhaps when the sufferers present an absurd appearance. The Brazilians called this disease the *polka fever;* the Spanish, *la pantomina;* the French, *bouquet* and *giraffe*, the latter because of the stiff manner in which those affected often carry the head. *Stiff-necked fever* and *broken-wing fever* are likewise terms suggested **by the posture of** the convalescent; while *break-bone fever*, **and the** Batavian designation, *Knockel Koorts* (bone fever), refer to the torturing **joint-pains** that attend it. In Spanish-American countries it has been known as *colorado*, on account of the red color of the eruptions. It is probable that this is also the derivation of the French term *bouquet*. *Toorhutia* and *three-day fever* are East Indian names for it; and *aburuka-bah* and *date fever*—the latter because it has been observed to prevail during the date-harvest, are Arabian folk-terms.

Medical observers have designated it by various **terms of** classical derivation, according to the views which they have entertained concerning its nature. Of such terms, those **based upon its fancied** relationship to other well-known affections, as scarlet fever and rheumatism, are inapplicable now that it is known to be a distinct affection; **others fall to the** ground, because they **are based upon the assumption that the disease is** peculiar to a country or locality, now **that it has become known as pandemic in tropical and subtropical climates; while others still fail of acceptation because they are not sufficiently distinctive.**

Historical Sketch.

Dengue first excited general attention by **its epidemic prevalence in the West India Islands in 1827.** Previous to that **date, however, it had occurred in less extended outbreaks in** tropical countries and elsewhere. The earliest **account of the disease,** according to De **Wilde,**[1] dates from the year 1779. **David Brylon, the chief** physician **of Java at** that time, briefly describes, under the **name of** *Knockel Koorts*, an epidemic disease which prevailed among the **natives and** colonists. Rush[2] published an account of an epidemic which **occurred in** Philadelphia **in** the following **year,** 1780. The disease was **then, as now, described in North America** as *break-bone fever*. At the same time it was observed **by the missionary Wise,** according to an anonymous French **writer,**[3] on the coasts of Coromandel, Africa, Arabia, Persia, and Thibet. It is said also to have oc-

---

[1] J. J. de Wilde : Deague in Fort William I., in Java. Niedl. Tijdschr., 1873, quoted by Zuelzer.
[2] Medical Enquirer and **Observer,** 1789.
[3] Zuelzer : Ziemssen's **Cyclopædia, vol. ii.**

curred at Lima, in January, 1818, and in the United States, at Savannah, in 1826.

Previous to the general epidemic which made its appearance first in September, 1827, at St. Thomas, there exists not the slightest trace in medical literature of this disease in the West Indies.[1]

From the Island of St. Thomas it spread in October to St. Croix. In these islands almost every inhabitant in a population of 12,000 suffered. It passed thence toward the northwest, over the great Antilles to the main-land of North America, and southward over the Caribbean Islands to Columbia.

Following its course toward our own country, we find that in the spring of 1828 it had reached Pensacola, and that it spread thence in June to Charleston in one direction, and in the other, to Mobile and New Orleans, where it prevailed early in the summer. It made its appearance in Savannah in September. In the same year sporadic cases were observed in Boston, New York, and Philadelphia, and, according to some accounts, in some of the cities of the West, although the evidence in regard to the last statement is not conclusive.

In the beginning of the year 1828, the disease prevailed in Columbia, and nearly at the same time in Porto Rico, Hayti, and Jamaica. It broke out in Cuba in March. In these islands and in Columbia it continued till September of the same year.

Two decades now passed without the occurrence of dengue in extended epidemics. It is true that in 1839 an outbreak took place at Iberville, in Louisiana, and one in 1844 at Mobile, but they appear to have been confined to the localities in which they first appeared.

In the summer of 1848, it again showed itself in New Orleans, and less extensively in Vicksburg and Natchez. In these outbreaks, dengue appeared simultaneously with the yellow fever. In the autumn of the same year it was again observed in Mobile. Cases occurred during the next two years, from time to time, in the cities along the Gulf Coast.

In 1850, a wide-spread epidemic visited the Southern States. Appearing in Charleston toward the end of July, it spread successively to Savannah, Augusta, Mobile, New Orleans, and intermediate points, and into Texas, in which state it became epidemic in October. The extent of prevalence of this disease in some of the localities visited by this epidemic is remarkable. Dickson informs us that, in Charleston, all the members of large households were attacked, without a single exception, and that of his own family, numbering eleven persons, he alone escaped. It is computed that ten thousand persons were ill in Charleston at one period, and that between seventy and eighty per cent. of the population suffered during the epidemic. The number of inhabitants in the town of New Iberia, Louisiana, in 1857, did not exceed two hundred and fifty;

---

[1] Hirsch: Handbuch der historisch-geographische Pathologie. Erlangen, 1860.

of these, two hundred and ten contracted the disease during a period of six weeks.

Less-extended epidemics occurred at various points in the same belt of country in 1861, and again in 1866.

In the summer of 1846, dengue appeared in Brazil, and prevailed widely. It reappeared, at the same season of the year, in the three following years. In 1852 it visited Peru. This visitation was followed by yellow fever.

In the eastern hemisphere, from the time of the epidemic reported by Wise, already alluded to, no outbreak is mentioned till 1799, when it broke out in Lower Egypt, and prevailed extensively in and around Cairo, under the name of the "knee evil." Pruner, who had seen the affliction on the coast of Arabia in 1835, again encountered it at Cairo, in August, 1845, and a little later in Alexandria. No further accounts are met with concerning the existence of the disease in tropical Africa until 1871. It has, at various periods, prevailed very extensively in India; in the year 1824, dengue made its appearance as an unknown disease, both to the physicians and the public, in the southern parts; it spread, in the rainy season, to Calcutta, and from there, along the Ganges, to Berhampoor, whence it extended over the southern portion of Bengal and a part of Madras. In March, 1825, it reappeared at Berhampoor, and became epidemic in the surrounding country in the rainy season. Again, in 1836, the disease visited Calcutta, and Pruner states that travellers, who came from India to Cairo in 1845, told of its epidemic prevalence in that country and along the borders of the Red Sea.

In 1860, it appeared among the ships at Martinique, and spread later to the garrison. Balbot states that, of four hundred men constituting the garrison, one hundred and twelve suffered from the disease.

The dengue prevailed in Spain from 1864 to 1868.

It appeared in Arabia in 1871, and was observed by Read, especially in Mecca, Medina, and Aden. In the last of these cities it was epidemic during a period of more than seven months; of the garrison of nine hundred men, seven hundred had the disease. Following the line of travel, it spread, in 1871, to Zanzibar and other points on the African coast. It was observed in the same year at Port Said, where it is said to have prevailed every year at the season of the date-harvest. In November, 1871, it broke out in Java. In 1872, it spread through all India, starting from Bombay and Cananore, and following, at first, the line of the railroads. Cases appeared at the same period in the English stations of China, Burmah, and Nepaul. This epidemic was as intense as it was wide-spread. In some localities scarcely an individual escaped. In Madras it prevailed so violently that not a house escaped. The epidemic reached its height in September and October, and subsided suddenly, after a heavy rain, about the middle of the latter month.

A mild epidemic of dengue prevailed at Charleston and in some of the neighboring localities, and at various points along the Gulf Coast, during the summer and autumn of 1880. It ceased with the advent of cold weather and frost. At the same time this disease was extensively prevalent in Northern Egypt.

Etiology.

1. Predisposing Causes.

There can be no doubt whatever that climate has a large influence in the development of dengue. It is a disease of tropical and subtropical lands. When it has occurred in colder countries, it has made its appearance almost exclusively in the summer or autumn, and upon the advent of cold weather has promptly disappeared. Its prevalence has also been restricted to sporadic cases or to circumscribed local epidemics. Arnold[1] declares that "this disease is undoubtedly affected by frost. The diminution of cases after a frost last fall, was as marked as the diminution of cases in our endemic climate fever usually is."

With this exception the *condition of the weather* has no direct influence either upon the origin or upon the epidemic spread of the disease. Within the tropics it has occurred alike in the hot, the cool and the rainy season of the year. In our Southern States it has prevailed in wet and dry, in cool and warm weather, indifferently, though it is to be remarked that it has almost always first broken out in summer and disappeared to a great extent, if not wholly, in the winter months. In the West Indies it prevailed continuously for a period of "nearly twelve months through a variety of seasons, and was neither perceptibly influenced by vernal nor autumnal equinoxes, by our strong, wintry north wind, nor by the scorching, fiery sea-breezes of June and July" (Maxwell).

The supposition that a peculiar condition of the atmosphere, combining a high degree of moisture and "stagnation of the air" with prolonged and very intense heat, are necessarily associated with the origin and transmission of the disease, or that its appearance as an epidemic is necessarily preceded by prolonged, heavy rains, falls to the ground, in view of the recorded fact of its prevalence at all seasons of the year in tropical regions, and its steady advance in the direction of the lines of human intercourse without regard to the dryness or moisture of the weather.

Dengue is in the strictest sense a pandemic disease. With the exception of influenza, no other disease has prevailed over so wide an extent of the surface of the globe, or attacked with such impartiality the inhabitants of the countries over which it has passed.

---

[1] Charleston Medical Journal, May, 1851.

Race and nationality have but little influence upon this disease. Observers in all countries where it has prevailed agree in the **statement that it** spreads equally among the white and colored of all nations. To this general remark must be made the exceptions, that in the South **the negro race** is attacked a little later and suffers less generally than the whites, and that, in the last epidemic in **India, Europeans recently arrived appeared** to suffer from the disease in **a milder form than residents already** acclimated.

The disease **spares** neither *age, sex,* **nor** *occupation.* Infants in arms and octogenarians **are** equally prone **to** it. All classes of society alike suffer. The physician enjoys no **immunity.** He is almost invariably attacked. Aitken suggests that **this is the reason that** the details of symptoms in epidemics of this disease are so **minute.**

It prevails, as a rule, chiefly in cities, less generally in the open country. To this there have been, however, notable exceptions.

II. THE EXCITING CAUSE.

The exciting cause of **the disease is unknown.** That it is specific is no longer **open to** doubt. Whether it **is** **capable** of being conveyed by **human intercourse,** and in most instances has **spread by a steady progress, in direct lines from the** points of early infection. **Dickson** and some of the observers of the later Indian epidemic look upon the **disease as contagious, but** adduce no direct proofs; others strongly **oppose this view.** Most physicians who have had the opportunity of personally **observing the disease express no** opinion upon this point. It is not generally regarded as contagious.

Its mode of invasion, its rapid march, the unsparing manner in which it attacks entire families, cities, and even districts within a brief space of time, **are opposed** to the assumption **that it** is propagated by contagion alone.

In some of the outbreaks, dengue **has** preceded or followed yellow fever. But **it has** so often occurred independently of any association with that disease, that the existence of **any** pathological relationship is in the highest degree improbable. Dengue **has** not only prevailed in the maritime countries subject to yellow fever, **but it has extended to the mountainous** back-country, in which the latter **fever is unknown. The** occasional **association** of dengue with scarlet fever **and whooping-cough in epidemics is** accidental, not causal.

No hypothesis adequate **to explain the** fierce **epidemic outbreak of this** disease in widely separated **localities, and at long intervals of time, has yet** been advanced.

Clinical History.

The period of incubation in about half the cases is extremely brief. At the commencement and at the maximum of an epidemic, the attack may follow exposure in a few hours, occurring without preliminary symptoms. Toward the close of an epidemic, the period which elapses between the exposure and the onset of the attack may be lengthened to several days.

The invasion of the disease is generally abrupt; there may in some cases, however, be a prodromal stage of from one to three days, characterized by lassitude, headache, a furred tongue, loss of appetite, muscular soreness, and chilliness.

Usually, however, the patient is seized, upon waking, with intense headache, burning pain in the temples, backache, and severe pain in the joints. Sometimes the first symptom is an acute pain in one of the joints—for example, one of the joints of the hand or foot; this may come on while the patient is engaged at his ordinary occupation, and apparently in full health.

The affected joints rapidly become swollen, and the skin of the face and neck is flushed and turgid. Painful stiffness of the muscles follows; the affected members are moved with great difficulty and suffering. The muscles of the eyes sometimes become stiff and immovable, the conjunctiva reddened, the eyelids swollen, so that the patient wears a staring expression, while the eyeballs feel too large for the sockets. There is intolerance of light and sound. At the same time symptoms of gastric disturbance occur; the tongue is coated; a burning pain is felt in the epigastrium, and there is nausea, followed by bilious vomiting. The irritability of the stomach is often so great that scarcely anything is retained. In most cases desire for food is wholly lost; but not infrequently, especially in children, the appetite is retained; thirst is not urgent; the bowels are constipated.

Fever makes its appearance at the onset of the attack, and reaches its height within the first twenty-four hours. A temperature of 41.5° C. or 42° C. (106.7° F. or 107.6° F.) is not infrequently observed in the axilla. The fever is now continuous; the pulse is full, hard, strong and exceedingly frequent, beating from 120 to 140, and even higher in children. The breathing is quickened, the skin hot and dry. Confusion of thought, and even delirium, particularly in children, also occur, and in young children the disease sometimes commences with convulsions. There are no other symptoms primarily referable to the nervous system.

In a great majority of the cases, an exanthem of variable character now shows itself. This eruption most frequently resembles the efflorescence of scarlatina, and for this reason dengue was regarded by many of the older observers as an epidemic scarlatinal rheumatism.

The duration of the first febrile paroxysm is variable, lasting from a few hours to several days. Its average duration is from two to three days. The fever generally abates suddenly, often with the occurrence of critical discharges, such as profuse sweats, epistaxis, or diarrhœa, the evacuations being dark, greenish, tawny, and foul-smelling.

Exceptionally the fever subsides slowly by lysis. The subsidence of the fever is marked by the disappearance of the eruption, if any be present, by the appearance of moisture on the skin, and an amelioration of the pains in the muscles and joints. In most cases the patient, although much relieved, is unable to leave his bed by reason of the great prostration following the fever. In other cases the relief is so great and the strength so well preserved, that the patients do not hesitate to arise from bed and even to leave the chamber.

This stage of the disease, which is analogous to the intermission of relapsing fever, is thought by recent observers to be in most cases a very marked remission, in which the temperature closely approaches, but does not reach, the normal; in others it amounts to a period of complete apyrexia. Its duration is from two to three days. In some cases it is wanting altogether, or of so short duration that it is overlooked. Notwithstanding the great amelioration of all the symptoms during this period, some headache, and more or less of the stiffness of the joints and muscles, remain. At the expiration of some hours, or of two or three days, as the case may be, acute symptoms reappear and the second febrile paroxysm sets in. Fever again arises, but it is not so intense as before, and its type is remittent rather than continued. The tongue again becomes coated, appetite is lost, nausea distresses the patient, but vomiting at this stage is rare. Headache attends it, and in many cases there is an exacerbation of the pains in the joints and some increase in the stiffness of the muscles, both these symptoms having continued in some degree throughout the remission. Coincidently with the reappearance of the fever, an eruption shows itself. This eruption, which may be looked upon as the distinguishing feature of the second paroxysm of the disease, has been variously described by different observers as erythematous, reseola-like, rubeolous, or as resembling urticaria. Appearing in many instances, first upon the palms of the hands or upon the soles of the feet, it extends over the greater part of the surface of the body. Or it may be localized in certain regions. It is attended by annoying itching, and, after an existence varying from some hours to two or three days, it vanishes and is followed by furfuraceous desquamation.

The duration of these symptoms is usually about two or three days. The fever gradually subsides, and the acute symptoms disappear, leaving the patient in an enfeebled state, often requiring months for the re-establishment of health. Besides the debility, which is often very great, emaciation, diarrhœa, and painful stiffness and swellings of the joints, protract the convalescence.

Complications do not occur, and there are no sequels. Relapses often take place, and occasionally repeated relapses befall the same patient. They run a milder course than the primary attack. The affection is scarcely ever fatal. Convulsions may occasion an unfavorable termination in infants. On the other hand, it is among the most painful of the epidemic diseases, and not seldom gives rise to serious impairment of health by the exhaustion which follows the high fever, the prolonged, severe pains, the sleeplessness, the inability to retain food, and the abundant critical discharges.

The course of the disease may be divided according to recent observers into—

a, the period of first febrile access, two to three days.
b, the intermission, some hours to two or three days.
c, the second febrile stage, two to three days.
Whole duration of the acute symptoms, about eight days.
The intermission may be altogether absent.
The duration of epidemics varies from two to seven months.

It remains to consider more in detail some of the prominent symptoms.

The affection of the joints and limbs, which accompanies the first paroxysm, gives rise to the peculiarities of gait and attitude which are expressed in so many of the popular names of the disease. It attacks large and small joints alike, often six or eight being affected at once. The joints of the hand, foot, and knee, the spine, the fingers, the toes, the elbow and shoulder, are ofttimes involved successively in the order given. In severe cases all the joints are implicated (Zuelzer). The joints are swollen, red, immobile, painful, and often exquisitely sensitive to the touch. The stiffness of the affected limbs is not wholly due to the condition of the joints. The muscles are likewise stiffened and sore, and there is an effusion of serum in the connective tissue surrounding certain of the tendons. The fingers are often stiff, and the hand cannot be closed. This is particularly the case in the morning, and constitutes an annoying circumstance of the convalescence.

The pains are described as rheumatic or rheumatoid, by most writers. De Wilde observed isolated painful spots in several instances, and in others found a single nerve-trunk, as the ulnar, to be affected. The pains in this disease, as in acute articular rheumatism, pass from one set of joints to another with remarkable rapidity. At the onset of the attack, only severe headache and pain in the hands may be complained of, yet in a few hours the joints of the feet and the knees may have become involved. Each new invasion of a part is accompanied in such instances by twitching of the muscles in the neighborhood of the joint affected (Aitken).[1] Patients describe the pains as of exceeding severity; they express them by such terms as "boring" and "breaking." Few can endure them without complaining. This affection, much less prominent during the second paroxysm of the fever than in the first, gradually disappears; but it may persist for several weeks, or even for some months, becoming fixed in one or more joints.

In three cases examined after death, serous infiltration of the connective tissue in the neighborhood of affected joints was found twice, and reddening of the crucial ligament of the knee once (Hirsch).

As has already been pointed out in the definition of the disease, a *primary* and a *secondary exanthem*, corresponding respectively to the first and second febrile paroxysms, occur in a majority of the cases.

The first, though present in a large proportion of the cases, is by no means constant. When present it appears and disappears coincidently with the fever.

The latter is, as a rule, always encountered. Much diversity of view as to the character of this eruption is found in the writings of those who have recorded their personal observations of the disease. The forms commonly assumed by the eruption have already been indicated. In some instances they are mixed, as, for example, erythema and urticaria may be present at the same time. Urticaria is common in children. Considerable swelling of the skin attends the appearance of this eruption in some regions, especially upon the palms and soles, at the lobe of the ear, and about the eyes, where it induces conjunctivitis and lachrymation. In severe cases the *mucous membrane* of the mouth and throat, and that of the nostrils, is inflamed. Aphthous ulcerations occur upon the tongue and buccal mucous membrane. The secretion of saliva is sometimes increased, and the salivary glands, and in particular the parotids, are swollen.

The *superficial lymphatics*, about the angle of the jaw and in the groin, are also in some instances transiently enlarged. Less commonly, *boils* occur during the convalescence, and some observers have recorded the occasional occurrence of extensive subcutaneous abscesses.

The desquamation is usually bran-like, but this is not always the case. The epidermis has been observed to peel off in large flakes, leaving a denuded, painful surface, which has sometimes resulted in superficial ulcerations. The *urine* during the access of fever is scanty and of dark color; its specific gravity is high; albumen has not been observed. With the crisis its quantity is augmented.

Restlessness, sleeplessness, headache, especially involving the forehead and temples, and sometimes nocturnal delirium, attend the fever.

---

[1] Reynolds' System of Medicine. Article on Dengue. Vol. i., 1868.

In children the fever is of shorter duration, and the course of the disease is modified by the convulsions by which its advent is not seldom heralded, and which sometimes persist, and even lead to a fatal issue. *Rapid emaciation,* and, as has been pointed out, an extreme debility, attend this disease. Weakness and loss of muscular power, in the legs especially, often continue far into the **convalescence.**

Hirsch informs us that **affections of** the heart appear **to have been in no case encountered.** His opinion is the result of a study of **the histories of the epidemics prevailing** previous to 1860. In the recent epidemics in India, **M. Sheriff and** Dunkley not seldom observed, after the fever, an affection **of the heart, which** was considered to be pericarditis. **In no case did it, however, result in death,** and after a time it disappeared (Zuelzer).

The respiratory organs are not implicated in the disease. In very rare instances pleurisy has been noted. It is probable that its association with dengue in such cases was accidental.

It is stated by observers of the West Indian epidemics, that females at various periods of pregnancy suffered the severer forms of the fever without any tendency to abortion. But, in the visitation in India in 1872, this accident not infrequently took place.

Dr. F. P. Porcher furnished to the *National Board of Health Bulletin,* September 25, 1880, the following account of the mild epidemic of break-bone fever which prevailed during the past summer at Charleston:

"It began, it appears, in the extreme northwestern portion of the city, above Calhoun, near Line and Columbus streets, in what was formerly called the 'Neck.' Afterward it seemed to progress into the lower or oldest part, and there is every indication that it is now diminishing.

"The earth had been disturbed in the paving of King street, an extensive thoroughfare running north and south the entire length of the city, and the special section of the city where the first cases were noticed was not in as good a condition as others, being near the marshes, and new streets having been opened there; but, though we were at first inclined to search for the causes of disease in these conditions, the simultaneous appearance of the fever in the West, and, as we learn, in Savannah and Augusta, must exclude such a supposition, and refer it to general and wide-prevailing atmospheric influences.

"Besides our own experience, which has been limited on account of temporary absence, we have made diligent inquiries of many persons, of physicians as well as the laity, and learn the following particulars, which we present in default of a more complete report, which will doubtless be made in the future.

"The symptoms vary exceedingly—some being present and some absent—as follows: the disease generally begins with a feeling of coldness, or by a chill, followed by fever; this, with a temperature ranging from 100° to 105°, lasts generally from twenty-four to forty-eight hours, occasionally extending to four or five days, and even in rare cases to seven. Relapses occasional, especially in those who have gone out too early. Headache frequent, generally frontal, from the beginning. Miliary eruptions, sometimes elevated and red, like measles, and the occasional presence of *sudamina over the face, neck, and body ;* sometimes the eruptions were confined to the body, and endured for days after recovery. We have seen some examples of slight desquamation—furfuraceous or branny in character. Sweating profuse in many persons, though *often absent.* Hence, some physicians are inclined to consider the disease to be *suette miliare* of a mild form. 'Break-bone' is the best name, because pain in the bones and limbs is the most constant symptom. There is often great restlessness during the fever, and in some a feeling of tightness or congestion about the throat, with bleeding in a few cases known to us. Catarrhal symptoms are rarely present, although cough has occasionally existed. Bleeding from the nose not unusual in children, and also increase in the menstrual molimen has been observed. Pain in the back and limbs markedly present, but no decided swelling of joints, no carbuncular enlargements or boils, as in the epidemic of dengue of forty years since, or in that of 'break-bone' which followed some years subsequently. Weakness and prostration have been very decided, but not nearly to such an extent as in previous epidemics. Some of the physicians consider that there has been a tendency to hepatic torpor or congestion—of no great severity, however. We have heard of no cases of decided jaundice. Nausea and vomiting seldom occur.

"The disease does not affect all the members of a household, oftentimes only one or two being seized, though we have known six to be taken in one house; in this respect differing from the dengue, as described by Prof. Dickson, and from the epidemic seen by us some thirty years since. Then ten thousand were down; no one was well enough or strong enough to help his neighbor, and one had to learn to walk over again.

"It is difficult to calculate the number who have suffered, as very many have not employed a physician; from two to three thousand, perhaps, approximates the number.

"Very little active treatment has been used—as far as we can learn, as follows: a mild laxative, saline or mercurial, hot teas, nitre, pediluvia, sinapisms, etc., and quinine during and after the attack, upon theoretical grounds, with occasionally mild stimulants. Several persons have recovered with no treatment whatever.

"It has prevailed among both races, perhaps equally, and *not a single death is ascribed to this disease,* as far as we can learn. The only disadvantage which accrues to those who take it is the time lost, and the temporary pain and weakness from which they suffer.

"Persons who were in the city and who visited the country had mild attacks. We know of four such; one of these had reached Asheville,

N. C., where we saw him. Cases of the fever have occurred in Summerville, thirty miles off, on the line of the South Carolina Railroad, among persons who had never visited the city; others sickened there who had paid flying visits, remaining a part of a day only."

DIAGNOSIS.

The diagnosis is not attended with difficulty. No other disease presenting analogous symptoms spreads with the same rapidity through a community. No other disease whatever, except influenza—with which dengue can by no possibility be confounded—attacks entire communities, sparing neither the young nor the old, the poor nor the rich, and, as has more than once been recorded, not a single individual in a district.

The natural history of dengue makes it unnecessary to point out the points of differential diagnosis between it and *acute articular rheumatism*, to which it presents, in the first febrile paroxysm, strong resemblances; or between it and *scarlet fever* or *measles*, which the eruptions of the second paroxysm are said, in certain instances, to resemble. Its likeness to *relapsing fever* is confined to its course, which is in fact that of a relapsing fever. In future outbreaks careful microscopic examinations of the blood are urgently called for, in view of this resemblance, and the discovery by Obermeier of a minute organism in the blood of relapsing fever patients.

TREATMENT.

Efficient methods of *prophylaxis*, as regards the individual in infected localities, are not known. As regards communities, it has been recommended that a rigid quarantine of the districts in which dengue prevails, and the isolation of the patients, may prevent its spreading. These measures, in view of the march of epidemics along the lines of human intercourse, the facility of its transportation in ships, and the enormous aggregate of human suffering which its unchecked progress occasions, will demand vigorous enforcement by the authorities of the city or region in which dengue shall next make its appearance.

There is no abortive treatment.

It is a specific disease, for which we possess no specific remedy. Nevertheless, much can be done by a judicious medication to mitigate the symptoms and abridge the period of convalescence. The treatment is to be conducted in accordance with general therapeutic principles, and is for the most part symptomatic.

Neither *general* nor *local blood letting* is of service. Either increases the tendency to debility and gives rise to vertigo during the convalescence, which is, at the same time, protracted.

Eliminative measures, in accordance with the practice of the tropics, have usually been employed in the beginning of the treatment.

Emetics are highly spoken of. In several epidemics, pushed to the production of free bilious vomiting, they have greatly relieved the head and eased the pains. For this purpose tartarized antimony and ipecacuanha were used. The latter is to be preferred.

Purgation is called for by the constipation which exists during the first period of fever, and by the dark green color and highly offensive character of the evacuations which commonly take place at its critical termination. It is desirable to anticipate elimination by the bowels by recourse to mild, but efficient purgatives. The disappearance of the green color and the occurrence of more natural fecal discharges, has coincided with a further amelioration of the symptoms. It is not necessary to push purgation to the bringing about of watery discharges. Rhubarb, aloes, magnesium sulphate, and the like, variously associated and combined, are proper remedies. The aggravation of the sufferings of the patient which attends the act of defecation cannot be regarded as a contraindication to their use, in view of the concurrent testimony of almost all observers that they are of undoubted service. The bowels should be kept open throughout the sickness by the occasional administration of mild laxatives.

With a view of *acting upon the skin*, the sweet spirits of nitre, neutral mixture, or the effervescing draughts may be regularly given at intervals of two or three hours. Warm baths have also been employed. Bartholow[1] suggests that, as the first paroxysm usually terminates by crisis and commonly with sweating, the "behavior of nature" may be imitated, and this stage possibly shortened by the administration of pilocarpine.

If necessary, *diuretics* are to be administered along with the foregoing remedies.

Opiates, to relieve the pain, restlessness, and inability to sleep, form an important part of the treatment. The subcutaneous injection of morphia will in most instances best fulfil this indication. Dover's powder may be given at night.

Belladonna in large doses has been highly extolled as favorably influencing the joint-pains. Its local application to the painful joints, in the form of a soft ointment, would probably constitute a valuable adjunct to the treatment. Salicylic acid and the salicylates, as yet untried in this disease, would also probably prove useful against the rheumatoid phenomena.

Alcoholic stimulants should be given from the decline of the initial fever, in carefully regulated doses, regard being had to the habits and mode of life of the individual patient.

---

[1] Practice of Medicine, 1880.

Quinine, combinations of iron with strychnia, and particularly the tincture of the chloride of iron, or that preparation of it known as Basham's mixture, are to be given upon the subsidence of the fever of the second paroxysm. The impaired appetite and enfeebled digestion are best managed by minute doses of strychnia, 0.0015—0.001 gramme (gr. $\frac{1}{48}$ — $\frac{1}{30}$ t. d.), either alone or in combination with dilute phosphoric acid, and with or without iron.

The itching which is so distressing a symptom in the second paroxysm, and during the desquamation which supervenes, may be in part relieved by the application of lotions of ammonium chloride and corrosive sublimate in almond emulsion:

℞. Ammonii chloridi . . . .   1—1.3 grm.   gr. xv.—xx.
   Hydrargiri chloridi corr... 0.008—0.016 grm. gr. $\frac{1}{8}$ — $\frac{1}{4}$.
   Misturæ amygdalæ. . . . . .   32 c.c.   fl. ℥ j.
M.

or a solution of carbolic acid, one-half of one per cent. to one per cent. The lingering stiffness, pain, and soreness of the muscles and joints are best treated by systematic hot douches and massage, and by mild galvanic currents.

INDEX.

ABSCESS, hepatic, in typhoid fever, 185
    of the spleen in relapsing fever, 333
Abscesses, in relapsing fever, 333
    in typhoid fever, 189
    multiple, in cerebro-spinal fever, 90
Adénopathie bronchique, in influenza, 34
Age, in etiology of cerebro-spinal fever, 59
    in etiology of typhoid fever, 118
    in etiology of typhus fever, 252
    influencing the mortality in typhoid fever, 219
Albuminuria, transitory, in typhoid fever, 188
Alcohol, excess of, in etiology of typhus, 252
    in typhoid fever, 226, 299
Alcoholism, diagnosticated from typhus fever, 297
Anæmia, after relapsing fever, 333
Anæsthesia, cutaneous and muscular, in typhoid fever, 167
    in cerebro-spinal fever, 77
Arachnoid, condition of, in cerebro-spinal fever, 91, 92
Arteries, fatty degeneration of, in typhoid fever, 208

BED-SORES, in typhoid fever, 189, 240
    in typhus fever, 286
Blood, changes in, in typhoid fever, 208
    condition of, in cerebro spinal fever, 90
Boils, in typhoid fever, 189
    in typhus fever, 285
Bowels, hemorrhage from, in typhoid fever, 174
Brain, changes in, in typhoid fever, 209

Brain, condition of, in cerebro-spinal fever, 93
    condition of the dura mater of, in cerebro-spinal fever, 91
    condition of the substance of, 93
Bronchitis, capillary, as a complication of influenza, 33
    in relapsing fever, 332
    in typhus fever, 281, 284

CALVARIUM, condition of, in cerebro-spinal fever, 91
Cancrum oris, in typhus fever, 287
Catarrh, bronchial, in typhoid fever, 176, 181
    the, of influenza, 30
Cellulitis, in typhus fever, 286
Cerebro-spinal fever (see Fever), 46
Chill, in cerebro-spinal fever, 72
Choroiditis, a sequel of relapsing fever, 334
Circulation, in influenza, 29
    in simple continued fever, 5
Circulatory system, symptoms referable to, in typhoid fever, 161
Climate, in etiology of cerebro-spinal fever, 57
    in etiology of relapsing fever, 309
    in etiology of typhoid fever, 116
    in etiology of typhus fever, 251
Cold, antipyretic use of, in typhoid fever, 229
    as an antipyretic in typhus fever, 298
Coma, in cerebro-spinal fever, 74
    in relapsing fever, 324
Coma vigil, 266
Congestion, hypostatic, in typhoid fever, 239

Constipation, in typhoid fever, 237
  in typhus fever, 283
Contagium vivum, of typhoid fever, 121
Convulsions, in cerebro-spinal fever, 75
  in typhoid fever, 165
  in typhus fever, 268
Cornea, perforation of, in typhus fever, 287
Cough, the, of influenza, 31
Countenance, in influenza, 30
Cutaneous lesions, in cerebro-spinal fever, 77
Cystitis, in typhoid fever, 188

DEAF-MUTISM, after cerebro-spinal fever, 89
Deafness, after cerebro-spinal fever, 89
  in typhoid fever, 191
  in typhus fever, 287
Death, mode of, in cerebro-spinal fever, 98
Debility, in influenza, 32
  in relapsing fever, 325
  in typhus fever, 267
Deglutition, difficult in typhoid fever, 184
  difficult in typhus fever, 282
Delirium, in cerebro-spinal fever, 73
  in relapsing fever, 324
  in typhoid fever, 164, 235
  in typhus fever, 265, 300
Dengue, 344
  clinical history of, 350
  diagnosis of, 356
  etiology of, 348
  exciting cause of, 349
  historical sketch of, 345
  treatment of, 356
Diarrhœa, a complication and sequel, of relapsing fever, 333
  in typhoid fever, 173, 236
Diet, in typhoid fever, 225
Dietetics, of typhoid fever, 222
Digestive system, condition of, in simple continued fever, 5
  derangement of the organs of, in cerebro-spinal fever, 84
  in influenza, 30
  symptoms due to disturbance of, in relapsing fever, 331
  symptoms referable to, in typhoid fever, 170
Diphtheritic processes, in typhoid fever, 184
Drinking-water, contamination of, in etiology of typhoid fever, 124

Duodenum, changes in, in typhoid fever, 203
Dura mater of the brain, condition of, in cerebro-spinal fever, 91
  condition of, in typhoid fever, 209
Dysentery, a sequel of relapsing fever, 334
  in typhus fever, 287
Dyspnœa, the, of influenza, 31

EAR, disorders of, in cerebro-spinal fever, 86
  disorders of, in influenza, 35
  disorders of, in typhoid fever, 166, 191
Ecchymoses, subpleural, in typhus fever, 294
Effusions of blood, in typhoid fever, 189
Emaciation, in dengue, 354
Endocarditis and pericarditis, in typhoid fever, 184
Endocarditis, with cerebro-spinal fever, 91
Enteric fever (see Fever), 107
  distinguished from cerebro-spinal fever, 95
Enteritis, diagnosticated from typhoid fever, 212
Epididymitis, in typhoid fever, 188
Epistaxis, in typhoid fever, 166, 169
Eruption, of cerebro-spinal fever, 77
  of typhoid fever, 167, 168
  of typhus fever, 169, 277
Erysipelas, after relapsing fever, 333
  facial, in typhoid fever, 189
  in cerebro-spinal fever, 78
  in typhus fever, 280, 285
Erythema, in cerebro-spinal fever, 78
Exanthem, in dengue, 353
Exanthems, diagnosticated from typhoid fever, 211
Excreta, decomposing, in etiology of typhoid fever, 124
Eye, condition of, in cerebro-spinal fever, 85
  disorders of, in typhoid fever, 167, 191
  disorders of, in typhus fever, 268

FEBRICULA, 1
Fever, ardent continued, 4
Fever, asthenic simple, 4
Fever, cerebro-spinal, 46
  analysis of the symptoms of, 72
  clinical history of, 64
  complications and sequels of, 87
  diagnosis of, 94

INDEX.

Fever, cerebro-spinal, disturbances of the organs of special sense in, 85
   etiology of, 56
   historical sketch of, 47
   pathology and morbid anatomy of, 89
   prognosis and mortality of, 97
   symptoms referable to the organs of respiration in, 84
   symptoms referable to the skin in, 77
   treatment of, 98
   varieties, 69, 71
Fever, enteric or typhoid (see Fever, typhoid), 107
Fever, herpetic, 6
Fever, **infantile remittent,** 195
Fever, **pernicious** intermittent, distinguished **from** cerebro-spinal fever, 95
Fever, relapsing, 302
   anatomical lesions of, 336
   analysis of the principal symptoms **of,** 324
   clinical history of, 320
   complications **and sequels of,** 332
   diagnosis of, 337
   diagnosticated **from typhoid, 211**
   etiology of, 309
   exciting cause **of,** 312
   following typhus, 319
   historical sketch of, 303
   prognosis and mortality **of,** 335
   symptoms due **to disturbance of the digestive organs in,** 331
   treatment of, 340
Fever, remittent, **diagnosticated from relapsing fever,** 339
   diagnosticated from typhoid, 211
   diagnosticated from typhus **fever,** 296
Fever, scarlet, **distinguished from cerebrospinal fever,** 95
Fever, simple continued, 1
   analysis of symptoms of, 5
   diagnosticated from typhoid, 211
   clinical history of, 3
   duration and **diagnosis of, 7**
   etiology of, 2
   prognosis, mortality, and treatment **of, 8**
Fever, typhoid, 107
   analysis **of the** chief **symptoms** of, 153
   anatomical lesions of, 202

Fever, typhoid, clinical history of, 147
   complications and sequels of, 178
   diagnosis of, 210
   differential diagnosis of, 95, 338
   etiology of, 116
   exciting cause **of,** 120
   expectant treatment of, 234
   geographical distribution, 115
   historical sketch of, 108
   management of the patient during convalescence from, 240
   prognosis and mortality of, 213
   relapses of, 196
   special forms of treatment of, 227
   symptoms referable to the circulatory system in, 161
   symptoms referable **to the** digestive tract in, 170
   symptoms referable **to the nervous** system in, 163
   symptoms **referable to the organs of respiration in,** 176
   skin, condition of, in, 167
   treatment of special symptoms, complications, and sequels of, 235
   treatment and prophylaxis of, 221
   urine, condition of, in, 177
   varieties of, 192
Fever, typho-malarial, 196
Fever, typhus, 241
   analysis of the principal symptoms of, 264
   anatomical lesions of, 293
   clinical history of, 260
   complications and sequels of, 284
   diagnosis of, 295
   diagnosticated **from** cerebro-spinal fever, 96
   diagnosticated **from** typhoid, 211
   differential diagnosis of, 338
   exciting cause of, 256
   etiology of, 251
   following relapsing **fever,** 319
   fomites of, 258
   historical sketch **of,** 242
   phenomena of the fever in, 269
   prognosis and mortality of, 290
   symptoms manifested by the **skin in,** 277
   symptoms referable to the respiratory and digestive systems in, 281
   treatment of, 297

Fever, typhus, varieties of, 288
Fomites, in typhoid fever, 130
    in typhus fever, 258

GANGRENE, in relapsing fever, 333
    of the lung in typhoid fever, 182
    of the lung in typhus fever, 285
Germ, of enteric fever, 121
    reproduction of, 136
    capable of reproducing itself outside of the human body, 143
    elimination of, with the fecal discharges, 140
    must undergo certain changes before it becomes capable of producing the disease, 141
    propagated by the atmosphere, 145
    remains in water and is conveyed by it, 144
    retains its activity for a long time, 142
Glossitis, in typhus fever, 287
Gurgling, in right iliac fossa in typhoid fever, 174

HAIR, falling of, in typhoid fever, 189
Hæmaturia, in typhoid fever, 188
Hæmoptysis, in typhus fever, 285
Hallucinations, in cerebro-spinal fever, 73
Headache, in cerebro-spinal fever, 72
    in influenza, 32
    in relapsing fever, 324
    in typhoid fever, 163, 235
    in typhus fever, 264, 300
Hearing, disturbances of, in cerebro-spinal fever, 86
Heart, condition of, in cerebro-spinal fever, 90
    changes in, in relapsing fever, 337
    changes in the muscle of, in typhoid fever, 208
    enfeeblement of, in typhoid fever, 162
Hemorrhage, from the bowels, in typhoid fever, 174, 238
Hemorrhages, cutaneous, in typhoid fever, 189
    in relapsing fever, 332
    in typhus fever, 285
Herpes, in cerebro-spinal fever, 77
    in influenza, 35
    in typhoid fever, 188
    in typhus fever, 280

Hiccough, in typhoid fever, 165
Hydrocephalus, chronic, after cerebro-spinal fever, 88
Hypostasis, in typhoid fever, 176
Hyperæsthesia, cutaneous, in typhoid fever, 167
    in cerebro-spinal fever, 76
Hysteria, 97

INFANTILE remittent fever, 195
Infarctions, hemorrhagic, in typhoid fever, 182
Influenza, 10
    analysis of the symptoms of, 29
    clinical history of, 26
    complications and sequels of, 33
    diagnosticated from typhoid fever, 212
    etiology of, 21
    historical sketch of, 12
    morbid anatomy and diagnosis of, 37
    pathology of, 36
    prognosis and mortality of, 38
    symptoms referable to the nervous system in, 32
    treatment of, 39
Insomnia, in relapsing fever, 325
Intelligence, feebleness of, after cerebro-spinal fever, 87
Intestines, changes in, in relapsing fever, 336
    changes in, in typhoid fever, 203
    perforation of, in typhoid fever, 185
Iritis, a sequel, of relapsing fever, 334

JAUNDICE, in cerebro-spinal fever, 84
    in relapsing fever, 321, 332
    in typhoid fever, 185
    in typhus fever, 287
Joints, affection of, in dengue, 352
    inflammation of, in cerebro-spinal fever, 85

KIDNEYS, condition of, in cerebro-spinal fever, 91
    changes in, in typhoid fever, 208
    changes in, in typhus fever, 293

LARYNGITIS, acute, in influenza, 35
    in typhoid fever, 179
    in typhus fever, 284
    in relapsing fever, 333

INDEX. 363

Liver, abscess of, in typhoid fever, 185
    condition of, in cerebro-spinal fever, 91
    changes in, in typhoid fever, 207
    changes in, in typhus fever, 294
    changes in, in relapsing fever, 337
    enlargement of, in relapsing fever, 331
Locality, in etiology of cerebro-spinal fever, 58
Lungs, changes in, in relapsing fever, 337
    changes in, in cerebro-spinal fever, 90
    gangrene of, in typhus fever, 285
    hypostatic congestion of, in typhoid fever, 181
Lymphatics, in dengue, 353

MANIA, in typhoid fever, 190
Measles, diagnosticated from typhus fever, 296
Meat, diseased, in etiology of typhoid fever, 132
Membranes of the spinal cord, condition of, in cerebro-spinal fever, 92
Memory, weak, after cerebro-spinal fever, 87
Meningitis, diagnosticated from typhoid fever, 212
    epidemic, 96, 97
    in typhoid fever, 180
    tuberculous basilar, distinguished from cerebro-spinal fever, 94
Menstruation, in typhoid fever, 188
Mesenteric glands, changes in, in typhoid fever, 206
Meteorism, in typhoid fever, 172
Mode of death, in cerebro-spinal fever, 98
Mucous membrane, gastro-intestinal changes in, in typhus fever, 293
    intestinal, in cerebro-spinal fever, 91
Muscles, changes in the voluntary, in typhoid fever, 203
    condition of, in cerebro-spinal fever, 90
    changes in, in typhus fever, 293
    disorders of, in typhoid fever, 165

NAILS, condition of, in typhoid fever, 189
Nausea, in relapsing fever, 331
    in typhoid fever, 171
    in typhus fever, 282
Neck, stiffness of, in cerebro-spinal fever, 74

Necrosis, in typhus fever, 287
Nervous system, condition of, in simple continued fever, 6
    symptoms pertaining to, in cerebro-spinal fever, 72
    symptoms referable to, in influenza, 32
    symptoms referable to, in typhoid fever, 163
Neuralgia, a sequel of relapsing fever, 334
    in influenza, 35
    in typhoid fever, 191
Noma, in typhus **fever, 287**
Nutrition, state of, in cerebro-spinal fever, 83

ŒDEMA, pulmonary, in typhoid fever, 181, 210
Occupation, in etiology of typhoid fever, 119
    in etiology of typhus, 252
Ophthalmia, post-febrile, 334
Opium, in cerebro-spinal fever, 102
Orchitis, in typhoid fever, 188
Otorrhœa, in typhoid fever, 191
Overcrowding, in the etiology of typhus, 253

PACHYMENINGITIS, hemorrhagic, a sequel of relapsing fever, 334
Pain, in influenza, 32
    in relapsing fever, 325
    in typhoid fever, 163
    abdominal, in typhoid fever, 171
    in typhus fever, 264
Pancreas, changes in, in typhoid fever, 210
    changes in, in typhus fever, 294
Paralysis, in cerebro-spinal fever, 76
    after cerebro-spinal fever, 87
    in typhoid fever, 190
    in typhus fever, 267, 287
    local, after relapsing fever, 334
Parotid gland, swelling of, in cerebro-spinal fever, 85
    inflammation of, in influenza, 35
    inflammation of, in typhus fever, 286
    swelling of, after relapsing fever, 333
    swelling of, in typhoid fever, 184
Perforation, intestinal, in typhoid fever, 185
Pericarditis and endocarditis in typhoid fever, 184

Pericardium, condition of, in cerebro-spinal fever, 91
Peritonitis, diagnosticated from typhoid fever, 212
   in typhoid fever, 239
   in typhus fever, 294
Petechiæ, in cerebro-spinal fever, 78
   in typhoid fever, 169
   in typhus fever, 261
Phlegmasia dolens, in typhus fever, 285
Phthisis, in typhus fever, 285
Physiognomy, of typhoid fever, 170
Pia mater, condition of, in cerebro-spinal fever, 91
   in typhoid fever, 209
Plague, diagnosticated from typhus fever, 296
Pleura, condition of, in cerebro-spinal fever, 91
Pleurisy, in cerebro-spinal fever, 87
   in influenza, 35
   in relapsing fever, 333
   in typhoid fever, 183
   in typhus fever, 285
Pleurosthotonos, in cerebro-spinal fever, 75
Pneumonia, catarrhal and croupous, as complications of influenza, 33
   in cerebro-spinal fever, 87
   in relapsing fever, 332
   in typhoid fever, 182
   in typhus fever, 284
Pregnancy, in influenza, 35
   with relapsing fever, 323
   with typhoid fever, 188
   with typhus fever, 288
   influencing the mortality in typhoid fever, 220
Prophylaxis, in cerebro-spinal fever, 98
   in typhoid fever, 221
Protomycetes, of relapsing fever, 312
Pulse, in relapsing fever, 328
   in typhoid fever, 161
   in typhus fever, 272
Pupil, condition of, in cerebro-spinal fever, 85
Pyæmia, in typhus fever, 285

RACHIALGIA, in cerebro-spinal fever, 76
Relapses, in dengue, 355
   in typhus fever, 290

Relapsing fever (see Fever), 302
Remittent fever, diagnosticated from relapsing fever, 339
Respiration, symptoms referable to the organs of, in cerebro-spinal fever, 84
   symptoms referable to the organs of, in typhoid fever, 176
Restlessness, in cerebro-spinal fever, 74
Retinitis, a sequel of relapsing fever, 334
Roseola, in cerebro-spinal fever, 78

SALIVARY glands, changes in, in typhoid fever, 209
Scarlatina, diagnosticated from cerebro-spinal fever, 95
Season of the year, in etiology of cerebro-spinal fever, 37
   in etiology of relapsing fever, 309
   in etiology of typhoid fever, 116
   in etiology of typhus fever, 251
Secretions, the, in influenza, 30
Sewage, in etiology of typhoid fever, 122
Sex, in etiology of typhoid fever, 119
   in etiology of typhus fever, 252
   influencing mortality in typhoid fever, 220
Simple continued fever (see Fever), 1
Skin, condition of, in cerebro-spinal fever, 90
   condition of, in simple continued fever, 5
   hyperæsthesia of, in cerebro-spinal fever, 76
   in relapsing fever, 330
   state of, in typhoid fever, 167
   symptoms referable to, in cerebro-spinal fever, 77
Sleeplessness, in cerebro-spinal fever, 74
   in typhoid fever, 235
   treatment of, in typhus fever, 300
Small-pox, diagnosticated from typhoid fever, 211
Sordes, in typhoid fever, 236
   in typhus fever, 282
Somnolence, in typhoid fever, 165, 235
Spinal cord, changes in, in cerebro-spinal fever, 93
   condition of the membranes of, in cerebro-spinal fever, 92
Spine, contraction of the erector muscles of, in cerebro-spinal fever, 74
Spirilli, of Obermeier, 322

INDEX. 365

Spleen, abscess of, in relapsing fever, 333
   condition of, in cerebro-spinal fever, 84, 91
   changes in, in relapsing fever, 337
   condition of, in typhoid fever, 172, 207
   changes in, in typhus fever, 293
   enlargement of, in relapsing fever, 332
   enlargement of, in typhus fever, 283
Splenization, in typhoid fever, 210
Stomach, changes in, in relapsing fever, 336
   changes in, in typhoid fever, 203
Sudamina, in cerebro-spinal fever, 78
   in typhoid fever, 169
Synocha, 1

TEMPERATURE, in cerebro-spinal fever, 78
   in etiology of typhoid fever, 117
   in influenza, 29
   in simple continued fever, 5
   in relapsing fever, 326
   in typhoid fever, 153
   in typhus fever, 269
Tenderness, abdominal, in typhoid fever, 171
Thirst, in cerebro-spinal fever, 84
   in typhoid fever, 171
   in typhus fever, 282
Thrombosis, venous, in typhoid fever, 183
Tinnitus aurium, in influenza, 16
Tongue, condition of, in cerebro-spinal fever, 84
   condition of, in relapsing fever, 331
   in typhoid fever, 170, 236
   in typhus fever, 282
Tremor, in typhoid fever, 236
Trichiniasis, diagnosticated from typhoid fever, 213

Trismus, in cerebro-spinal fever, 75
Tuberculosis, acute, diagnosticated from typhoid fever, 212
   acute miliary, after typhoid fever, 182
Tympany, in typhoid fever, 237
   in typhus fever, 283
Typhoid fever, 107
Typho-malarial fever, 196
Typhus fever, 241

URINE, condition of, in cerebro-spinal fever, 85
   condition of, in relapsing fever, 321
   condition of, in simple continued fever, 5
   condition of, in typhoid fever, 177
   in dengue, 353
   in relapsing fever, 330
   in typhus fever, 274, 280
   retention of, in typhoid fever, 165
Urticaria, in cerebro-spinal fever, 78
   in dengue, 353

VENTRICLES, cardiac, dilatation of, in typhoid fever, 183
Vertigo, in cerebro-spinal fever, 73
   in relapsing fever, 324
   in typhoid fever, 163
   in typhus fever, 264
Vibices, in typhus fever, 279
Vomiting, in cerebro-spinal fever, 73
   in relapsing fever, 331
   in typhoid fever, 171
   in typhus fever, 282

WAKEFULNESS, in typhoid fever, 165
Weather, state of, in etiology of typhoid fever, 117

www.ingramcontent.com/pod-product-compliance
Lightning Source LLC
Chambersburg PA
CBHW030349230426
43664CB00007BB/581